IX International Congress Motor Vehicles and Motors (MVM 2022)

IOP Conference Series: Materials Science and Engineering
Volume 1271

Kragujevac, Serbia
13 – 14 October 2022

ISBN: 978-1-7138-6517-9
ISSN: 1757-8981

Printed from e-media with permission by:

Curran Associates, Inc.
57 Morehouse Lane
Red Hook, NY 12571

Some format issues inherent in the e-media version may also appear in this print version.

Institute of Physics
Dirac House, Temple Back
Bristol BS1 6BE UK

Phone: 44 1 17 929 7481
Fax: 44 1 17 920 0979

techtracking@iop.org

Additional copies of this publication are available from:

Curran Associates, Inc.
57 Morehouse Lane
Red Hook, NY 12571 USA
Phone: 845-758-0400
Fax: 845-758-2633
Email: curran@proceedings.com
Web: www.proceedings.com

TABLE OF CONTENTS

IOP Conf. Series: Materials Science and Engineering 1271 (2022) 011001 doi:10.1088/1757-899X/1271/1/011001

9th International Congress
Motor Vehicles & Motors 2022
Kragujevac, Serbia
October 13th-14th, 2022
Motor Vehicles & Motors

Preface

In October, the Faculty of Engineering traditionally holds gatherings of researchers and academics who study motor vehicles, engines and road traffic. From 1979 to 2004, thirteen, biennal MVM Symposiums have been held and they grew into an International Congress MVM in 2006. Since then, eight MVM Congresses have been held and in October 2022, the Faculty of Engineering of the University of Kragujevac organized the ninth International Congress MVM from 13th to 14th October 2022.

With the traditional support of the Ministry of Educetion, Science, and Technological Development of Republic of Serbia, University of Kragujevac, Faculty of Engineering and the International Journal "Mobility and Vehicle Mechanics", the organizers of the Congress decided to organize the event on the Internet network (online) due to COVID-19 pandemic situation, through ZOOM platform, for the second time.

The theme of the Congress MVM 2022 was "Ecology - Vehicle and Road Safety - Efficiency". Along this journey we learned from the various sessions, which broadly cover a wide range of topics related to automotive engineering from fundamental research to industrial applications, highlight the interaction between the driver, vehicle and environment and stimulate scientific interactions and collaborations.

The International Scientific Committee in collaboration with the Organising Committee built up a stimulating scientific program. The program offered over 49 presentations, including key-note speakers and paper sessions. The presentations to this conference covered current research in motor vehicle and motors conducted in 15 countries from all over the world. Maunscripts were submitted and peer review were done through Morressier platform.

We were pleased to have professors Giovanni Belingardi and Alessandro Scattina (from Politecnico di Torino, Italy) and professor Ralph Pütz (from Hochschule Landshut, Germany) as the keynote speakers, addressing Challenges and solutions in the design of battery pack for battery electric vehicle as well as European policy on future road mobility.

We are sure this program will trigger lively discussion and will project researchers to new developments.

The 9th Congress MVM 2022 was finacially supported by Ministry of Education, Science and Technological Development of the Republic of Serbia.

We would like to thank experienced and young researchers, for attending and bringing their expertise and innovative ideas to our conference.

Special thanks are due to the International Scientific Board Members and all reviewers for their significant contribution in the high level of the conference.

Scientific and Organising Committee of Congress MVM2022

Editors:

Dr Jovanka Lukić, prof.
University of Kragujevac
Faculty of Engineerng

Dr Jasna Glišović, prof.
University of Kragujevac
Faculty of Engineerng

University of Kragujevac, Faculty of Engineering, Sestre Janjic 6, 34000 Kragujevac,
e-mail: congress.mvm@kg.ac.rs; lukicj@kg.ac.rs; http://congressmvm.fink.rs/

IOP Conf. Series: Materials Science and Engineering 1271 (2022) 011001 doi:10.1088/1757-899X/1271/1/011001

Legend

- FE – Faculty of Engineering
- FTTE - Faculty of Transport and Traffic Engineering
- FTS – Faculty of Technical Sciences
- FME – Faculty of Mechanical Engineering

IOP Conf. Series: Materials Science and Engineering 1271 (2022) 011002 doi:10.1088/1757-899X/1271/1/011002

All papers published in this volume have been reviewed through processes administered by the Editors. Reviews were conducted by expert referees to the professional and scientific standards expected of a proceedings journal published by IOP Publishing Publishing.

Type of peer review: Single anonymous

Conference submission management system: Morressier

Number of submissions received: 68

Number of submissions sent for review: 49

Number of submissions accepted: 30

Acceptance Rate (Submissions Accepted / Submissions Received × 100): 44.1 %

Average number of reviews per paper: 1.07

Total number of reviewers involved: 11

Contact person for queries:

Name: Jovnka Lukic

Email: congress.mvm@kg.ac.rs

Affiliation: University of Kragujevac, Faculty of Engineering

A contribution to investigation of oscillatory loads of driving axles in order to create conditions for laboratory tests of trucks

Miroslav Demić1,2, Aleksandar Đurić [3], Aleksandar Grkić [4], Momir Drakulić [3], Slavko Muždeka [3]

[2] Engineering science academy of Serbia, Belgrade, Serbia,
[3]Military academy, University of defense, Belgrade, Serbia
[4]Auto Moto Asociation of Serbia, Center for motor vehicles, Belgrade, Serbia

Abstract: Motor vehicles are complex dynamic systems due to spatial vibratory displacements during movement, changes in the characteristics of components during life cycle, a large number of influences and disturbances, the occurrence of clearance, friction, hysteresis, etc. The mentioned dynamic phenomena, especially vibrations, cause fatigue of the driver and passengers, reduce the life cycle of the vehicle and its systems, etc. In the general case, the movement of motor vehicles is done on uneven roads (terrain) and curvilinear paths in flat roads (terrains). Oscillatory movements cause loading of vehicle parts, but they also negatively affect human health. Therefore, even in the design phase of the vehicle, special attention must be paid to the harmonization of the mutual movement of the vehicle subsystem, and in particular, the suspension system. Theoretical, experimental or combined methods can be used for this purpose, and it is very useful to have experimental results of vibration of the vehicle subsystems in real operational conditions. Therefore, it is very useful to have experimental results of vibration of vehicle subsystems in operational conditions. Bearing that in mind, the aim of this wpaper was to use the movement of FAP 1118, 4x4 wheel formula, in operational conditions (due to higher speeds - in road conditions) to define laboratory conditions for testing. This is made possible by registering and identifying statistical parameters of registered quantities.

Keywords: *Motor vehicle, axles, oscillatory loads, laboratory tests*.

1. Introduction

Motor vehicles are complex dynamic systems due to spatial vibratory displacements during movement, changes in the characteristics of components during life cycle, a large number of influences and disturbances, the occurrence of clearance, friction, hysteresis, etc. [1-12]. The mentioned dynamic phenomena, especially vibrations, cause fatigue of the driver and passengers, reduce the life cycle of the vehicle and its systems, etc.

In the general case, the movement of motor vehicles is done on uneven roads (terrain) and curvilinear paths in flat roads (terrains). Oscillatory movements cause loading of vehicle parts, but they also negatively affect human health [1-3, 11]. Therefore, even in the design phase of the vehicle, special attention must be paid to the harmonization of the mutual movement of the vehicle subsystem, and in particular, the suspension system [12]. Theoretical, experimental or combined methods can be used for

[1] ***Miroslav Demić, PhD, Engineering science academy of Serbia, Belgrade,*** demic@kg.ac.rs
Ljubiše Jovanovića 7, 34110 Kragujevac

this purpose, and it is very useful to have experimental results of vibration of the vehicle subsystems in real operational conditions.

The road (terrain) can be identified on the basis of its spatial geometry (macro relief) and micro-roughness (micro relief) [13-16]. The movements of the vehicle subsystem are conditioned, in first, by the shape and size of the unevenness, as an external factor and oscillatory-inertial characteristics, engine torque and vehicle velocity, as phenomena related to the vehicle itself. Based on that, it can be stated that careful research and definition of micro-roughness characteristics of roads on which vehicles move, both in terms of periodicity characteristics and in terms of energy levels, development and automation of roughness measurement and mathematical apparatus for processing data, contribute to reliability, optimality and safety of the construction of the vehicle itself. As the description of the road parameters and their identification are described in detail in [1-16], they will not be elaborated here…

As it is known [17], in laboratory conditions, reproductions of signals registered in operation can be performed on shakers. Therefore, the aim of this work was to establish the oscillatory movements of axles in operating conditions (when driving in road conditions), FAP 1118 vehicle, in order to create conditions for laboratory testing.

2. Experiment

Desing of experiment is a complex problem [17]. In this particular case, the subject of research is a vehicle FAP 1118, with 4x4 drive formula and load capacity of 4t. The maximum mass of the test vehicle is 11000 kg, and during the test the vehicle was partially loaded (total mass 7800 kg - static load of the front axle was 4200, and the rear 2850, daN). The test truck vehicle is equipped with hydraulic servo steer and classic suspension systems with leaf springs. The roll axe is located, approximately, between vehicle axles and the pins of leaf springs (high 435, mm from ground).

The measuring chain for measuring the dynamic parameters of the vehicle consisted of the following elements:

- Kistler Correvit S-350 sensors for direct, slip-free measurement of longitudinal and transverse vehicle dynamic,
- HBM Quantum MX 840B universal measuring acquisition system connected to a QuantumX CX22B-W computer,
- B-12 acceleration sensor, located in the center of gravity of the rear truck bridge,
- SST 810 dynamic inclinometer, placed in the center of gravity of vehicle axles. With it, angle, velocity and acceleration was measured around the X, Y and Z axes.

All measured quantities are being processed by using HMB Catman easy software.
The scheme of measuring points is given in Figure 1.

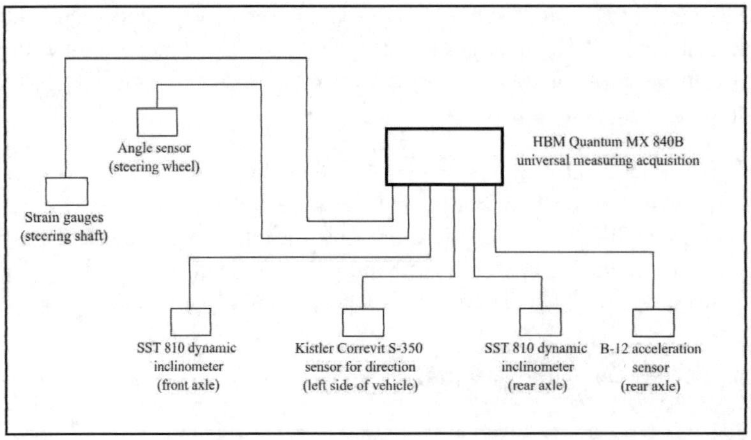

Figure 1. The scheme of measuring points during the FAP 1118 road test

Based on previous experiences at the Military Academy, it was considered expedient to test the vehicle while driving in real operational conditions, on an asphalt regional road near Belgrade (maximum speed, 67 km/h – design vehicle limited speed 80 km/h). During the experiment, the weather was sunny and the surface was dry. The length of the time records was 260 s (13000 points, and the sampling step was 0.02 s).

For illustration, Figures 2 - 4 show the time histories of vehicle velocity and the oscillatory movements of the front drive axle.

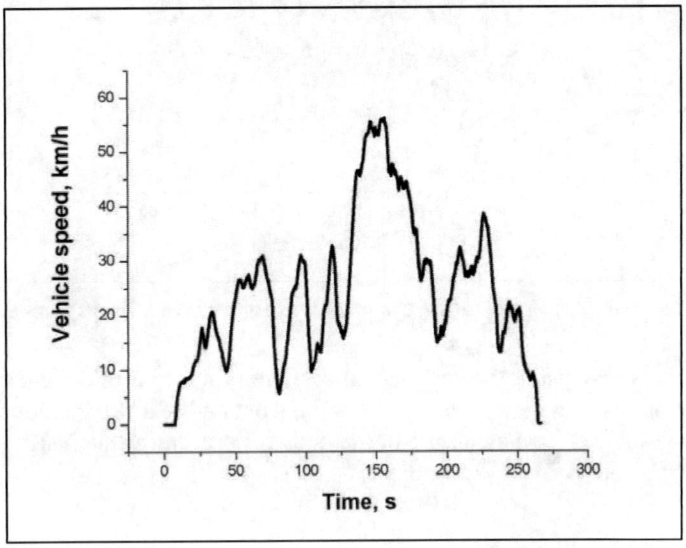

Figure 2. Vehicle velocity change during drive

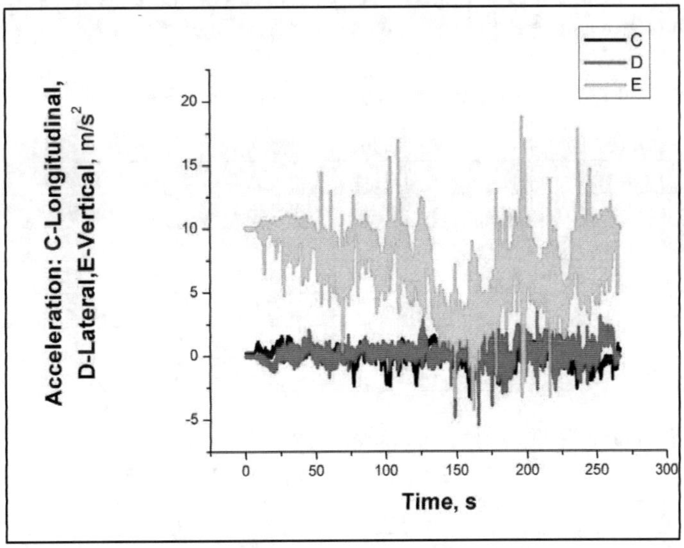

Figure 3. Longitudinal, lateral and vertical acceleration of the vehicle's front drive axle

IOP Conf. Series: Materials Science and Engineering 1271 (2022) 012001 doi:10.1088/1757-899X/1271/1/012001

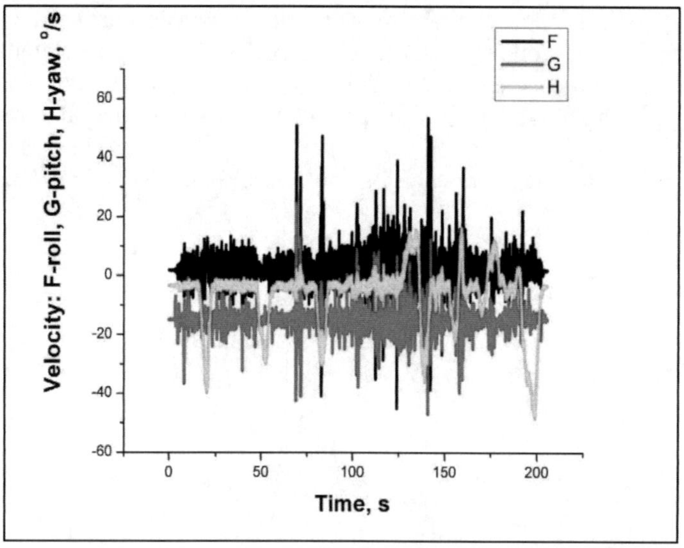

Figure 4. Roll, pitch and yaw velocity of vehicle's rear drive axle

Figures 2 – 4 shows that the registered parameters of oscillatory motions of drive axles depend on time. The registered accelerations and angular velocities of the front and rear drive axles belong to the group of random processes [18]. There are a number of methods for processing such signals [18] and we will use some of them in this paper.

3. Signal processing and data analysis

Having in mind the random character of the registered values, it was considered expedient to calculate: mean, RMS, minimum and maximum value of all registered oscillatory values. For this purpose, the software Analsigdem [19] was used, and the values given in Tables 1 and 2.

Table 1. The statistical parameters of front axle vibration

Measured parameters	Min.value	Max.value	Mean value	RMS
Vehicle velocity, km/h	0.0	61.7	31.2	34.9
Longitudinal acceleration, m/s^2	-3.2	1.64	-0.0039	0.67
Lateral accelerations, m/s^2	-5.45	5.61	0.15	0.70
Vertical acceleration, m/s^2	-4.24	18.7	7.34	7.86
Roll velocity, o/s	-48.1	54.1	2.2	4.1
Pitch velocity, o/s	-16.6	18.6	1.0	2.3
Yaw velocity, o/s	-32.6	20.2	-2.3	7.4

Table 2. The statistical parameters of rear axle vibration

Measured parameters	Min.value	Max.value	Mean value	RMS
Vehicle velocity, km/h	0.0	61.7	31.2	34.9
Longitudinal acceleration, m/s^2	-8.8	5.0	0.3	1.1
Lateral acceleration, m/s^2	-4.7	7.5	0.2	0.9
Vertical acceleration, m/s^2	-10.1	31.9	9.65	9.8
Roll velocity, °/s	-45.1	53.4	1.9	5.8
Pitch velocity, °/s	-47.0	24.5	-14.8	15.2
Yaw velocity, °/s	-48.4	16.3	-6.4	11.9

By analyzing the data from Tables 1 and 2, it can be concluded that there are differences in the oscillatory movements of the front and rear axles in operating conditions. As the differences are not unambiguous, additional analyzes are needed.

In order to determine the character of the registered signals, using the same software [19], autocorrelation functions were calculated and shown in Figures 5-8.

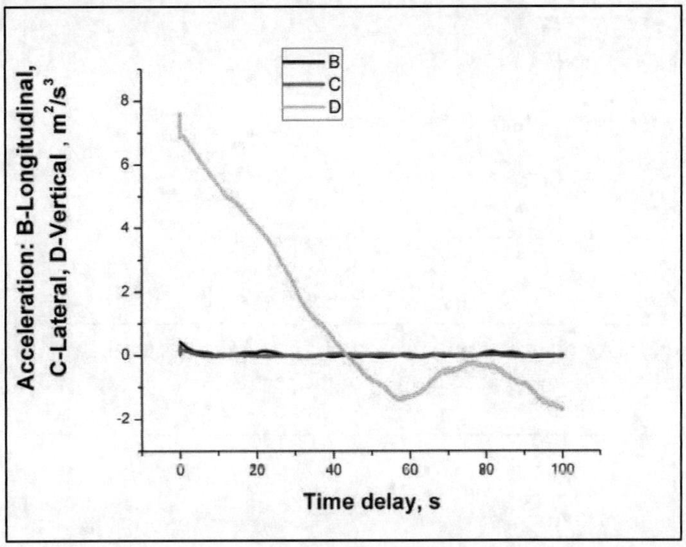

Figure 5. Autocorrelation function of longitudinal, lateral and vertical acceleration of the front axle

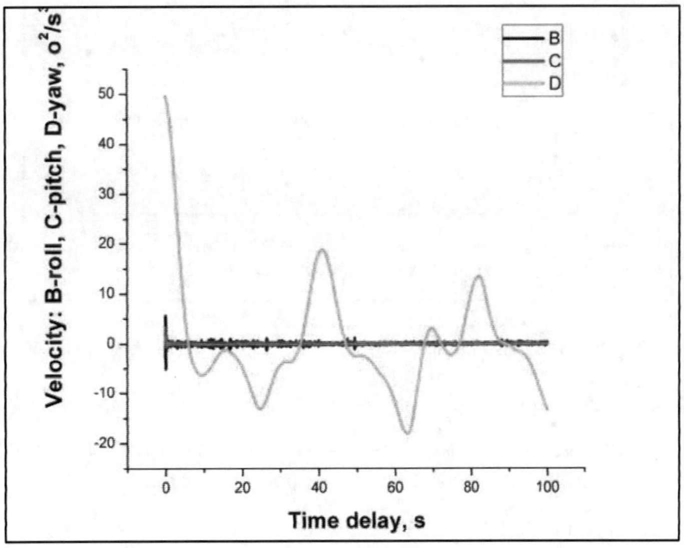

Figure 6. Autocorrelation function of roll, pitch and yaw velocity of vehicle's front drive axle

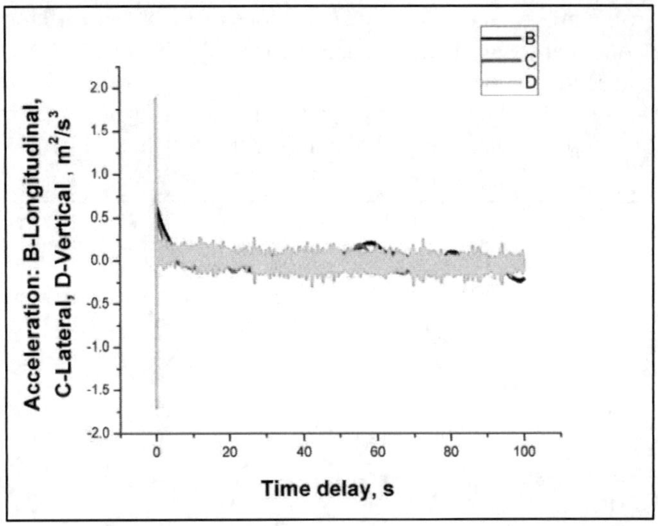

Figure 7. Autocorrelation function of longitudinal, lateral and vertical acceleration of the rear axle

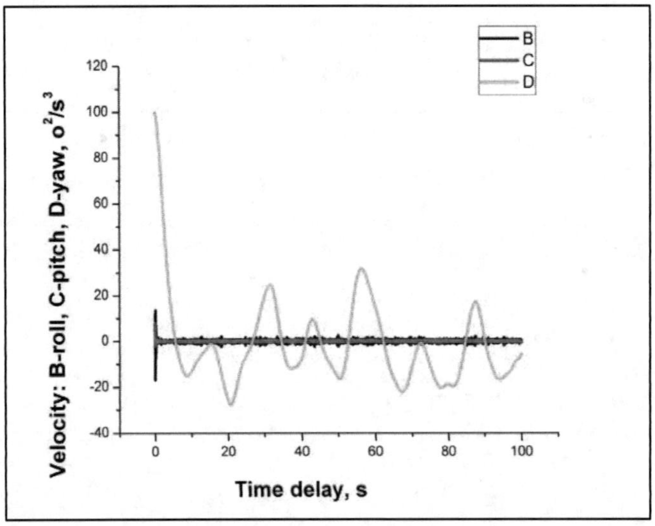

Figure 8. Autocorrelation function of roll, pitch and yaw velocity of vehicle's rear drive axle

Analysis of all data, partially given at Figures 5-8. shows that autocorrelation functions decrease with increasing time delay and that they oscillate slightly around zero. This points to the fact that registered oscillatory values can be treated as stationary [18].

IOP Conf. Series: Materials Science and Engineering 1271 (2022) 012001 doi:10.1088/1757-899X/1271/1/012001

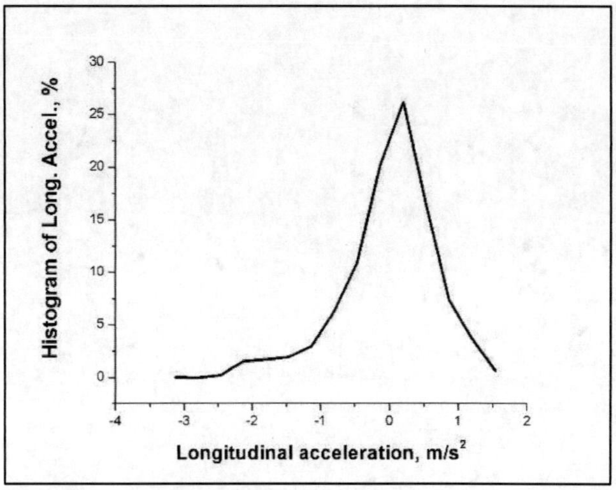

Figure 9. Front axle longitudinal acceleration distribution histogram

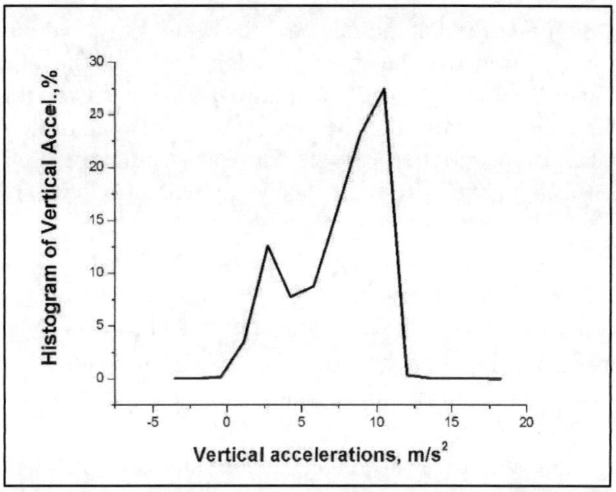

Figure 10. Front axle vertical acceleration distribution histogram

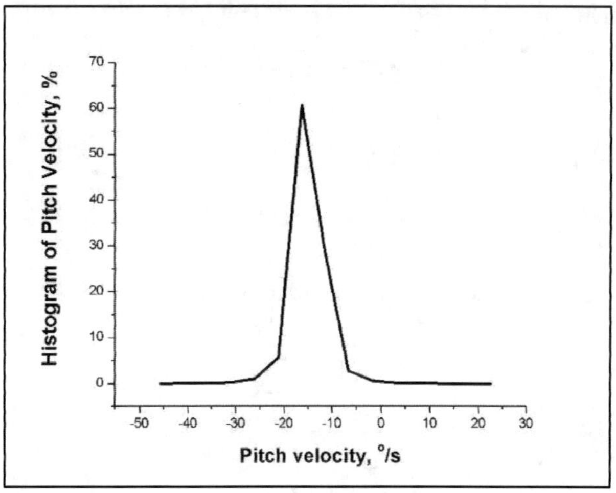

Figure 11. Rear axle distribution of the pitch velocity histogram

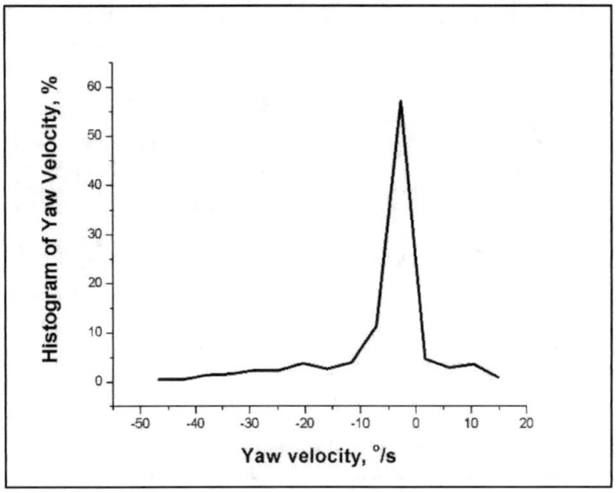

Figure 12. Rear axle yaw velocity distribution histogram

It was considered interesting to perform amplitude analysis of all registered parameters [18-20]. More precisely, the probability of occurrence of the observed values by levels (histogram, %) was calculated, using the software [19]. First, the minimum and maximum value of each variable, in 13000 points, was calculated using the aforementioned software. Then the interval between those values was divided into 500 fields and the number of occurrences of the variable within the limits for each field was automatically determined (histogram, %). The calculated values are, for the sake of illustration, partially shown in figures 9-12.

By analyzing the histogram of all registered values, it was observed that the maximum values are close to the domain of mean values. Bearing that in mind, it was assessed as relevant to test all data for normality, using the so-called „Null hypotheses" (from statistics known as „Normality test") [20].

The „Normality test" consists of comparing the mean value of the distribution of the registered data with a data that has a normal distribution and an infinite number of members. In this case, the mentioned test was implemented with a risk of 5%, using the Origin 8.5 software [21].

More precisely, using that software, the analysis of the correctness of the adopted null hypothesis (in this case for the mean value of 0) was performed for all measured values. It was found that there is a significant difference in the mean values, so the hypothesis cannot be accepted unconditionally.

Having that in mind, it was estimate useful to use the so-called $\chi 2$ test [20], which is defined as:

$$\chi^2 = \sum_{1}^{N} \frac{(f_i - f_{ti})^2}{f_{ti}} \quad (1)$$

where are:

f_i - frequencies of i-th class;

f_{ti} – theoretical frequency of i-th class;

N-number of classes.

In [20], a simple and reliable test, known as the „Romanovsky test", was especially recommended. The criterion is defined by expression:

$$R = \frac{|\chi^2 - k|}{\sqrt{2k}} \quad (2)$$

where:

$$k = N - L - 1 \quad (3)$$

In expression (3), N is the number of additions in (1), and l is the number of unknown parameters in the assumed probability distribution. If R<3 the hypothesis is accepted, and for R>3 the hypothesis is rejected.

Bearing in mind that the mean values are not always positive, it was considered useful to perform hypothesis tests with Gaussian, Laplace and Cauchy distributions [20]. Previously, on the basis of experimental and theoretical distribution functions, using the method of minimizing the square of the difference, the parameters of the territorial distributions were identified (this procedure is known from [12], so it will not be discussed further). Analyzes have shown that the assumed probability density functions do not meet the Romanovsky criterion, so they do not belong to any of the observed distributions.

With this in mind, it was considered useful to show in Tables 3 and 4 the mean values of the observed values and the maximum density of probability of occurrence by levels.

Table 3. Front axle

Measured parameters	Mean value	Maximum prob. density, %
Longitudinal acceleration, m/s^2	0.20	26.24
Lateral acceleration, m/s^2	0.10	52.71
Vertical acceleration, m/s^2	8.90	23.13
Roll velocity, °/s	3,01	87.32
Pitch velocity, °/s	1.01	63.30
Yaw velocity, °/s	-2.41	59.90

Table 4. Rear axle

Measured parameters	Mean value	Maximum prob. density, %
Longitudinal acceleration, m/s^2	0.90	35.11
Lateral acceleration, m/s^2	-0.25	37.92
Vertical acceleration, m/s^2	10.75	55.75
Roll velocity, °/s	4.01	53.87
Pitch velocity, °/s	-16.36	60.74
Yaw velocity, °/s	-2.80	57.23

The data from Tables 3 and 4 can be used when generating schaker excitations, during vehicle testing in laboratory conditions.

For practical reasons, it can be argued that, in the vehicle design phase, it is acceptable that the registered quantities belong to the Gaussian distribution.

Frequency analysis was performed using the Analsigdem software [19] with 8192 points and a sampling step of 0.02 s, which enabled reliable analysis in the range 0.061 to 25 Hz [18] (the DC components are turned of for frequency analysis).

Analysis of random and bias errors, for the number of data used, showed that a sufficient number of averaging of 100 for one signal and 138 for two signals, which achieves a minimum reliable frequency of 0.049 Hz (this is acceptable in this case, because it is lower than those obtained on the basis of signal length [18]). Having in mind that the phases of the calculated spectra do not enable the analysis of the energy carried by the signal, it was considered expedient to observe only the magnitudes of the calculated spectra, which are shown in Figures 13-16.

IOP Conf. Series: Materials Science and Engineering 1271 (2022) 012001 doi:10.1088/1757-899X/1271/1/012001

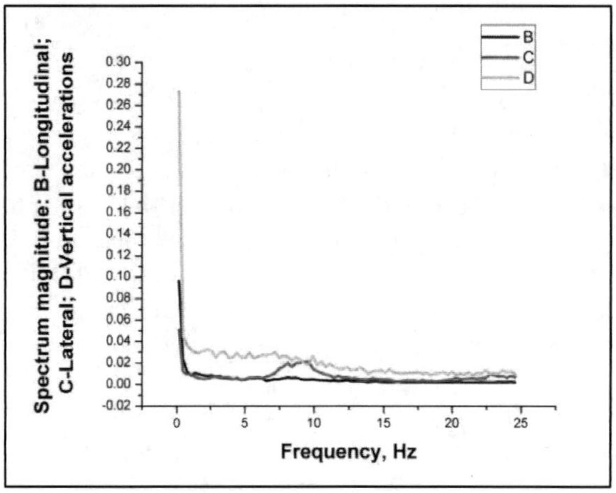

Figure 13. The magnitudes of the spectrum of the longitudinal, lateral and vertical acceleration of the front axle

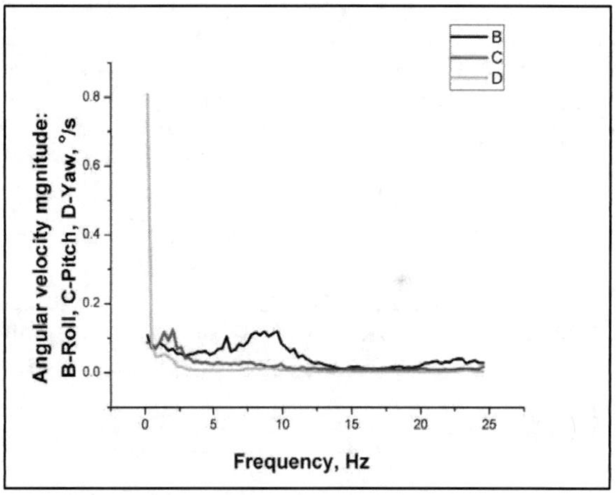

Figure 14. The magnitudes of the spectrum of the roll, pitch and yaw velocity of front axle

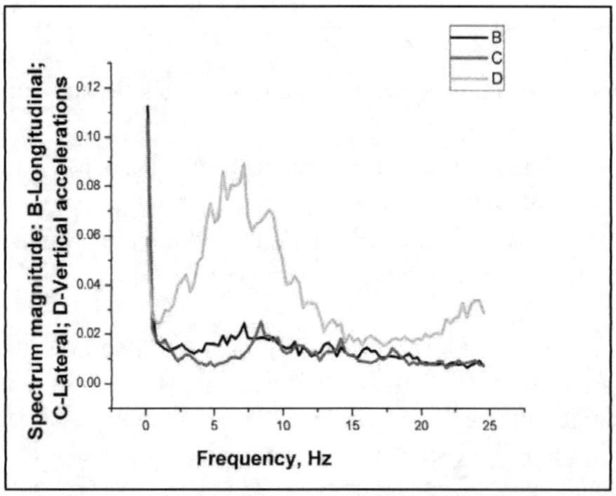

Figure 15 The magnitudes of the spectrum of the longitudinal, lateral and vertical of the rear axle

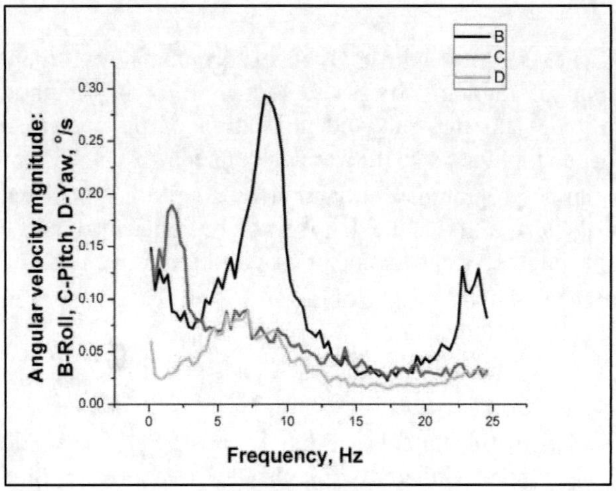

Figure 16. The magnitudes of the spectrum of the roll, pitch and yaw velocity of rear axle

Analyzes of the calculated spectrum magnitudes have shown that the largest amplitudes are not unambiguous: they depend on the axle (front, rear), as well as on the registered value. Namely, with the front axle (all observed values) and with the rear axle (linear accelerations), the maximum values occur at very low frequencies. However, when it comes to the angular velocities of the rear axle, this is not the case. Therefore, it was found useful to determine the magnitudes of the spectra at resonant frequencies. It is noted that the analysis does not include resonances originating from the sprunged mass, drive group, etc. Resonance parameters are shown in Table 5, and can be used when programming laboratory tests of trucks.

Table 5. Resonant parameters

Measured size	Amplitude/frequency front axle	Amplitude/frequency rear axle
Longitudinal acceleration, m/s^2	0.0577/3.59	0.016/4.42
Lateral acceleration, m/s^2	0.015/7.47	0.080/4.73
Vertical acceleration, m/s^2	0.029/5.34	0.083/5.64
Roll velocity, °/s	0.119/9.61	0.119/5.03
Pitch velocity, °/s	0.015/9.00	0.089/7.17
Yaw velocity, °/s	0.068/5.95	0.091/7.18

It can be seen from Table 5 that the corresponding resonance parameters of the front and rear axle are not equal to each other [11], so this should be taken into account when designing the suspension system for cargo vehicles. It is noted that, in practice, the vehicle suspension system is usually designed according to the resonance of vertical vibrations of the sprunged and unsprunged vehicle masses.

Based on the results of the performed analyzes (parameters of time identification-mean values and autocorrelation function, parameters of amplitude identification-probability density and parameters of frequency identification-spectra) it is possible to design research in laboratory conditions-on schakers. Depending on the available type of schaker, we need to select the values that will be reproduced. Most often, these are vertical vibration, but it can also be some other vibration movements (it is noted that there are rare schakers that can generate six excitations at the same time).

As the values have been registered in the middle of the axles during the measurements, in some potential laboratory test this should be taken into the account.

IOP Conf. Series: Materials Science and Engineering 1271 (2022) 012001 doi:10.1088/1757-899X/1271/1/012001

4. Conclusion

Tests performed on the FAP 1118 truck with 4x4 wheel formula showed that the observed measured values belong to the group of random processes, so the following methods were used for their identification: time, amplitude and frequency identification of parameters. Namely, mean values, autocorrelation functions and amplitude spectra were calculated.

Based on them, research can be programmed in laboratory conditions, on schakers. Depending on the available type of schaker, the size of the value that is to be reproduced should be selected. Most often, these are vertical vibrations, but they can be some other vibration movements.

As the values have been registered in the middle of the axles during the measurements, in some potential laboratory test this should be taken into account.

REFERENCE

[1]. Demić M. *Cybernetic system: Human-Vehicle-Environment*, Serbian Academy of Sciences and Art Research Centre and Kragujevac University (in Serbian, monograph), 2008.
[2]. Hachaturov A. A. at all, *Dynamics of system: Road-Tire-Vehicle-Driver*. Mašinostrojenije, Moscow (in Russian), 1976.
[3]. Fiala E., *Mensch und Fahrzeug*, ATZ/MTZ Fachbuch, 2006.
[4]. Abe M. and Manning W., *Vehicle handling Dynamics*, Elsevier, 2009.
[5]. Ellis J. R., *Vehicle Dynamics*, Business Books, London, 1973.
[6]. Genta A., *Motor Vehicle Dynamics – Modeling and Simulation*, Singapore: World Scientific Publishing, 2003.
[7]. Gillespie T., *Fundamentals of vehicle Dynamics*, SAE, 1992.
[8]. Milliken W.: Milliken, D., *Race car dynamics*, SAE, 1995.
[9]. Mitschke M., *Dynamik der Kraftfahrzeuge*, Springer Verlag, 1972.
[10].Rajamani R., *Vehicle Dynamics and Control*. Springer Verlag, 2006.
[11].Simić, D., *Motor vehicle dynamics*, Naučna knjiga, Belgrade (in Serbian), 1980.
[12].Demić, M., *The Vehicle Oscillatory Systems Parametric Optimization*, monograph, Faculty of Mechanical Engineering from Kragujevac, 1997.
[13].Preleov V. Vl., *A fast method for measuring of the density of the road spectrum*, Bulletin of the Moscow University, Mat. Mech. Number 6, Pages 74-76, 1989.
[14].Demić M. at all, *The Road Surface Profile Investigation Perspective*, The 30' Session «Modern technologies in XXI Century», Bucharest, Romania, CD, p.p. 35-42, 2003.
[15].Demić M., *Dynamic excitation of vehicles*, Institute for nuclear sciences Vinča – Centre for engines and vehicles, 2006.
[16].ISO 8608, *Mechanical vibration – road surface profile – Reporting of measured data*, 1995.
[17].Cox D. R. at all, *The Theory of the Design of Experiments*, Chapman and Hall/CRC, Boca Raton, London, New York, Washington, 2000.
[18].Bendat J.S., Piersol AG, *Random Data Analysis and Measurement Procedures*. John Wiley and Sons, London, 2000.
[19].Demić M., Software-“Analsigdem”, www.ptt.rs/korisnici/i/m/imizm034, 2001.
[20].Vukadinovic S., Basic of Probabily and Statistics (in Serbian), Privredni pregled, 1973.
[21].OriginLab corporation, Origin 8.5, USA.

3D printing of hydrodynamic coupling for school education

V Šušteršič[1], V Vukašinović[1], D Gordić[1], M Josijević[1]

[1] University of Kragujevac, Faculty of Engineering, Sestre Janjić 6, 34000 Kragujevac, Serbia

E-mail: vanjas@kg.ac.rs

Abstract. The application of 3D printing technologies in the industry is growing as new possibilities for their application are discovered, which take advantage of their functionalities. Also, today more and more 3D technology is used for educational purposes. Incorporating of 3D printing in the form of production of three dimensional shapes into the teaching process improved the understanding of geometry.
Hydrodynamic couplings works on the hydrodynamic principle, which takes place based of the indirect principle of operation. The pump wheel transfers the introduced mechanical energy into the kinetic energy through the fluid flow. The fluid of higher energy flows centrically from the pump wheel to the turbine wheel, where re-conversion to mechanical energy takes place. In this paper, the description and principles of operation of hydrodynamic couplings are presented, as well as their basic characteristics and advantages. Also, the calculation of one hydrodynamic coupling and its 3D model that would be used in the education system are given.

1. Introduction

3D printing technology in developed countries has already entered the field of education, in primary and secondary school, as well as at the university, and has become a powerful tool for "intelligent manufacturing". The application of 3D printing at universities enables the acquisition of knowledge through the creation of 3D models during the creation of project tasks. Integration of 3D printing skills development into curricula is done through inclusion in existing courses. 3D printing is very important for learning at the university level. The application of 3D printing in education has several goals: training students and teachers about 3D printing, supporting new technologies in teaching, creating models that contribute to better learning, skill development and increased engagement of students and teachers in the subject. At technical faculties, almost all areas of engineering can be understood better through 3D printed models. Also, 3D printers in Universities have greater applicability and possibility for use in engineering and applied science [1]. Students can print models to simplify and understand complex theories and learn about new technology, while teachers can create models to teach subjects that are difficult to explain in 2D. Also, the equipment used for 3D printing within the teaching purposes is relatively cheap [2].

3D printing is the process of making 3D solid objects from a digital file by building up layer by layer of material. This type of manufacturing process has a many of advantages over traditional manufacturing: 3D printed designs do not become more expensive due to complexity, it is very cheap to customize the design, and shapes that are impossible to produce through manufacturing techniques can be printed [3]. Also, 3D printers are more environmentally sustainable than other manufacturing techniques because they use fewer raw materials, create fewer by-products, and represent a less energy-intensive process [4, 5].

The 3D printing process begins with loading the 3D printer with the appropriate material for printing, making sure that the material is selected to meet the needs of the printed object [2, 3].

2. Hydrodynamic coupling

There are two types of hydraulic power transmission, namely hydrostatic (displacement) and hydrodynamic (turbo transmissions). The advantage of hydraulic power transmission compared to others is found in small dimensions and low specific weight in relation to power, as well as in a small moment of inertia of moving parts. For example, a comparison of aircraft hydraulic transmissions with electric transmissions with an analogous purpose shows that the volume of the hydraulic motor and pump is about 12 % of the volume and the weight is 10 to 20% of the weight of the electric motor and generator.

Fluid couplings are designed in two types: constant-fill couplings and variable-speed (fill-controlled) couplings. Constant-fill couplings (figure 1) are mainly used for start-up (to limit torque) and to cushion the torsional vibration of the drive chain when variable-speed couplings are used to control or regulate the speed of the driven machine over a wide range below the drive speed [6].

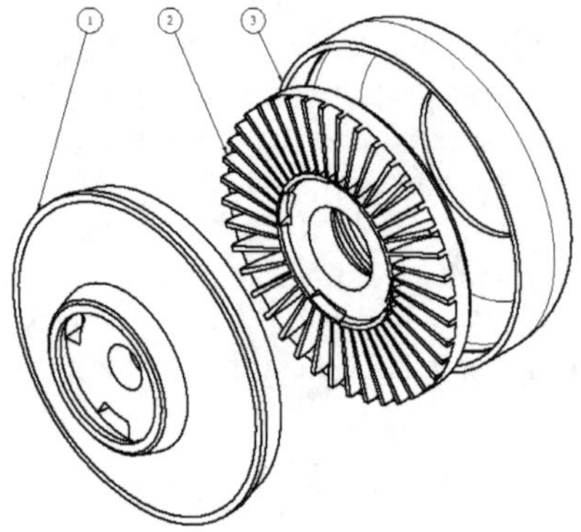

Figure 1. Hydrodynamic coupling assembly elements: 1 – pump wheel, 2 – turbine wheel, 3 – housing.

Hydrodynamic couplings work on the hydrodynamic principle. In drives where there are hydrodynamic couplings, there is no mechanical contact between the driving and driven machines, and the power is transmitted using fluids. Due to the mechanical separation between the driving and the driven machine, the hydrodynamic coupling enables the achievement of two separate acceleration values: a high acceleration value for the driving machine and, at the same time, a low acceleration value for the driven machine. Hydrodynamic couplings are often used to drive large inertial machines in combination with cage motors. They allow the engine to accelerate without load, and consequently, with increased oil filling, enable a soft/slightly quasi-stationary start of the machine [6].

The advantages of application of hydrodynamic couplings are:

- No load on the engine at start,
- Smooth start-up of driven machine,
- Avoid of shocks and overload protection,
- Dampens torsional vibration in the drive chain,
- Protects system components, thereby increasing service life,
- Power transmission is performed without wear; maintenance is easier,
- There is a possibility of load sharing in multi-motor drives [7].

IOP Conf. Series: Materials Science and Engineering 1271 (2022) 012002 doi:10.1088/1757-899X/1271/1/012002

3. Calculation of hydrodynamic coupling

The calculation of the geometric parameters of the working spaces of the hydrodynamic coupling is based on a one-dimensional flow model and the similarity theory. This method implies the averaging of current parameters by flow sections, taking as reference the values of these quantities in the meridian surface.

Its main task is to provide the possibility of designing a hydrodynamic coupling that meets the requirements of the design task, i.e. a coupling with pre-defined working characteristics. The project task defined the initial data used to calculate the geometrical parameters of the working areas of the coupling. These parameters are:

P_e - the power transmitted by the driving machine to the coupling input shaft [kW],
n_e - number of revolutions of the drive shaft [min^{-1}],
η - the desired degree of utility in the nominal mode of operation [/],
ρ - density of the working fluid [kg/m^3],
p - working pressure of the coupling [Pa].

The process of calculation is given in figure 2 and the calculation was done in the MathCAD software.

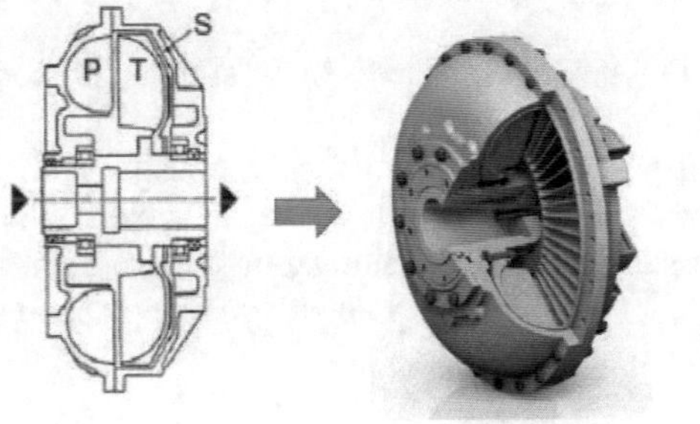

Euler's turbine equation

$$M = \dot{m} \cdot (r_e \cdot c_{u,e} - r_i \cdot c_{u,i}) \longrightarrow \dot{m} = f(V, v_n, \beta, v_{FL}), \quad Re = \frac{\omega_p \cdot D_a^2}{v_{FL}}$$

Hydrodynamic model equations

$$M_p = \lambda_M \cdot \rho \cdot D_a^5 \cdot \omega_p^2, \quad P_p = \lambda_P \cdot \rho \cdot D_a^5 \cdot \omega_p^3 \longrightarrow \lambda = f(V, v_n)$$

Figure 2. Calculation based on hydrodynamic principles, *adapted from* [8].

The calculation, 3D model and 3D printing were done at the Faculty of Engineering within the course Fluid Power Transmission. Initial and the calculated data are given in table 1. The data was adjusted to create 3D model.

Table 1. Input and calculated data.

Input data	
P_e (kW)	11
n_{emax} (min^{-1})	2770
ρ(kg/m^3)	860
Δt (K)	50
p (Pa)	$3.2 \cdot 10^5$
η (/)	0.97
Calculated data	
D_o (mm)	40
D_a (mm)	200

Q (m^3/s)	0.023
Y (J/kg)	588.6
z_p (/)	38
z_t (/)	40

After the calculation, the 3D model of the hydrodynamic coupling was created in the Autodesk Inventor software package, and after that, 3D print is accessed (figure 3). 3D printer supports .STL files which are obtained in any 3D modelling program.

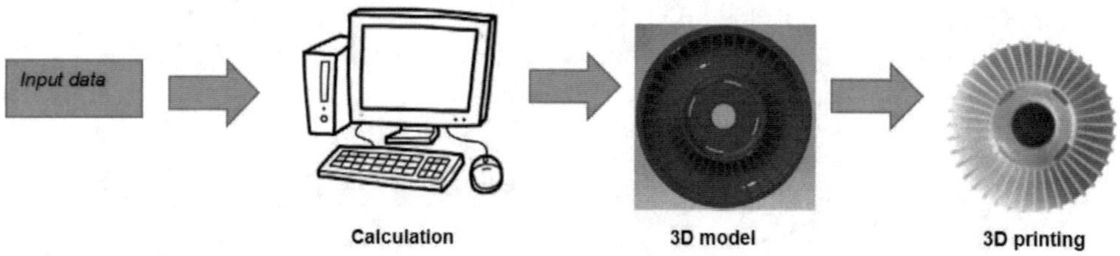

Figure 3. The process of creating a 3D model.

Figure 4 shows 3D model of pump and turbine wheels, housing and also assembly of hydrodynamic couplings, and figure 5 shows 3D printed models of the same elements. Before printing, it is necessary to set a number of parameters such as: plastic melting temperature, substrate temperature, printing speed, speed of movement of the printer head when passing from one place to another, percentage of filling of objects, support material etc.

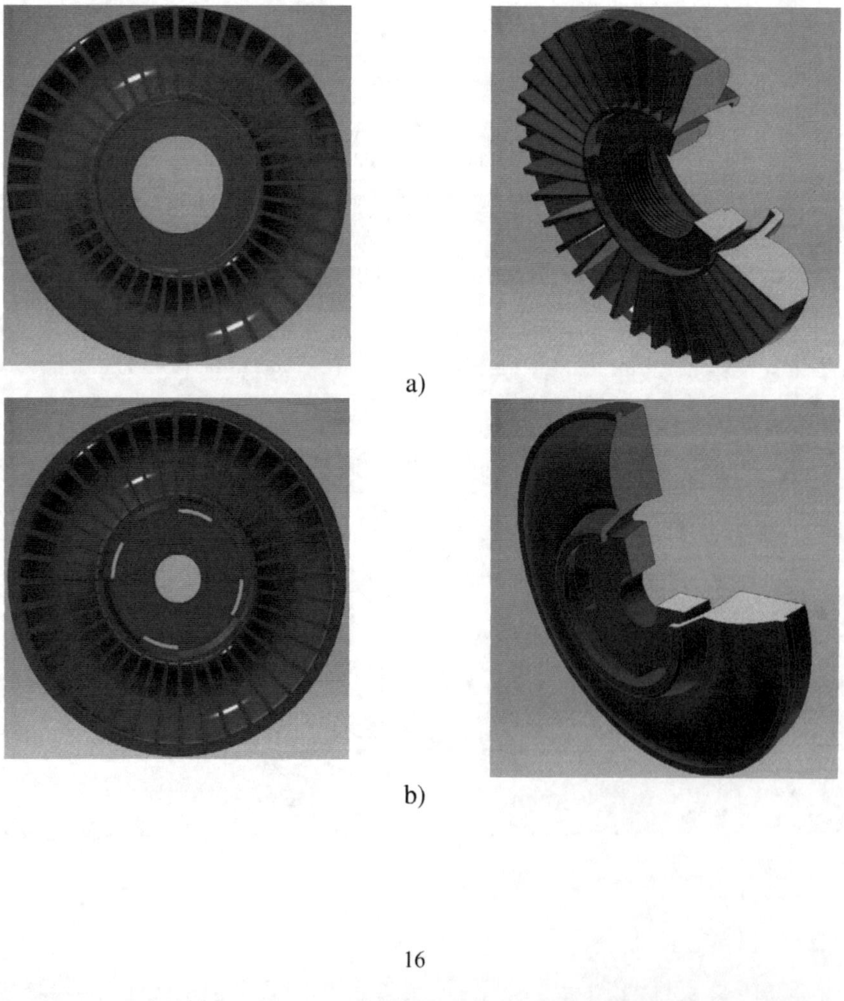

IOP Conf. Series: Materials Science and Engineering 1271 (2022) 012002 doi:10.1088/1757-899X/1271/1/012002

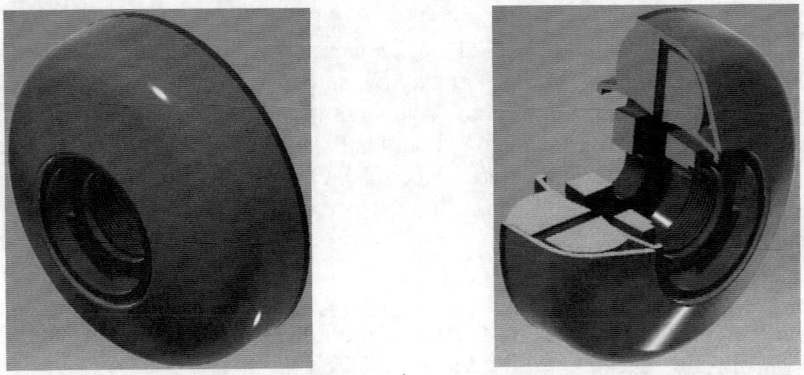

c)

Figure 4. 3D model of hydrodynamic coupling a) pump wheel, b) turbine wheel and c) assembly.

a)

b)

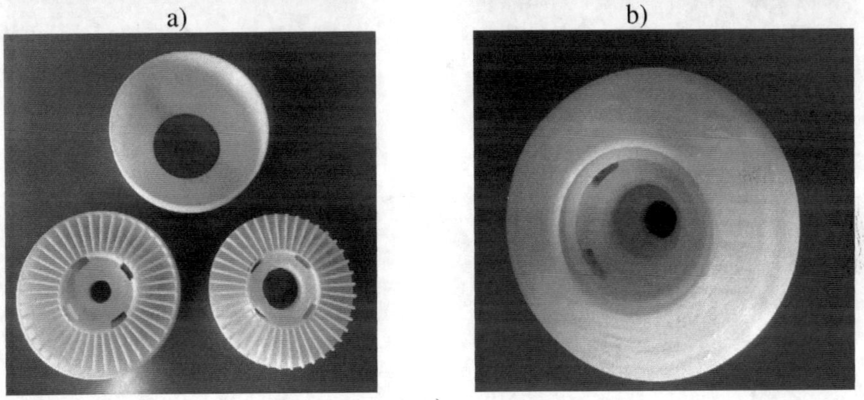

c)

Figure 5. 3D printing of hydrodynamic coupling a) pump wheel, b) turbine wheel and c) assembly.

4. Conclusion

Today, sustainable development is most often associated with raising awareness of the preservation of the environmental and the preservation of natural resources. 3D printing is often considered additive manufacturing (AM). It represents a manufacturing process of object through layer by layer with the selective fusion, sintering or polymerization of materials. The additive manufacturing process begins by taking a 3D computer generated file and slicing it into thin slices (typically ranging from 0.01mm to 0.25mm per slice depending on the technology used).

Today, 3D printers are used for various applications, from the auto industry, the food industry, the construction industry, for educational purposes etc. In education, 3D printing technologies facilitate learning, skills development, and increased engagement of student and teacher in the subject. Furthermore, 3D printing encourages greater creativity and collaboration in solving problems.

Hydrodynamic couplings are often used to drive large inertial machines in combination with cage motors. They allow the engine to accelerate without load, and consequently, with increased oil filling, enable a soft / slightly quasi-stationary start of the machine. As the hydrodynamic coupling enables quick acceleration of the motor and a short duration of high-value starting current, it results in an economical design.

References

[1] Jakovljević P, Dihovični Đ, Bijelić I, Kreculj D and Ratković Kovačević N 2022 Experiences in 3D printing applied in education *Struct. Integr. Life* **22** 1 43–7

[2] Ford S and Minshall T 2019 Where and how 3D printing is used in teaching and education *Addit. Manuf.* **25** 131-50

[3] Arvanitidi E, Drosos C, Theocharis E and Papoutsidakis M 2019 3D Printing and Education *Int. J. Comput. Appl.* **177** 24 55 - 9

[4] Diegel O, Singamneni S, Reay S and Withell A 2010 Tools for sustainable product design: additive manufacturing *J. Sustain. Dev*. **3** 3 68–75

[5] Gebler M, Schoot Uiterkamp A J M and Visser C 2014 A global sustainability perspective on 3D printing technologies *Energy Policy* **74** 158–167

[6] Shah K. P 2018 Construction, Working, Operation and Maintenance of Fluid Couplings https://practicalmaintenance.net/wp-content/uploads/Construction-Working-Operation-and-Maintenance-of-Fluid-Couplings.pdf (access 27.06.2022.)

[7] Voith Hydrodynamic coupling: Principles, Features, Benefits https://d2euiryrvxi8z1.cloudfront.net/asset/445934742530/d810385a0e1d0df78e04218fac8441f3 (access 27.06.2022.)

[8] Höller H, Selection of Operating Fluids for Hydrodynamic Power Transmitting Equipment Voith Turbo, http://voith.com/cn/297_e_ca_cr593_en.pdf (access 27.06.2022.)

Analysis of powertrain's workload during the turning process of a high-speed tracked vehicle

Luka Ponorac[1*], **Ivan Blagojević**[2] **and Aleksandar Grkić**[1]

[1] AMSS-Center for Motor Vehicles Ltd., Kneginje Zorke 58, 11000 Belgrade
[2] Faculty of Mechanical Engineering, University of Belgrade, Kraljice Marije 16, 11000 Belgrade

[*]luka.ponorac@amss-cmv.co.rs

Abstract. In order to analyze the efficiency of a high-speed tracked vehicle powertrain, it is necessary to understand the powertrain workloads in specific working regimes. Experimental tests are conducted on a specific high-speed tracked vehicle, in various terrain conditions, in order to obtain accurate workload data and isolate the maximum workload regimes. From the power balance point of view, the most demanding working regime is the vehicle turning process, so corresponding test results are analyzed with special attention. The obtained results indicate that the slip of the turning mechanism friction elements greatly influences the character of the power balance change.

1. Introduction

High-speed tracked vehicles have a great advantage over wheeled vehicles when it comes to overcoming hard condition terrains. In order to maintain high mobility in low adhesion and high moving resistance conditions, these vehicles have robust powertrain components and complex turning mechanisms, which can withstand high torque, while producing a respectful velocity at the same time [1]. This means that the power source, IC engine, has to produce enough power to satisfy all of these requests. Unlike with wheeled vehicles, the selection of the IC engine for a high-speed tracked vehicle is calculated according to the power requirements for the lowest turning radius achieved with the highest vehicle velocity, turning towards the hill of a maximum gradient [2], [3]. Such turning scenario is reflected by extremely high turning resistance, significantly higher than the moving resistance of the rectilinear movement. Given that, selection and design of the adequate powertrain and turning mechanism is no easy task.

When it comes to mechanical turning mechanisms with friction elements, the powertrain efficiency greatly depends on the type of the turning mechanism. For example, asymmetric turning mechanisms provide better power balance during the turning process and have less power loss due to slip than symmetric turning mechanisms. On the other hand, during the turning process, the velocity of the vehicle with asymmetric turning mechanism is reduced, thus interrupting general vehicle performances, while the vehicle with symmetric turning mechanism maintains the velocity of mass center, increasing the velocity of the outer track and reducing the velocity of the inner track [4].

Experimental testing of the high-speed tracked vehicle powertrain in dynamic conditions requires great resources, while the repeatability of the results and testing conditions can be hard to achieve [5]. Nevertheless, this kind of testing is inevitable in order to acquire precise data of the complex processes which are the result of numerous variable factors and phenomena. Having an accurate image of the high-speed tracked vehicle power balance for every moving scenario gives us the possibility to analyze the powertrain efficiency, which can then be the subject of optimization, modernization and further improvement. Also, data collected through experimental testing is essential for simulation model development, which significantly reduces the required resources for further vehicle testing and development.

2. Methods

2.1. Powertrain construction

The analysed vehicle powertrain is an integrated powertrain, consisting of the conventional, five speed gearbox and an asymmetric turning mechanism, shown in figure 1.

Figure 1. Subject vehicle powertrain construction

Asymmetric turning mechanisms provide the vehicle turning by reducing the velocity of the inner track and the vehicle mass center, while the outer track maintains the velocity of the rectilinear movement. In the analysed powertrain, the turning mechanism consists of the summarizing planetary gear sets and friction clutches and brakes.

The analysed powertrain has two power flows: main power flow and auxiliary power flow shown in figure 2 [6].

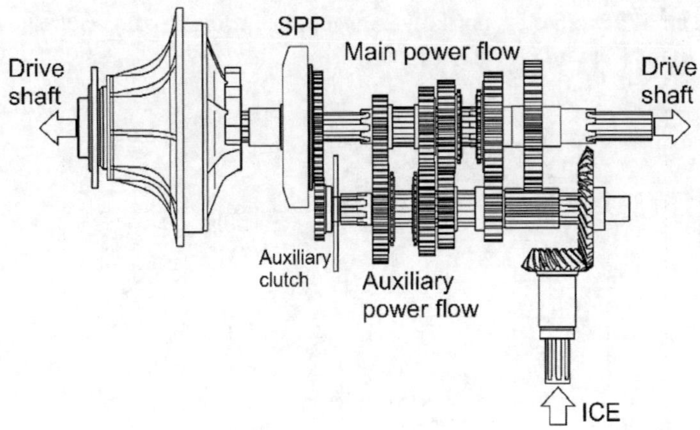

Figure 2. Powertrain conception

The main power flow permanently transfers power parameters from the IC engine and main friction clutch, through the gearbox, summarizing planetary gear sets (*SPP,* figure 2), and to the drive wheels and tracks. During the rectilinear movement, the summarizing planetary gear sets, on both output shafts, act as basic reducers and transfer power from the gearbox to the drive wheels and tracks. Auxiliary power flow is activated only during the turning process. A portion of the IC engine power is transferred through the auxiliary power flow to one of the summarizing planetary gear sets, by means of *auxiliary clutch* (figure 2). During the turning process, the power from the main and auxiliary power flow is summarized in one of the summarizing planetary gear sets. The power from the main power flow is delivered to the planet carrier, while the power from the auxiliary power flow is delivered to the sun gear of the planetary gear set. This results in ring gear speed reduction, which is connected to the powertrain output shaft, thus reducing the velocity of one track (inner track) which causes the vehicle to turn.

The state of the auxiliary clutch is of great importance for the vehicle turning kinematics. When the auxiliary clutch is fully activated, the vehicle is turning with a defined turning radius, frequently referred to as the *calculated turning radius* [7]. This means that the power from the IC engine is fully transferred through the auxiliary clutch, and the turning radius can be calculated, whatever the powertrain input values are. Partially activating the auxiliary clutch results in clutch slip, so a portion of the IC engine power, transferred through the auxiliary power flow, is lost.

Figure 3 shows the kinematic analysis of the difference between the velocity of the planetary gear components during the turning process with fully engaged auxiliary clutch and with partially engaged auxiliary clutch. The angular velocities of the sun gear, ring gear, and planet carrier are labeled ω_a, ω_b and ω_H, respectfully, while their peripheral velocities are labeled V_a, V_b and V_H. The values with the index "*s*" represent the same values, for the same powertrain input, when the auxiliary clutch is partially engaged (clutch slip). When the auxiliary clutch is fully activated, the sun gear angular velocity ω_a increases to the value proportional to the ICE rpm. Given the fact that the input speed from the gearbox ω_H is constant, it is obvious that the output speed ω_b is decreasing. It can easily be concluded that, when the auxiliary clutch is fully activated, the angular velocity of the ring gear, thus the output shaft, has the lowest value of all the planetary gear components.

When the auxiliary clutch is partially engaged, the input angular velocity is not fully transferred to the sun gear of the planetary gear set, due to clutch slip. This results in lower sun gear speed, so the braking effect on the ring gear, thus the output shaft, is reduced to minimum. It can be concluded that, the greater the sun gear speed, the lower the output speed [8]. This means that the vehicle turning radius is lower when the auxiliary clutch fully transfers power parameters via auxiliary power flow. On the other hand, when the auxiliary clutch is partially engaged, the turning radius is greater, and the power is transferred with power loss, due to clutch slip.

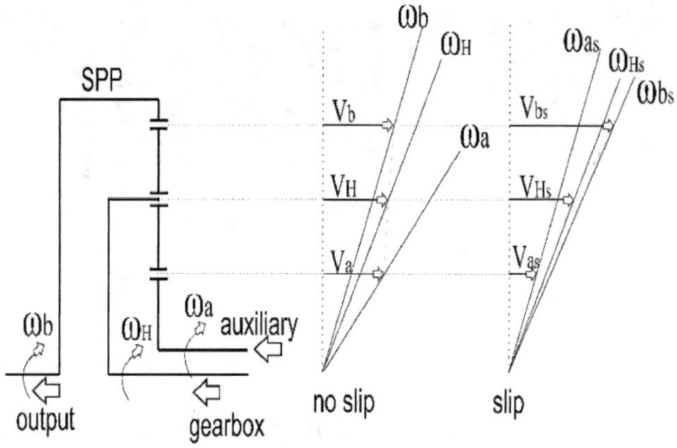

Figure 3. Sumarizing planetary gear set kinematic scheme-slip vs no slip

When the analyzed vehicle is turning with auxiliary clutch activated, a useful circulating power appears in the powertrain, also referred to as *recuperation power* [9]. The recuperation power appears when both of the following conditions are met: the drive wheels and tracks are mutually kinematically connected and a "negative" force (braking force) is acting on the inner track. When both of these conditions are met, the power in the powertrain is distributed in such a way, that a certain portion of power from the inner track is transmitted, through the powertrain, to the outer track, thus decreasing the engine power needed for the vehicle turning process, as shown in figure 4.

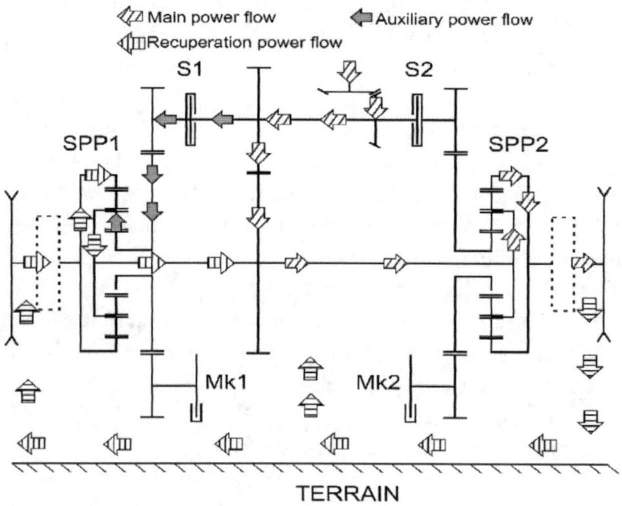

Figure 4. Powertrain turning process kinematic scheme

2.2. *Experimental testing*

The vehicle is driven on soft soil and in dry conditions, varying turning and rectilinear moving scenarios, with gear and vehicle velocity changes. In order to analyse the power losses during the turning mechanism slip, a scenario where the vehicle is driven with a relatively constant engine input velocity, in specific gear, with field condition change reduced to minimum is selected, varying only the activation pressure of the auxiliary clutch.

IOP Conf. Series: Materials Science and Engineering 1271 (2022) 012003 doi:10.1088/1757-899X/1271/1/012003

For the analysed part of the test, the vehicle was driven in second gear, with a relatively constant IC engine speed of approximately 800 rpm. For the first part of the test, by engaging the turning mechanism command, the auxiliary clutch activation pressure is increased to the maximum value of 6,5 bar. The vehicle is now turning with a calculated radius, with auxiliary clutch fully engaged. For the second part of the test, the auxiliary clutch was partially engaged with the activation pressure of 2 bar, which was, after one full turn of the vehicle, increased to the value of 4 bar. The results obtained during the first part of the experimental test, with the auxiliary clutch fully activated, are compared with the results obtained during the second part of the experimental test, while the auxiliary clutch was partially activated, thus transferring only a portion of the engine power, due to auxiliary clutch slip.

2.3. Measuring system

In order to determine the powertrain efficiency and power losses due to auxiliary clutch slip, an accurate measuring system had to be established. Since the measuring of the main parameters is done mostly on rotational parts of the powertrain, wireless transducers are used for signal transfer.

Torque and angular velocity are the main measured values, from which the power is derived from. Torque is measured with strain gauges on powertrain output shafts, after the summarizing planetary gear sets, as shown in figure 5. Wireless signal transducer is connected to each strain gauge, providing the gauges the power supply, and transmits measured signal to the signal receiver. Signal receiver is connected to the data acquisition unit, while the measured signal is displayed in data acquisition software. Measured torque is a significant value for perceiving the resisting forces during the turning process.

Figure 5. Position of strain gauges for torque measurement

Angular velocity is measured on powertrain input shaft, as well as on both output shafts. Angular velocity is measured by inductive gauges in positions shown in figure 6. Input angular velocity is measured on the powertrain entrance, while the output angular velocities are measured on the output shafts, which connect the ring gears of the summarizing planetary gear sets and the track drive wheels. By obtaining the angular velocity values, intensity of the auxiliary clutch slip can be perceived.

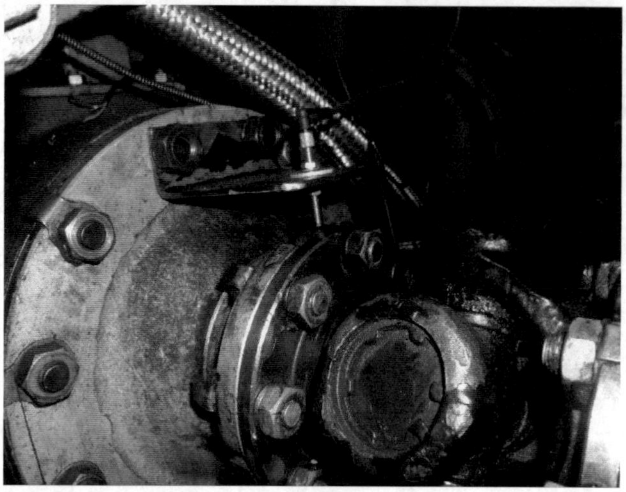

Figure 6. Position of rpm sensor

For the purpose of obtaining information about auxiliary clutch state, a pressure gauge is installed in the hydraulic control block, which controls the activation of the turning mechanism friction elements, as shown in figure 7. Knowing the activation pressure of the auxiliary clutch, it can easily be determined if the auxiliary clutch is fully or partially activated, thus perceiving the vehicle moving regime.

Figure 7. Pressure sensor position

The data acquired from the gauges is collected and processed in the QuantumX acquisition unit, while the results are analyzed in the Catman software.

3. Results

The first part of the experiment, where the auxiliary clutch is fully activated, increasing the activation pressure to the maximum of approximately 6,5 bar at $t \approx 5\ s$, is represented by pS1 in figure 8. For the second part of the experiment, the auxiliary clutch is partially activated, with a low activation pressure of 2 bar, which was increased to the value of approximately 4 bar at $t \approx 22\ s$, as represented by pS1s in figure 8.

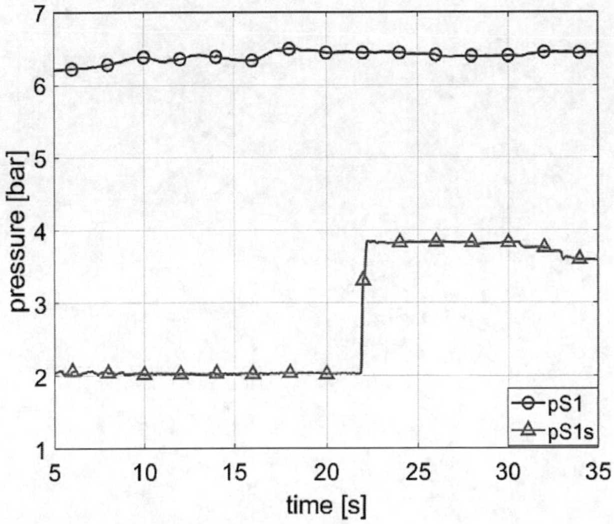

Figure 8. Auxiliary clutch activation pressure-slip vs no slip

The influence of the activation pressure difference on the vehicle turning process is primarily analyzed through turning kinematics. Powertrain input and output velocities, for both slip and no slip turning scenario, is shown in figure 9. The value n_{ul}, represents the powertrain input rpm with the auxiliary clutch fully activated, while the values n_1 and n_2 represent the rpm of the inner and outer track drive shaft, respectively. Values n_{uls}, n_{1s} and n_{2s}, represent the same values, but with the auxiliary clutch partially activated

The vehicle is driven at relatively constant powertrain input speed of approximately 800 rpm, in both scenarios. Since the auxiliary clutch on the outer track is disengaged, and the outer track velocity maintains the velocity of rectilinear movement, the outer track velocity for both fully and partially activated auxiliary clutch is of the relatively same value. The discrepancies between n_2 and n_{2s} come only from the incapability of the driver to maintain the exact same IC engine rpm during both experiments. Even though, it is clear from the vehicle construction and the experimental test results that auxiliary clutch slip does not affect the velocity of the outer track.

The impact of the auxiliary clutch slip is obvious when the inner track velocities are analyzed. For the time interval of $t \approx 5 \div 22\ s$, the value n_{1s} is greater than n_1. This is because, when the auxiliary clutch is partially activated, the IC engine velocity is not fully transferred to the sun gear of the planetary gear set, so the velocity of the ring gear, thus the velocity of the inner track, is not reduced as it would be if the auxiliary clutch was fully activated. Since the inner track velocity is greater in this case, so is the vehicle turning radius. When the activation pressure is increased to the value of approximately 4 bar, the auxiliary clutch is almost fully activated, so the curves n_{1s} and n_1 are almost overlapped.

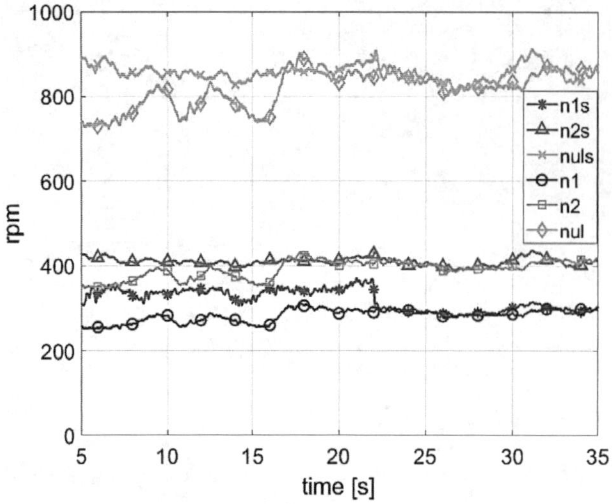

Figure 9. Input and output rpm-slip vs no slip

Dynamic analysis of the turning process in these two scenarios, shown in figure 10, is a little more complicated, because of the adhesion conditions of the track-soil contact. Values M_2 and M_1 represent the measured torque on outer and inner powertrain output shafts, respectively, while the auxiliary clutch was fully activated. M_{2s} and M_{1s} represent the same values, but with auxiliary clutch slip.

For the initial period of $t \approx 5 \div 22\ s$, the measured torque on the outer track is greater when the auxiliary clutch is fully activated then when it is partially activated. Fully activated auxiliary clutch means smaller turning radius, but greater turning resistance [10], thus greater measured torque on the outer track.

The same principal can be observed on the inner track. For the interval of approximately $t \approx 5 \div 22\ s$, the values of M_1 are greater than M_{1s}, as it is expected. When the auxiliary clutch pressure is increased at 4 bar, and the auxiliary clutch was almost fully activated, the measured torque values with the clutch slip are almost overlapping with the same measured values without slipping, which is expected due to turning radius decrease, thus the turning resistance increase.

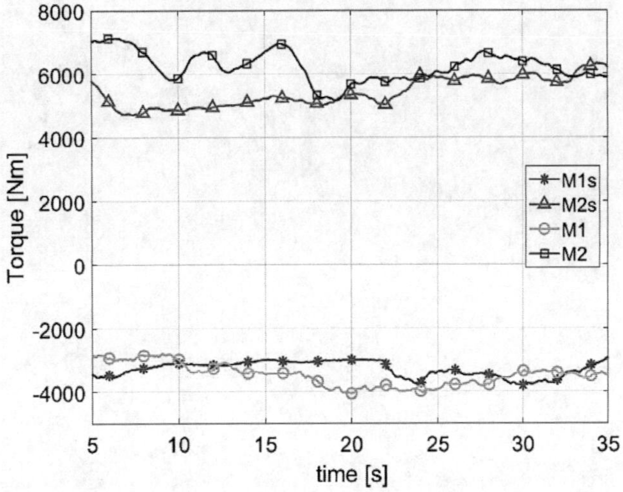

Figure 10. Powertrain output torque-slip vs no slip

In order to assess the efficiency of the turning process, it is useful to analyze the powertrain output power. The authors choose to analyze the power delivered on the outer track, given the fact that it is the sum of the ICE power and the recuperation power, coming from the inner track, as shown in equation (1) [11].

$$P_2 = P_{mz} \cdot \eta_t + P_1 \cdot \eta_p - P_s \cdot \eta_t \tag{1}$$

where

- P_2, P_1-power delivered on the outer and inner track
- P_{mz}-IC engine power required for the turning process
- P_s-power lost due to auxiliary clutch slip
- η_t, η_p-losses in powertrain mechanical components

Figure 11 shows the power delivered on the outer track drive shaft P_2 with auxiliary clutch fully engaged, while P_{2s} represents the same power with auxiliary clutch partially engaged. For the time interval of $t \approx 5 \div 22\ s$, the power values P_{2s} are significantly lower than P_2 because of the increased power loss due to auxiliary clutch slip. This confirms the theory behind the equation (1), where it is obvious that the greater the power loss P_s in the powertrain, less power is delivered to the outer track. When the auxiliary clutch activation pressure is increased to the value of 4 bar at $t \approx 22 \div 35\ s$, the power loss on the auxiliary clutch decreases, thus the power delivered on the outer track increases, tending to reach the value of the power delivered while the auxiliary clutch is fully engaged.

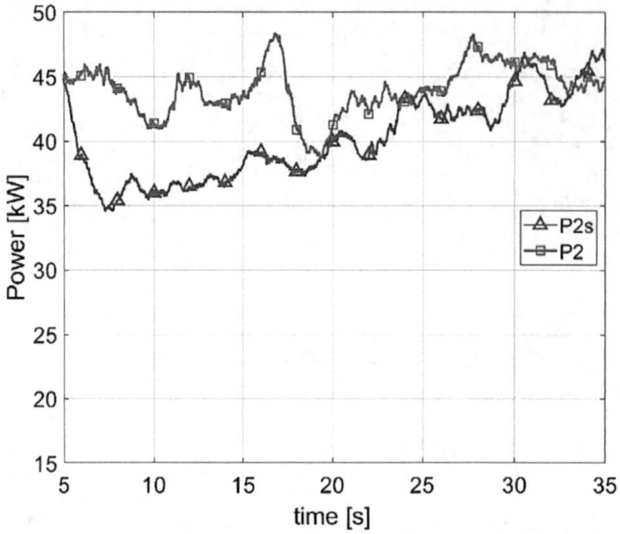

Figure 11. Power delivered on the outer track-slip vs no slip

The part of the equation (1) defined by the term $P_1 \cdot \eta_p - P_s \cdot \eta_t$ actually represents the recuperation power [12]. This is basically the circulating power useful to the turning process, which increases the power delivered to the outer track from the IC engine, that is, the power that would have to be compensated by the IC engine if there was no power recuperation. The difference between the two powers displayed in figure 11 is actually the difference in recuperation power value, which is clearly influenced by auxiliary clutch slip.

In order to quantify the power loss with and without auxiliary clutch slip, the power result values are expressed as a relative error shown in figure 12. Representing the power loss through a relative error, it can be observed that, during highest value of auxiliary clutch slip, up to approximately 25% less power is delivered to the outer track. When the activation pressure is increased to the value of 4 bar, this difference is significantly reduced, but still approximately 10% less power is delivered to the outer track, comparing to when the auxiliary clutch is fully engaged.

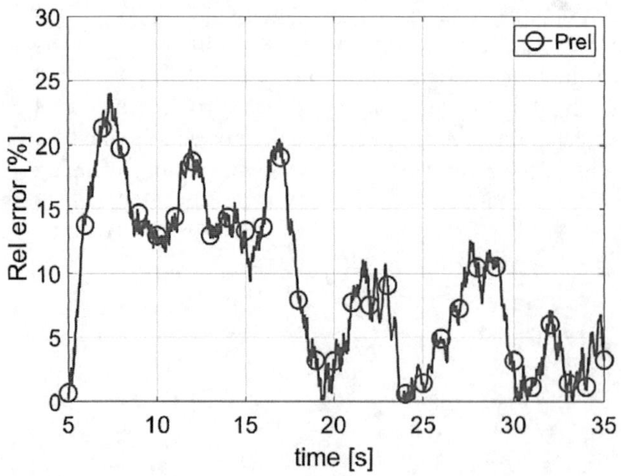

Figure 12. Power loss expressed through relative error

4. Conclusion

The turning process of a high-speed tracked vehicle is evidently the most complex moving scenario, influenced by a large number of variables such as vehicle mass and dimensions, construction of the powertrain and turning mechanism, vehicle velocity, engagement state of the powertrain components, adhesion conditions, moving resistance etc. With all these variables, dynamic analysis, such as one presented in this paper, is no easy task, but is necessary in order to assess the efficiency of the vehicle powertrain. One of the most efficient ways of analyzing complex powertrains, such as the one in this paper, is to analyze the power flow and power requirements for the most demanding working regime.

Analyzing the power delivered on the subject vehicle outer track, the authors conclude that, when the auxiliary clutch is partially activated, the power on the outer track is approximately up to 25% lower, meaning that 25% of the IC engine power is lost due to auxiliary clutch slip. This means that, in order to achieve the turning parameters as when the auxiliary clutch is fully engaged, the IC engine has to deliver more power in order to compensate this power loss.

It is obvious that, the lower the activation pressure of the auxiliary clutch, the greater the power loss due to slip, so the logical conclusion is to avoid these working regimes and turn the vehicle with auxiliary clutch fully activated. Unfortunately, these vehicles are most commonly driven in such working regimes, where the vehicle is slightly turned just for trajectory correction. Rarely is the vehicle turned with the lowest values of the turning radius (calculated radius). This means that, for the most of the vehicle usage, the vehicle is turned with great power losses in turning mechanism friction elements.

The problem of power loss on friction elements can be overcome by modifying the powertrain, in such a way, that the existing turning mechanisms with friction elements are replaced with hydrostatic or electric components, which would provide a continuous trajectory change, with an infinite number of calculated turning radii and no power loss.

References

[1] Hryciów Z and Rybak P 2019 Numerical research of the high-speed military vehicle track *AIP Conf. Proc. (Jora Wielka, Poland)* vol 2078 doi: 10.1063/1.5092032.

[2] Zhang Y, Mianhao Q, Xixia L, Li J, Song H, Zhai Y and Hu H 2021 Research on characteristics of tracked vehicle steering on slope *Math. Probl in Eng.* 2021 doi: 10.1155/2021/3592902.

[3] Lazarević D and Popović Z 1997 Specificities in research of the load regimes of fast tracked vehicles *Vojnoteh. Glas.* 45 300–312 doi: 10.5937/VojTehG9703300L.

[4] Vesic M and Muždeka S 2007 Analysis of influence of turning system kinematic scheme on turning power balance for high speed tracked vehicles *Vojnoteh. Glas.* 55 149–168 doi: 10.5937/vojtehg0702149V.

[5] Janarthanan B, Padmanabhan C and Sujatha C 2012 Longitudinal dynamics of a tracked vehicle: simulation and experiment *J. of Terramech.* 49 63–72, doi: https://doi.org/10.1016/j.jterra.2011.11.001.

[6] Muždeka S and Perić S 2012 *Osnovi borbenih vozila, praktikum za vežbe* (Beograd:Medija centar Odbrana).

[7] Stojkovic V and Mikulić D 2002 The impact of a fixed kinematic turning radius of a tracked vehicle on the engine power required in a turn *Stroj. Vestnik* 48 459–466.

[8] Muždeka S 2005 Strukturna analiza složenih planetarnih prenosnika snage / Structural analysis of compound planetary gear trains *Vojnoteh. Glas.* 53 doi: 10.5937/vojtehg0502178M.

[9] Ogorkiewicz R 1991 *Technology of Tanks* (Couldson, Surrey: Jane's Information Group).

[10] Taratorkin I, Derzhanskii V and Taratorkin A 2016 Experimental determination of kinematic and power parameters at the tracked vehicle turning *Procedia Eng.* 150 1368–1377 doi: https://doi.org/10.1016/j.proeng.2016.07.331.

[11] Nikitin A O and Sergeev L V 1962 *Teoriya tanka* (Moscow: Voennaja akademija bronetankovyh vojsk).

[12] Muždeka S 2012 *Osnovi borbenih vozila, udžbenik* (Beograd: Medija centar Odbrana).

The influence of the fuel stratification on the IC engine working cycle with the compressed natural gas

I Grujic[1], A Davinic[1], N Stojanovic[1], J Doric[2], S Narayan[3]

[1] University of Kragujevac, Faculty of Engineering, 6 Sestre Janjić Str., 34000 Kragujevac, Serbia
[2] University of Novi Sad, Faculty of Technical Sciences, 21000 Novi Sad, Trg Dositeja Obradovica 6, Serbia
[3] Qassim University, 52571 Buraydah, Saudi Arabia

E-mail: ivan.grujic@kg.ac.rs

Abstract. The gaseous fuels are very convenient from the economic aspect as well as from the ecology aspect. Fuel stratification allows the engine work with the globally lean mixture and in this way increase the engine efficiency and reduces the pump losses. The separation of the engine working space, on the area where the mixture will be always rich, and on the area where the mixture will be always lean, allows the spark ignition engine work with the globally lean mixture. In the paper it was investigated, how the fuel stratification ratio influences on the combustion process of the compressed natural gas in the spark ignition engine. It was found that the stratification ratio has a great influence on the combustion process, as well as on the engine performances and the efficiency. Based on the research it was identified the optimal combustion process from the aspect of the performances and the efficiency.

1. Introduction

The air pollution is global problem which negatively influences on the environment and people health. Traffic contributes with 25% to greenhouse gas emission, and with this it takes the second place, right after the energy sector [1]. If observe the CO2, the 95% originates from the road traffic, from which the 61% originates from passenger cars [2]. The new produced vehicles have less fuel consumption, and because of this emit less harmful combustion products. However, the great number of people still drive older vehicles, and because of this it is necessary to focus on the measures which will provide the reduction of the harmful products from the combustion process, and one of the measures can be the adaptation of vehicles, to use gas fuels [3].

In Italy, in period between 2005 and 2010 was noticed the reduction of PM, NOx and VOC emissions because of the usage of compressed natural gas (CNG) instead of the gasoline [4].Vehicles that use the CNG have less emission of harmful products compare to the vehicles that use conventional fuels, which leads to the significant improvement of the air quality in the cities [5]. The experimental research conducted by Aosaf et al. [6] for vehicles with four different fuels at idle, show that the highest emission of CO and NO has the vehicle which use diesel, while the highest emission of HC has the vehicle with CNG because of the slow combustion, Figure 1.

IOP Conf. Series: Materials Science and Engineering 1271 (2022) 012004 doi:10.1088/1757-899X/1271/1/012004

How the CNG can be considered as one of the fuels of the future, Garg and Ravikrishna [7] investigated how the mixture stratification influences on the emission and have show that the emission of the NOx reduces for 30% for the same load and engine speed. The same conclusions have come and from the research of Reynolds et al. [8]. The mixture stratification provides the possibility of engine work with globally lean mixture, and besides that it increases the efficiency. Besides that, the mixture stratification reduces the emission of HC and CO2. However, it is important to not violate the engine performances, and this was investigated by Yamato et al. [9], and they found that if the stratification is correct, it will not come to the engine performances violation, and to the same conclusions have come and the Salehi et al. [10].

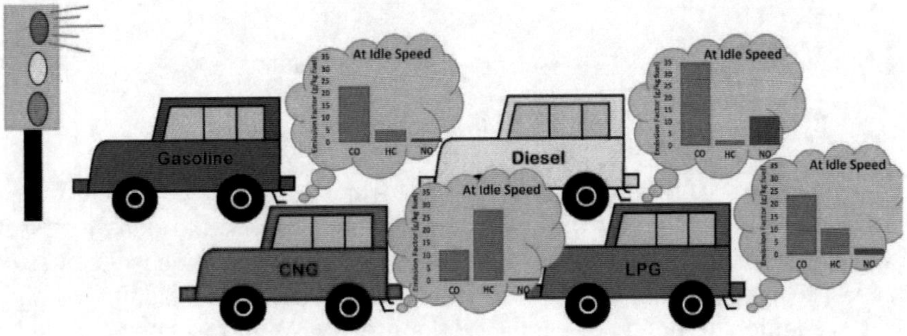

Figure 1. The emission for different fuels [6].

The main of this paper was to investigate how the mixture stratification influences on the combustion process as the main parameter of the working cycle, from which depend performances, efficiency and raw emission.

2. Experimental investigation

The experimental investigation of the IC engine working cycle was conducted on the experimental test engine, in the laboratory for the IC engines and fuels, at the Faculty of engineering. The used test engine, in its basic conception is a one-cylinder air cooled diesel engine, which characteristics are given in Table 1.

Table 1. Test engine characteristics.

Engine displacement, cm^3	454
Piston diameter/stroke, mm	85/80
Compression ratio, -	17,5
Valve timing	Inlet valve opens 16° BTDC
	Inlet valve closes 40° ABTC
	Exhaust valve opens 40° BBDC
	Exhaust valve closes 16° ATDC
Engine power/engine speed, kW/rpm	7,3/3000
Engine torque/engine speed, Nm/rpm	28/1700

In order to conduct the experimental investigation of the IC engine working cycle with the CNG (Compressed Natural Gas), the existing experimental installation it was upgraded with the gas installation for the IC engine supply, in which is also implemented and the measurement of the fuel consumption. The scheme of the gas installation is shown on the Figure 2.

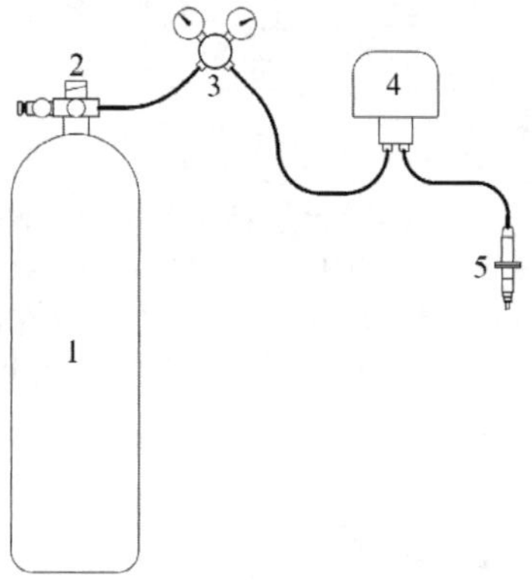

Figure 2. Gas installation [11].

The gas installation consists from the gas reservoir 1, which is equipped with electromagnetic valve 2. After that it goes the pressure regulator 3, which allows injection with different pressures. The fuel consumption was measured by flowmeter 4. At the end of the gas installation is the injector mounted in the prechamber 5. The main idea was to achieve the mixture stratification by addition of the prechamber. The prechamber is additional volume of engine working space. With this aproach, the mixture in the prechamber will be always rich, while the mixture in the main space will be always lean. The scheme of the engine reconstruction, by the prechamber addition is shown on the Figure 3.

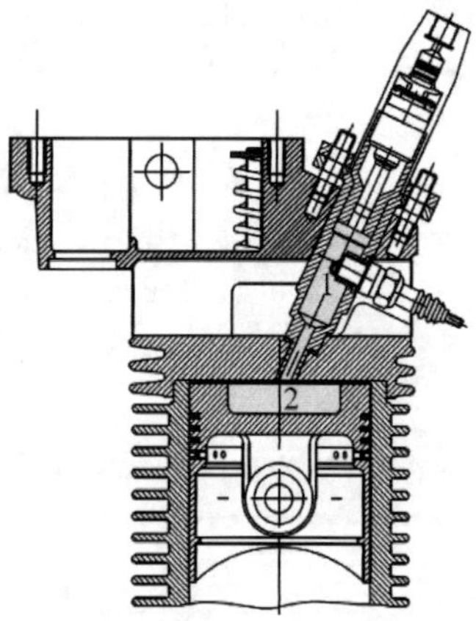

Figure 3. The engine working space [11].

By addition of the prechamber, it was achieved engine work with fully open throttle, and by this by work with globally lean mixture. This is good because provides, the reduction of pump losses, and the increment of the efficiency.

3. Results and discussion

In order to see how the fuel stratification influences on the IC engine work, it was created a mathematical model based on the ideal gas law, equations (1) and (2).

$$p_p \cdot V_p = m_p \cdot R_p \cdot T_p \tag{1}$$

$$p_{ms} \cdot V_{ms} = m_{ms} \cdot R_{ms} \cdot T_{ms} \tag{2}$$

The equation one is the ideal gas law for the prechamber, while the equation 2 is the ideal gas law for the main space. How the amount of the fuel which stayed into the prechamber and which exit to the main space, cannot be measured, the calculation was the only way. It should say, that the mathematical model is based on the assumptions, and simplifications. The calculation was conducted for two moments of the working cycle. The first moment is the moment when closes the inlet valve, or when the injection is finished, all depend, what happens latter. Second moment of calculation is the moment when the ignition is defined. In these calculation moments, the main space and prechamber were observed as two separate volumes, with the assumption that the pressures in the main space and prechamber are equalized. According to the mathematical model, were calculated the mass of air, and CNG, in main space and prechamber, and thanks to this are calculated the global air-fuel ratio, air-fuel ratio for the main space and air-fuel ratio for the prechamber (Table 2).

Table 2. Air fuel ratio for different working regimes.

Engine speed, rpm	Engine load, %	Global air-fuel ratio	Main space air-fuel ratio	Prechamber air-fuel ratio
1960	25%	3,35	6,24	1,04
1960	50%	2,79	4,63	0,97
1960	75%	2,24	3,23	0,91
1960	100%	1,86	2,42	0,80

The combustion process for four regimes which were investigated is given on Figure 4.

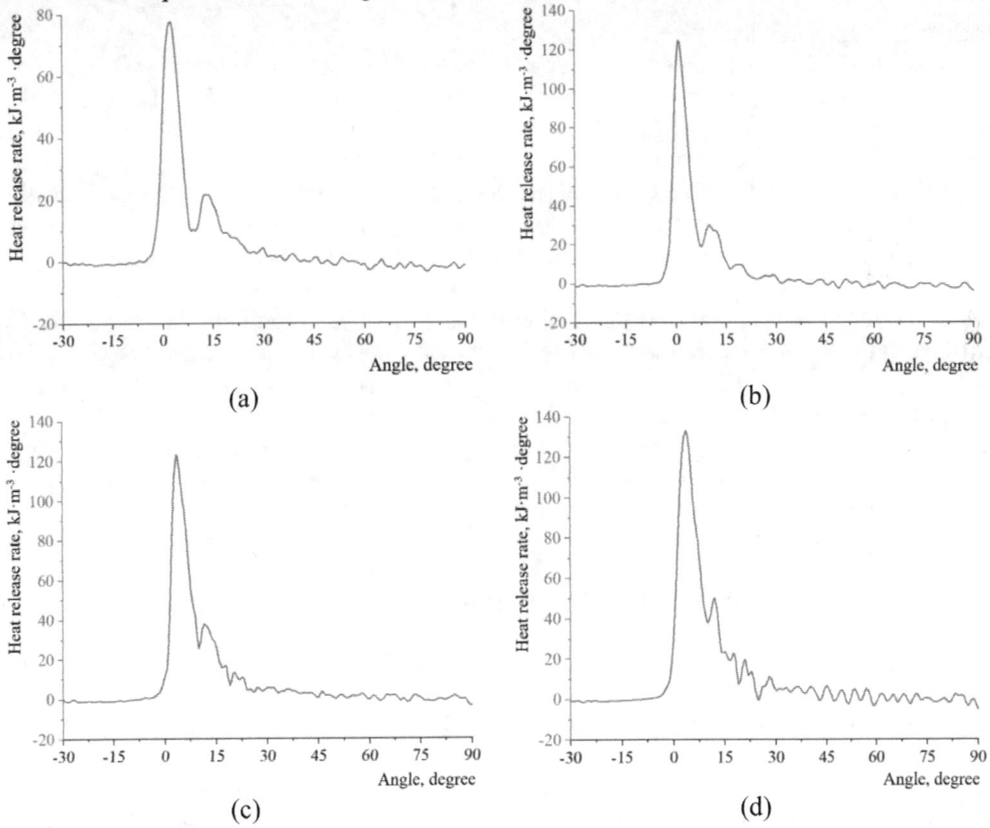

Figure 4. Combustion law for (a) 25% of engine load, (b) 50% of engine load, (c) 75% of engine load and (d) 100% of engine load.

From the combustion law it can be seen the different behaviour between the investigated regimes. The first part of the combustion (the highest intensity combustion) is highly related to the combustion which happened into the prechamber, where the mixture is ignited. That is, this part of the combustion illustrates the flow of gases from the prechamber to the main space.

Interesting part of the combustion, which is related to the fuel stratification, is the second part of the combustion which looks like the late combustion at diesel engine. This part of the combustion, is actually the combustion which continues from the prechamber into the main space. It can be seen that in this part of the combustion appears in some cases specific shape of the curve which reminds to the hump. This hump is more expressed on the regimes with lower load of the engine. Actually, this is highly connected with the air-fuel ratio into the main space. If the mixture is very lean in the main space (main space air-fuel ratio over the 2.5), combustion process almost stops in on moment and then continues. The very lean mixture in the main space is specific for the regimes with low load. With the load increment, the mentioned hump is less expressed. The important parameter which can be observed, and from which can lead some conclusions what is best is the indicating efficiency for these regimes, given in Table 3.

Table 3. Indicating efficiency for different working regimes.

Engine load	25%	50%	75%	100%
Indicating efficiency	0.30	0.35	0.35	0.37

It can be seen, that the indicating efficiency grows, with the load increment, that is with the mixture enrichment in in the main space.

4. Conclusions and remarks

In this paper was conducted the investigation, how the mixture stratification influences on the combustion process, as the main parameter of the working cycle of the IC engine. The engine worked with globally lean mixture, and with fully open throttle, which is good from the aspect of the efficiency increment, and pump losses decrement. The most important parameter which influences on the combustion process, and by that on the engine performances, is the air-fuel ratio in the main space. That is, for better performances, it is recommended for the air-fuel ratio in main space to be between the 1.5 and 2. To lean mixture in the main space, in this case, has negative influence on the continuity of the combustion process, between the prechamber and the main space. For future researches, it is good to think about the usage of prechamber with smaller volume, in order to provide the exit of the greater amount of gas into the main space, and in this way to enrich the mixture in the main space.

Acknowledgment

This paper was realized within the framework of the project "The research of vehicle safety as part of a cybernetic system: Driver-Vehicle-Environment", ref. no. TR35041, funded by the Ministry of Education, Science and Technological Development of the Republic of Serbia.

References

[1] Available: https://www.eea.europa.eu/data-and-maps/data/data-viewers/greenhouse-gases-viewer. [Accessed 03 08 2022]
[2] Artur Jaworski HK, Ustrzycki A, Balawender K, Lejda K and Woś P 2018 Environ. Sci. Pollut. Res. **25** 17862-17877
[3] Lejda K, Jaworski A, Mądziel M, Balawender K, Ustrzycki A and Savostin-Kosiak D 2021 *Energies* **14** 1631
[4] Pastorello C, Dilara P and Martini G 2011 Transp. Res. D: Transp. Environ. **16** 121-128.
[5] Fiscon G, Sori M and Meloni G 1997 Proceedings of the International Conference on Solid Waste Technology and Management (Philadelphia, USA)
[6] Aosaf MR, Wang Y and Du K 2022 Environ. Pollut. *305* 119296
[7] Garg M and Ravikrishna R 2019 *Appl. Therm. Eng.* **149** 425-438

[8] Reynolds C Evans R, Andreassi L, Cordiner S and Mulone V 2005 *SAE Technical Paper* 2005-01-0247
[9] Yamato T, Sekino H, Ninomiya T and Hayashida M 2001 *SAE Technical Paper* 2001-01-0610
[10] Salehi M, Kakaee A and Chitsaz I 2019 *Modares Mechanical Engineering* **19** 2329-2338
[11] Grujić I, 2021 *Identification of gaseous fuels combustion process in spark ignition engine*, Faculty of Engineering of the University of Kragujevac, Serbia, Doctoral Thesis

Internal combustion engine test plan execution order optimization using Travelling Salesman Problem heuristics approach

Predrag Mrdja[1], Marko Kitanovic[1], Nenad Miljic[1], Slobodan Popovic[1]

[1] Department of Internal Combustion Engines (ICED), Faculty of Mechanical Engineering, University of Belgrade, Kraljice Marije 16, 11120 Belgrade 35, Serbia

pmrdja@mas.bg.ac.rs

Abstract. For a given internal combustion (IC) engine stationary test plan, significant time savings during its realization could be achieved if the sequence of test points execution is adequately determined. The criterion for stabilizing the engine operating point is determined by the magnitude of the change in the most inert parameter, and in the general case, it is the temperature of the engine exhaust gases. A certain level of prior knowledge about the examined object is necessary to conduct such an analysis. If there are no results from previous tests, simulation models, or experiences, the Slow Dynamic Slope (SDS) tests are a great way to quickly gather the necessary information. The task of finding the optimal sequence for a stationary engine testing plan can be set as Travelling Salesman Problem (TSP). This paper will present the application of one of the heuristic methods for solving the TSP on the example of testing the IC engine, which is a very complex dynamic system. Following this model, it is possible to optimize the stationary test plan for any other dynamic system. The basic idea is to find such a sequence of stationary operating points, during the realization of which a minor deviation of the engine exhaust gas temperature is obtained, resulting in the operating point's shortest stabilization time.

1. Introduction

The tightening of legal regulations and increasingly strict market requirements in terms of performance and comfort demand lead to a continuous increase in the complexity of the vehicle's drive system. Although the powertrain may consist of various subsystems, the central point where all information intersects is the engine control unit (ECU). Modern vehicles are equipped with numerous additional systems that can be managed by separate control units, but a certain level of communication and information exchange between the peripherals and the central ECU must exist, so that the functionality and the possibility of conducting service diagnostics are satisfied. For this reason, modern ECU counts several thousand parameters based on which the management of the drive system and other vehicle systems will be formed.

To optimally adjust the control parameters based on which the ECU will form proper system control, the so-called process of calibration of the parameters of the control unit is carried out. As a result of the increased complexity, nonlinearity, and multidimensionality of the object, the calibration of the control parameters of the ECU has become a process that requires the most significant financial and time resources during internal combustion (IC) engine and drivetrain development. The algorithm of the control unit consists of numerous functions [1] and corresponding models that can be of different complexity. Mathematical models used for these purposes are divided into complex submodels of the physical process of a given subsystem and simplified submodels in the form of control maps or curves.

The calibration and verification of mathematical models and parameters that will later be used on the ECU cannot be successfully realized without data gathered from in-laboratory engine testing. Considering the complexity of the object, the efficiency of the testing process is of crucial importance, and one of the methods that can be used for testing efficiency improvement is the Slow Dynamic Slope (SDS) method [2,3], which was the subject of earlier research [4,5].

The methods of dynamic engine testing can contribute to stationary test plan improvement in terms of determining the operating points in which the tests make the most sense [6]. However, this paper will focus on the further use of this information and determining the most efficient experiment plan execution sequence. The method that will be explained in detail consists of configuring an efficient genetic algorithm (GA) that solves this problem extremely well and reduces the time needed for engine test bench operation.

2. Dynamic Engine testing using Slow Dynamic Slope Method

The idea of the SDS test is based on a slow continuous change of the control parameter. The continuous change of control parameters over time results in deviation (offset) of the measured output of the system concerning the stationary values that will occur if the test was carried out as a quasi-stationary. The values and position of the offsets will depend on the system's characteristics (system gain, time constants) and the slope of the input. In the ideal case, when increasing the value of the control parameter during a test, we expect this offset to have some constant value, and when decreasing this parameter by the same intensity in the opposite direction, we expect the offset to be symmetrical. By determining the mean value of the response during the ascending and descending control slopes, an approximation is obtained that adequately corresponds to the results of the quasi-stationary test.

During dynamic test sequences, the engine load parameter (represented as Effective Torque M_e) was varied while the Engine Speed (n) was kept at a constant level. After the test was completed, the Engine Speed was successively changed until the entire global operation field of the engine was covered.

Two concepts of this type of examination were carried out:

- Tests with stationary preparation (Figure 1) during which the dynamic sequence begins with the stationary engine operation (curve index S),
- Tests with dynamic preparation (Figure 2) during which no parts with a stationary operation are applied (curve index D).

Also, each dynamic engine load pass is configured with a different total ramp time, i.e., a different gradient of the ascending and descending slope function. To see the effect of the speed of change of the control parameter on the final results (thus system response), the duration of the experiment was varied from 120 s to 600 s for each examined Engine Speed. The revolutions were kept constant in the range of 1500 to 3000 min^{-1}.

Each of these tests has certain specific advantages and disadvantages [6], but at this point, it should be noted that the SDS data were used to quickly collect information and generate approximate stationary results on a wide engine operation domain, and later that results will serve to optimize the detailed stationary test plan.

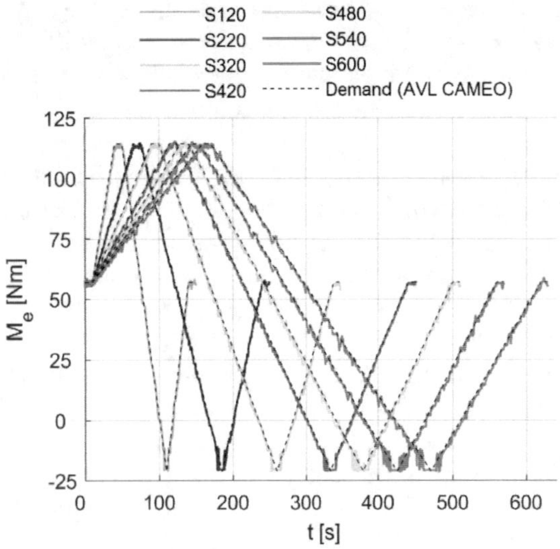

Figure 1. Engine load change for SDS tests with stationary system preparation at 1500 min^{-1}.

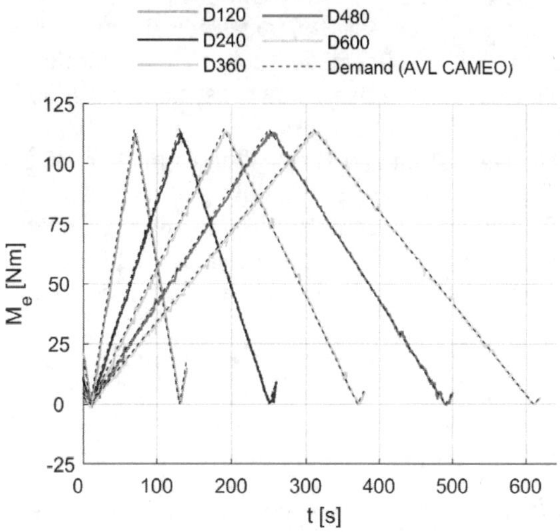

Figure 2. Engine load change for SDS tests with dynamic system preparation at 1500 min^{-1}.

3. Experimental Setup

The highest level of the engine test bench automation implies centralized supervision and control of all subsystems. The commercially available platform AVL PUMA Open [7] incorporates dyno control, control of other measuring devices, including a system for determining the composition of exhaust gases and analysis of the combustion process by engine indication, implementation of automated test plans with the possibility of calibrating the control parameters of the control unit. The use and maintenance of such a platform require exceptional financial and engineering resources and are justified in the companies of the global automotive industry. Automated engine test plans are implemented using commercially available software to define them; in this particular case, the AVL CAMEO software was used [8]. The AVL PUMA Open software's basic functionality in communication between the AVL CAMEO software and the Rotronics [9] brake controller, which does not support an interface for AVL CAMEO, is implemented within the NI LabVIEW programming environment. The AVL CAMEO software is an engineering tool that enables complete management of the test bench for engine testing, calibration of control parameters, and formation of simple mathematical models of parameters that characterize the engine working process. This software is primarily intended for testing the IC engines, but newer versions also provide support for testing hybrid powertrains and engines with a gearbox assembly. For this research purpose, the software module Test & Measure [8,10] was used and could be characterized by the following capabilities:

- Formation and execution of the IC engine test plan based on the factorial design concept and other Design of Experiment (DoE) methods;
- The possibility of optimizing the test plan by iterative means (direct model-based calibration);
- Connecting to application systems of engine control units (ECUs) through standardized interfaces, with the most significant support for the ETAS INCA software environment [11];
- Possibility of integration with the AVL PUMA Open system. By automating the test table, it is possible to send centralized commands for acquisition control and collect information about the statistical measurement values at the currently tested engine operation point;
- If the engine test plan is too extensive, there is support for networking test benches.

Thanks to the automated engine test bench and the SDS test method, Figure 3 shows the approximate results of exhaust gas temperature (T_31) measurements at the inlet section of the turbine for a wide operation field of the IC engine obtained in only 20 minutes of testing. Figure 4 shows the relative deviations compared to reference stationary tests of the same parameter (T_31). It is noted that the deviation is in the range of ±15%, but the trend of the change of T_31 has been obtained, and it will serve for further analysis and determination of the "almost" optimal sequence of execution of the stationary test plan. The engine under test was automotive PSA 1.4 HDI (40 kW at 4000 min^{-1}).

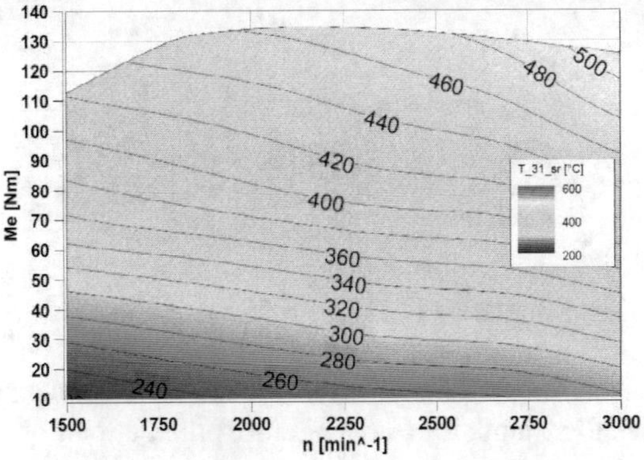

Figure 3. The exhaust gas temperature (T_31) during SDS with stationary preparation and 120 s sweep duration (S120).

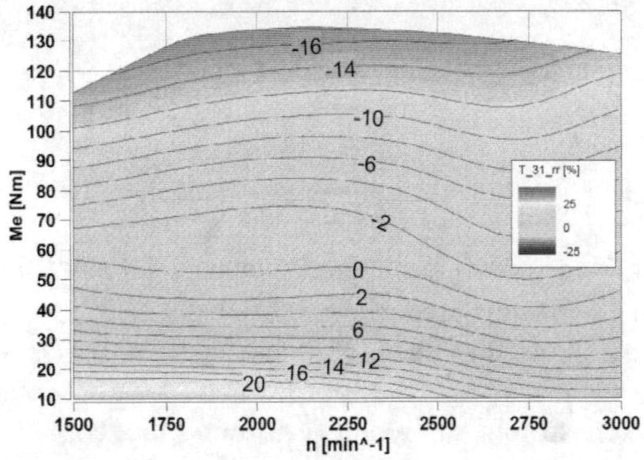

Figure 4. Relative change of T_31 gathered by SDS S120 test compared with stationary testing results.

4. Determining the optimal execution sequence for a stationary test plan

Assuming that a sufficiently good test plan exists, further time savings in its execution can be achieved if a favorable sequence of execution of the operating points is determined. As was emphasized, the criterion of stabilization of the operating point is determined by the magnitude of the change of the most inert parameter, and in the general case, it is the temperature of the engine exhaust gases. The basic idea is to find such a sequence of stationary operating modes, during the implementation of which a minor deviation of the engine exhaust gas temperature is obtained. This will result in the shortest stabilization time for a given series of operating points.

According to some predefined criteria, determining the favorable sequence of execution of operating points belongs to the category of optimization problems called "Traveling Salesman Problem" (TSP). The basic wording reads:

- For a given list of cities and their mutual distances, what would be the shortest possible route, such that each city is visited only once and after completing the tour of all cities, the merchant returns to the city from which he started?

The TSP mathematical formulation represents one of the most frequently considered numerical optimisation problems [12]. Algorithms were developed to determine possible solutions, and it was concluded that there is no way to solve this problem correctly except for an exhaustive search. The problem is that it is impossible to define a rule that would reduce the number of attempts to find a solution to a number smaller than the total number of solutions, that is, the total number of path combinations.

There are several different mathematical formulations of TSP, and the basic formulation is Dantzig-Fulkerson-Johnson [13] of the following form:

$$\min \sum_{i=1}^{p} \sum_{j=1}^{p} d_{ij} x_{ij} \tag{4.1}$$

$$\sum_{i=1}^{p} x_{ij} = 1, \qquad i = 1,2, \ldots, p \tag{4.2}$$

$$\sum_{j=1}^{p} x_{ij} = 1, \qquad j = 1,2, \ldots, p \tag{4.3}$$

$$\sum_{i \in S, j \in S}^{p} x_{ij} \leq |S| - 1, \qquad S \subset \{1,2, \ldots, p\} \tag{4.4}$$

$$x_{ij} = 0 \vee 1, \qquad \forall i,j \tag{4.5}$$

for nodes i and j , mutual distances $d(i,j)$ and for coefficient x_{ij} which takes the values $x_{ij} = 0 \vee 1$ depending on whether the route between nodes i and j belongs to the final optimal path. The first equation defines that it is necessary to minimize the total distance. The following two equations limit the number of branches of each node to one incoming and one outgoing branch. Equation 4.4 prevents the creation of sub-routes, i.e., routes that contain a smaller number of nodes (cities) than a total of p nodes.

TSP has numerous applications in designing logistics and optimizing production processes. Just as the number of points (cities) is one of the parameters in defining the problem, the values that quantify the relationships between them (distance) can be expressed through money, time, or some other parameter. In the case considered here, there are p engine operating points that need to be tested, and the relation between them is represented by the approximate value of exhaust gas temperature T_31. Given that during the stationary engine testing, there is no need to repeat the operating point by returning to the initial one, the case of an open TSP will be considered. Also, the repetition of one or more operating points is not of interest (nodes or branching), so the total number of potential sequences for executing the test plan can be described by the equation:

$$N = \frac{p!}{2} \tag{4.6}$$

where p is the number of points of the considered test plan, and dividing by 2 is the consequence of the formation of identical paths, only in the opposite order. For an examination plan of $p = 45$ points, an approximate $N \cong 6 \cdot 10^{55}$ potential combinations of execution order. Solving TSP by exhaustive numerical search belongs to the category of algorithms of factorial dependence of execution duration as a function of the number of parameters. Due to many such combinations, it is impossible to realize (execute) a program that will find the optimal solution through an exhaustive search on an ordinary computer.

The simplest algorithms are based on local path optimization. For this very reason, they are not able to globally optimize the problem [14], and the well-known among them are the Nearest Neighbor Algorithm, Cheapest Insertion Algorithm, Nearest Insertion Algorithm, Christofides Algorithm, and Replacement of pairs of routes (2-OPT, 3-OPT, 4 -OPT, k-OPT).

5. Genetic Algorithm configuration for stationary engine testing optimization

Genetic algorithms are based on the principles of natural processes of selection and evolution [15, 16]. Unlike simpler algorithms for solving TSP, where only the current solution and its quality are considered in the observed iteration, genetic algorithms (GA) perform a simultaneous search of a larger population of potential solutions. For this reason, GAs belong to the category of heuristic search algorithms. Such algorithms are often called metaheuristic because they are not based on an exclusively random search for the best solution, but rather the following series of steps are carried out:

- Upon initialization, a certain number of random sequences (vectors) are generated. Vector elements represent the indices of the operating points for which the minimum is to be determined by the criteria. One such array represents a potential solution to the TSP and is called a path vector;
- After forming a given number of such vectors, the value of the criterion function is determined, i.e., the total path length for each path vector is determined. This identifies better solutions that will be considered in the further steps of the algorithm, while worse solutions are rejected;
- A reduced number of path vectors that have undergone selection (parents) are subjected to so-called mutations, i.e., random changes, which form new path vectors (offspring). In addition to the changed path vectors, the previous path vectors (parents) are also passed into the next iteration because it can happen that the mutation had a negative effect even though the initial vector (parent) has potential, i.e., a quality genetic code in the sense of having parts of the route that are an integral part of the optimal solution (or are close to the optimal solution);
- In the iterative process, the selection is performed according to the value of the criterion function and certain mutations, which ensures the transfer of the best genetic material, i.e. parts of the path within the TSP, to the next iteration. The condition for stopping the GA can be the control of the change of the final solution or the total number of iterations. If the algorithm is well-adjusted, a sufficiently accurate solution (almost optimal) should be obtained after a certain number of iterations.

Genetic algorithms enable a high level of flexibility in defining the way of changing the path vector, i.e., the previously mentioned mutations. Within the used genetic algorithm [17], mutations of the selected path vector are based on two of the randomly selected indices (points in the path) on which GA:

- Performs full rotation of path vector elements between specified indices (points);
- Replaces the value of the path vector at the given indices;
- Performs a phase shift of part of the path vector with an adequate displacement of one element.

In this way, for every parent path, three modified paths are formed, for which the criterion function is determined, and the process of further selection is carried out. All mentioned random choices are characterized by a uniform distribution of the probability function.

6. Optimal execution sequence based on GA and SDS results

In the following chapter, the DoE test plan of 45 operating points will be analyzed. Figure 5 shows the sequence of execution according to the DoE, and an analogy with the random selection of the sequence can be observed, i.e. it is impossible to establish any rule on which the given sequence was formed. If the condition of minimizing the total Euclidean distance between the points is set for the same set of points, and the approximately optimal solution is determined using GA, the results shown in Figure 6 will be obtained. To get appropriate results, the total span in terms of engine speed and engine load would be at the same absolute level.

Determining the shortest path within the n-Me domain is justified if there is demand for the smallest possible deviations of the control parameters during the experiment, but from the aspect of the necessary stabilization time, it is not possible to determine which is closer, a change in engine speed or a change in the engine load. To identify the impact of execution order, feedback on system performance is required. The approximate models of exhaust gas temperature T_31 for stationary and tests with SDS will be discussed in this sense. The criterion function, i.e., the matrix of cross distances, can be formed based on the value of the absolute deviation of the exhaust gas temperature between the modes of the given test plan according to:

$$\boldsymbol{M}_{D(T\_31)}(i,j) = \left|\hat{y}_{T\_31}\big(DoE(i)\big) - \hat{y}_{T\_31}\big(DoE(j)\big)\right|, \qquad i,j = [1,p] \tag{6.1}$$

where DoE indicates the operation points contained in the test plan, and $\hat{y}$ represents certain mathematical models formed over different groups of data. Given that GA successfully finds (almost) optimal execution sequences for a given $\boldsymbol{M}_D$, it is necessary to compare the results obtained based on SDS approximations and the ideal case of execution sequences, thus for stationary test results. In this sense, Figure 7 shows the cumulative exhaust temperature growth formed based on optimal execution sequences and different models $\hat{y}_{T\_31}$ according to:

$$\sum |\Delta T\_31| = \sum_{i=1}^{p-1} \boldsymbol{M}_{D(T\_31)}\,(i, i+1) \tag{6.2}$$

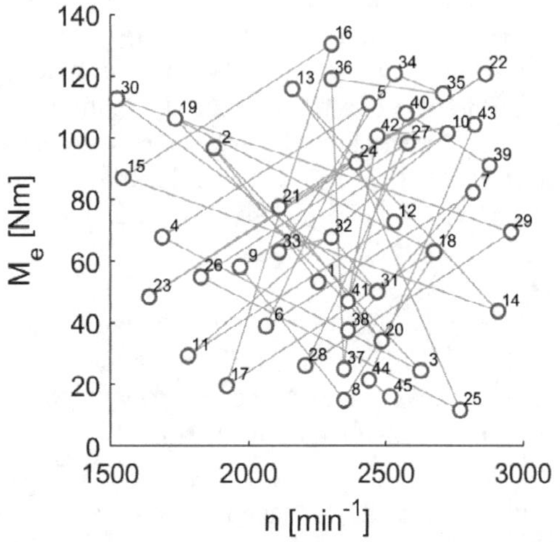

Figure 5. Initial sequence of execution DoE operating points for stationary engine testing.

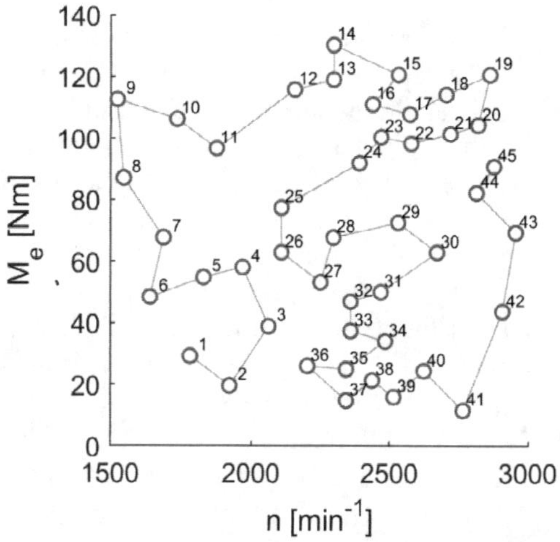

Figure 6. The shortest execution path within n-Me operation domain.

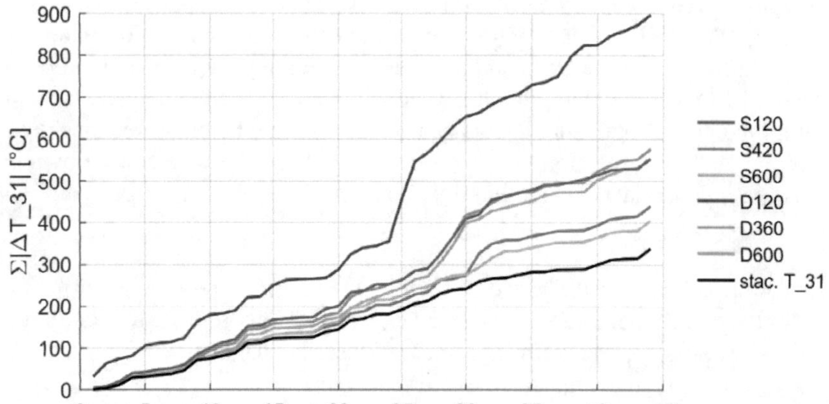

Figure 7. Cumulative deviation of exhaust gas temperature T_31 for optimal execution sequences based on different stationary approximation models.

In the worst case, if the sequence of execution were random (Figure 5), the average deviation of the exhaust gas temperature would be about 140 °C. If the optimized execution order is determined based on the minimum distance criteria in the n-Me space (Figure 6), the mean temperature deviation of T_31 would be 34 °C.

In the ideal case, for the known stationary test results, the mean value of the exhaust gas temperature deviation is 7.7 °C per operating point. If the models obtained from SDS testing were used to determine the optimal sequence, the mean values of the deviations per operating point would range from 20.5 °C to 9.2 °C, depending on the type and duration of the SDS experiment, which is very close to the ideal case.

Figure 8 and 9 shows the sequences obtained based on the minimization of the total distance of the operating points in terms of the absolute temperature curves T_31. The approximately optimal execution sequences for the $\boldsymbol{M_D}$ based on the approximate reference stationary model and the stationary approximation of the exhaust gas temperature determined by the dynamic test S120, are shown. Given that the principal trend of exhaust gas temperature values is very similar in both models, the results of GA application in the execution sequence are also very similar.

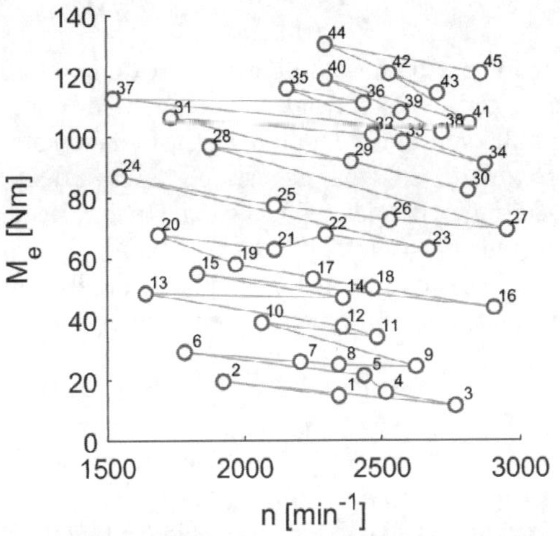

Figure 8. The execution sequence of the operating regimes determined by GA for the reference stationary model T_31.

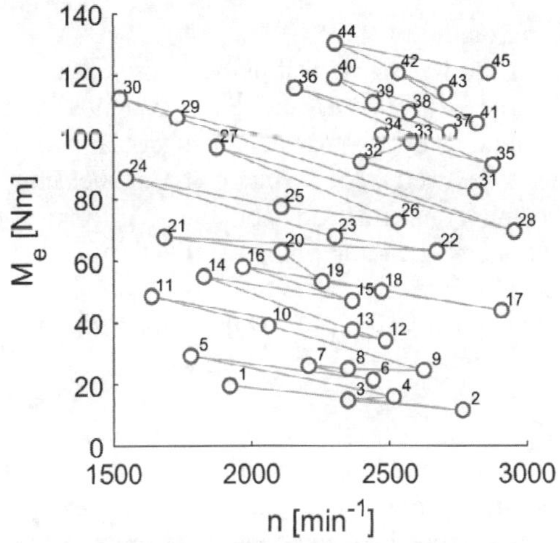

Figure 9. The execution sequence of the operating regimes determined by GA for the model is based on SDS (S120).

The initial matrix of the mutual distances of the test plan points in terms of temperature deviation T_31, for the model, formed based on the SDS test S120, is given in Figure 10. After determining the approximately optimal path according to the criterion of minimizing the deviation of the exhaust gas temperature, a matrix of cross distances is shown in Figure 11. If the execution sequence is optimal, this diagram should indicate monotonically increasing (or decreasing) differences in exhaust gas temperature according to the execution sequence of operating points. Given that GA cannot provide a correct execution sequence of operating points based on the minimization of the criterion function, a slight irregularity is observed in the central part of the diagram in Figure 11. Considering the principle flow of the temperature difference shown in this diagram and the speed of execution of GA, of only 6 seconds, the presented results are more than satisfactory. For the execution sequence shown in Figure 9, the mean deviation of the temperature of the exhaust gases according to the operating point during the stationary test of the engine would be about 12.5 °C.

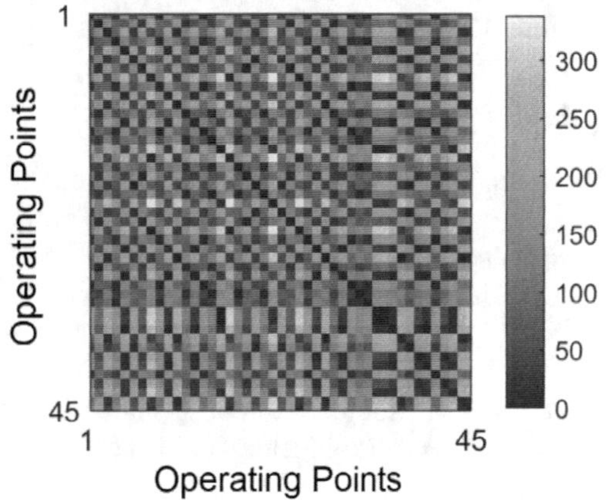

Figure 10. Initial matrix of distances $\boldsymbol{M}_D$ for DoE 45 test plan and SDS S120 T_31 model.

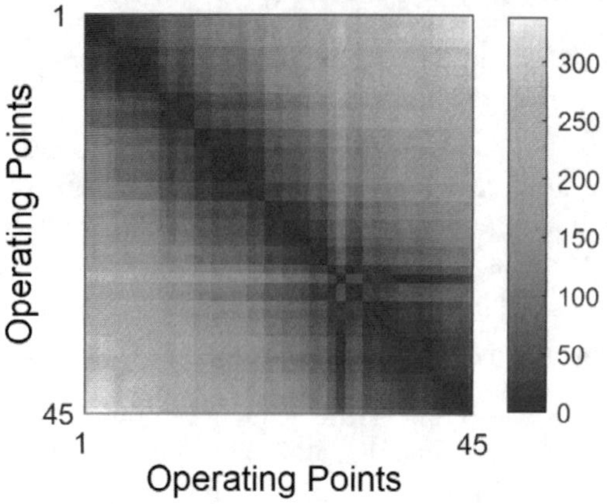

Figure 11. GA resulting matrix of distances $\boldsymbol{M}_D$ for DoE 45 test plan and SDS S120 T_31 model.

7. Conclusions

Forming the IC engine test plan requires prior knowledge about the tested object. For example, prior knowledge is needed to define the system's boundaries in terms of allowed combinations of control parameters, or prior knowledge is necessary for choosing the appropriate architecture of mathematical models to be used in the formation of the mentioned approximations. Dynamic testing is a very effective way to collect information about the investigated object. During dynamic engine testing, the time needed for stationary operating point stabilization is saved, but it is necessary to define an appropriate dynamic test plan and develop a method for interpreting the results of such a test. In this paper, the dynamic method is presented on IC engine tests during which one control parameter changes relatively slowly - the SDS method.

The IC engine exhaust gas temperature belongs to the category of parameters characterized by longer stabilization time. With approximate exhaust gas temperature characteristics obtained based on SDS testing, an approximately optimal test plan execution sequence is formed to obtain the shortest possible stabilization time between operating points of the given test plan. The analysis was carried out by setting the problem as the open path TSP and forming the corresponding genetic algorithm for solving it.

8. References

[1] K. Mollenhauer and H. Tschöke, Eds., *Handbook of Diesel Engines*. Berlin Heidelberg: Springer-Verlag, 2010. Accessed: Aug. 05, 2019. [Online]. Available: https://www.springer.com/gp/book/9783540890829

[2] J. Godburn, D. Brown, and R. Case, *Computer-controlled non-steady-state engine testing*, International Journal of Vehicle Design, vol. 12, no. 1, pp. 50–60, 1991, DOI: 10.1504/IJVD.1991.061671.

[3] M. C. Ward, C. J. Brace, N. D. Vaughan, R. Ceen, T. Hale, and G. Kennedy, *Investigation of 'Sweep' Mapping Approach on Engine Testbed*, Mar. 2002. DOI: 10.4271/2002-01-0615.

[4] N. Miljić, S. Popović, P. Mrđa, and M. Kitanović, *Slow Dynamic Slope method in I.C. engine benchmarking*, Thermal Science, Vinča Institute of Nuclear Sciences, Belgrade, vol. 22, no. 3, pp. 1271–1283, 2018, DOI: 10.2298/TSCI170921226M.

[5] P. Mrđa, N. Miljić, S. Popović, and M. Kitanović, *Continuous slow dynamic slope approach for stationary base internal combustion engine mapping*, Thermal Science, Vinča Institute of Nuclear Sciences, Belgrade, vol. 24, no. 1, pp. 147–158, 2020, DOI: 10.2298/TSCI190308171M.

[6] P. Mrđa, N. Miljić, S. Popović, and M. Kitanović, *Stationary test plan optimisation using Slow Dynamic Slope engine screening*, in Proceedings of the 8th International Congress Motor Vehicles & Motors 2020, Kragujevac, Serbia, 2020, pp. 77–84. [Online]. Available: http://www.mvm.fink.rs/congress.htm

[7] AVL List GmbH, *AVL PUMA Open*, AVL LIST GmbH Graz, 2018. [Online]. Available: https://www.avl.com/web/guest/-/avl-puma-open-2-

[8] AVL List GmbH, *AVL CAMEO 4*, AVL LIST GmbH Graz, 2018. [Online]. Available: https://www.avl.com/-/avl-cameo-4-

[9] Dynosens-Rotronics, *Dyno Control Module PGB301 - User Manual*, La Roche sur Foron, France, 2014.

[10] AVL List GmbH, *Smart Calibration - CAMEO Test & Measure*, AVL LIST GmbH Graz, 2018. [Online]. Available: https://www.avl.com/documents/10138/885965/CAMEO+TEST+and+MEASURE.pdf/979039e4-c9a8-4855-830c-e2b88a7dbbe6

[11] ETAS GmbH, *INCA Software Products - Software Products & Systems - ETAS*, 2019. https://www.etas.com/en/products/inca_software_products.php (accessed Jul. 28, 2019).

[12] W. J. Cook, *In Pursuit of the Traveling Salesman: Mathematics at the Limits of Computation*, First Edition edition. Princeton, N.J: Princeton University Press, 2012. [Online]. Available: https://press.princeton.edu/books/paperback/9780691163529/in-pursuit-of-the-traveling-salesman

[13] G. Dantzig, R. Fulkerson, and S. Johnson, *Solution of a Large-Scale Traveling-Salesman Problem*, Journal of the Operations Research Society of America, vol. 2, no. 4, pp. 393–410, 1954.

[14] D. J. Rosenkrantz, R. E. Stearns, and P. M. Lewis, *An Analysis of Several Heuristics for the Traveling Salesman Problem*, SIAM Journal on Computing, vol. 6, no. 3, pp. 563–581, 1977, DOI: https://doi.org/10.1137/0206041.

[15] John H. Holland, *Adaptation in Natural and Artificial Systems*, M.A., United States: The MIT Press, 1992. Accessed: Jan. 22, 2020. [Online]. Available: https://mitpress.mit.edu/books/adaptation-natural-and-artificial-systems

[16] D. E. Goldberg, *Genetic Algorithms in Search, Optimisation, and Machine Learning*. M.A., United States: Addison-Wesley Longman Publishing Co., Inc, 1989.

[17] J. Kirk, *Open Traveling Salesman Problem - Genetic Algorithm - MATLAB Central*, 2014. Accessed: Sep. 17, 2019. [Online]. Available: https://www.mathworks.com/matlabcentral/fileexchange/21196-open-traveling-salesman-problem-genetic-algorithm

IOP Conf. Series: Materials Science and Engineering 1271 (2022) 012006 doi:10.1088/1757-899X/1271/1/012006

A comparison of main bearings load of two-cylinder "V" and boxer motorcycle engines

N Nikolić[1], N Crnogorac[2], J Dorić[1], D Feher[1], S Galamboš[1]

[1] Faculty of Technical Sciences, University of Novi Sad, Trg Dositeja Obradovića 6, 21000 Novi Sad, Serbia
[2] Continental automotive Serbia DOO, Narodnih heroja 3, 21000 Novi Sad, Serbia

E-mail: nebnik@uns.ac.rs

Abstract. The paper deals with determining the load of the crankshaft main bearings in two-cylinder motorcycle engines. Two cylinder arrangements, used in two-cylinder motorcycle engines, are considered - the boxer-engine and V-engine. The aim of the research is to determine the influence of cylinder arrangement on the magnitude and circumferential distribution of the bearing load of the engines considered. For this purpose, models of a two-cylinder boxer engine and a corresponding V-engine crankshaft drives were developed using multibody dynamics simulation software. Previously, by using a one-dimensional gas analysis software the engine cycles were simulated in order to determine the cylinder pressure required to simulate piston dynamics. As a result of the dynamics simulations, bearing load diagrams of the boxer- and V-engine were obtained, which were then compared and analyzed.

1. Introduction

The first idea of a motorized bicycle appeared in 1867, when Ernest Michaud, the son of a French motorcycle manufacturer, came up with the idea of installing a steam engine on his father's bicycle [1]. A couple of decades later, the first motorcycles get basic features, which are not much different from modern motorcycles. Unlike automobiles where some type of in-line engine was installed in most vehicles, the concept of design in motorcycles has varied greatly throughout history. Some of earlier design concepts, which are still in use, are two-cylinder boxer and V engines [2].

In 1889, Gottlieb Daimler patented the first 2-cylinder engine with the cylinder axes forming a Latin letter "V" (V-2 engine). The implementation of V-2 engine in motorcycles followed very soon, so that as early as 1903, many manufacturers power their motorcycles with V-2 units. The main advantage of a V-2 engine was a relatively narrow profile (resulting in a small front surface of the motorcycle) and easy positioning of the rider's legs as well. The disadvantage of these engines was uneven firing interval, depending on the angle between the cylinder banks. The bank angle was usually 90 degrees so these engines were also called "L-2" engines. An additional problem of these engines was uneven heating of the rear and front cylinders, as well as large heat transfer to the driver's legs, due to the position of the rear cylinder. Nowadays, many manufacturers such as Honda, Suzuki, KTM, Ducati, use this configuration [3].

In 1896 Karl Benz patented the first two-cylinder boxer engine (boxer-2 engine). Shortly thereafter, boxer engines found widespread use in aircraft engines. In 1905, the Light Motors company, in cooperation with the Douglas company, patented a motorcycle with a boxer-2 engine. However, this

engine was placed transversely, which made it difficult to cool the rear cylinder and required a large wheelbase between the front and rear wheels [4].
In 1923, BMW produced the BMW R32 motorcycle, with a boxer engine. The Boxer engine with cylinders mounted transverse to the direction of motion remains a characteristic feature of BMW motorcycles to this day, alongside the cardan shaft. Advantages of such a design are reflected in low position of the centre of gravity and uniform cooling of the cylinders. The problem of this engine is the increased mass, given the shape and design of the crankshaft and the engine block. Since the engine cylinders are located directly in front of the rider's legs, they can be a nuisance, but, on the other side, they provide an additional protection to the rider in case of an accident or a fall from the motorcycle [4, 5, 6].
Not many researchers dealt with load of crankshaft drive components in motorcycle engines. Regulsky et al. [7] presented a procedure of studying the fatigue strength of crankshafts of a two-cylinder motorcycle engine. They analyzed the results of comparative fatigue tests of some series of crankshafts of various design and technological versions. De He [8] analyzed force of the single cylinder motorcycle engine connecting rod bending and fracture by engine linkage dynamic model. The author compared four connecting rod types based on the rod bend parameters by calculating the connecting rod compression, combustion chamber total pressure and reciprocating movement of the inertia. Gao [9] analyzed the crankshaft of motorcycle engine and solved the Reynolds and energy equations to show that the load capacity of the crankshaft journal is quite large, but at high eccentricity ratio the bearing will be damaged.
The authors of this paper have not found literature references dealing with the influence of cylinder position on the main bearings load in two-cylinder motorcycle engines, so they decided to investigate the topic. In this sense, they developed models of a two-cylinder boxer and a „V" engine, in order to simulate the crankshaft drive operation so as to determine the main bearings load. The research procedure, as well as the results obtained, is presented below.

2. Research method

The schematic view of the calculus procedure leading to the bearing forces is shown in figure 1. The procedure is applied to both boxer-2 and V-2 engines.

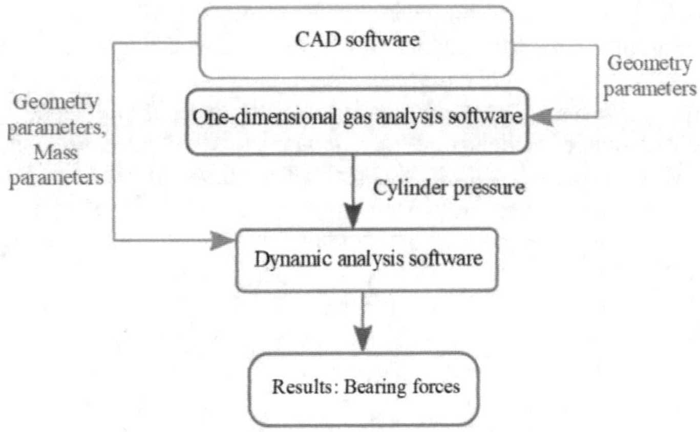

Figure 1. Flow chart of the calculus procedure.

First, the crankshaft drives have been modelled by using the CAD software "Autodesk Inventor". In order to obtain the geometry and mass parameters of the mechanisms, BMW GS1200R and Ducati Multistrada engines were taken as examples for boxer-2 and V-2 engines, respectively. The CAD models are shown in figure 2.
In order to make the comparative analysis of the bearings load as realistic as possible, the piston assemblies and the connecting rods of the CAD models are designed to be of equal dimensions and shapes. The only difference between the CAD models was in the shape of the crankshafts because of

different cylinder arrangements. For simplicity of the subsequent dynamic modelling, the piston rings, wrist pins, circlips and the connecting rod bearings have not been modelled as separate elements. Instead, the masses of the pistons and the connecting rods have been increased accordingly.
Knowing the materials density of the crankshaft drive components, their mass parameters have been calculated by using the Autodesk Inventor. These data as well as the cylinder pressure are necessary for the subsequent dynamic analysis. Due to the equal basic geometric parameters of both engines, it is approximated that identical cycles take place in each engine cylinder. Therefore, it is enough to model engine cycle in one engine cylinder only.

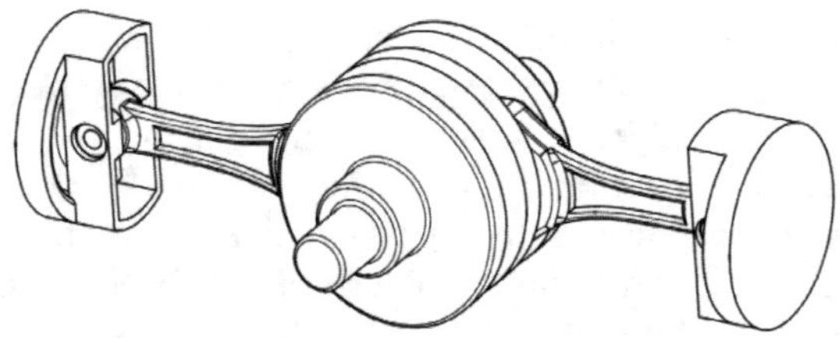

a)

b)

Figure 2. CAD models of crankshaft drives: a) boxer-2 engine, b) V-2 engine.

In order to calculate the cylinder pressure during the engine cycle, appropriate one-dimensional gas analysis software was used. After entering the required parameters obtained from the engines CAD models and by using the parameter values recommended by other researchers for engines with similar characteristics [10, 11], one-dimensional gas analysis software results are obtained. More details about the parameters adopted are given in [12].
The main result of the engine cycle modelling is the p-α (pressure – crank angle) graph, shown in figure 3. The analysis was performed for 6 engine speeds, representing the entire engine operating range in a satisfactory manner. The engine speeds simulated are 2000, 4000, 6000, 8000, 10000 and 12000 rpm. In order to reduce the number of virtual experiments, only the full engine load has been taken into account, as this is the most unfavourable case regarding the forces on the crankshaft drive components.
Knowing the geometric and mass parameters of the crankshaft drive, as well as the cylinder pressure, it was possible to create dynamic models of both engines. For that purpose, a multibody dynamics simulation software, MSC "ADAMS View", was used. The complete geometry of the both 3D CAD engines models has been transferred to the ADAMS View as it can be seen in figure 4. The mass parameters (masses and moments of inertia), obtained by Autodesk Inventor, are assigned to each part in the ADAMS View models. In the process of dynamic engine modelling the crankshaft is modelled first and then connected to the Ground by a revolute joint. The big and small ends of the connecting rod are connected to the crankshaft and the piston, respectively, also by revolute joints. The connection of the piston to the Ground has been achieved by a slider joint, in order to enable translational motion of the piston in the cylinder (the cylinder is here represented by the Ground). The same principle of modelling has been applied to the other cylinder and the other engine model as well.
To provide a motion of the mechanism a rotational motion is applied to the revolute joint connecting the crankshaft to the Ground, with 6 engine speeds desired. As a result of the motion the inertial forces appear, which are calculated automatically by "ADAMS View" during the simulation. Gas forces are applied to the piston crown taking into account the firing interval between the two cylinders. These forces are simply calculated multiplying the cylinder pressure by the piston crown area.

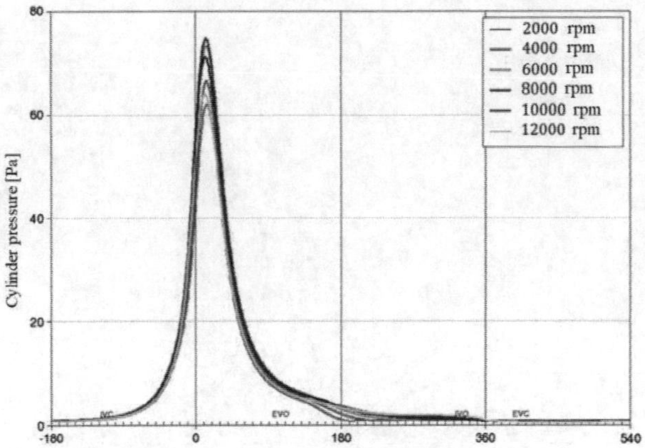

Figure 3. Engine cycle cylinder pressure at full engine load.

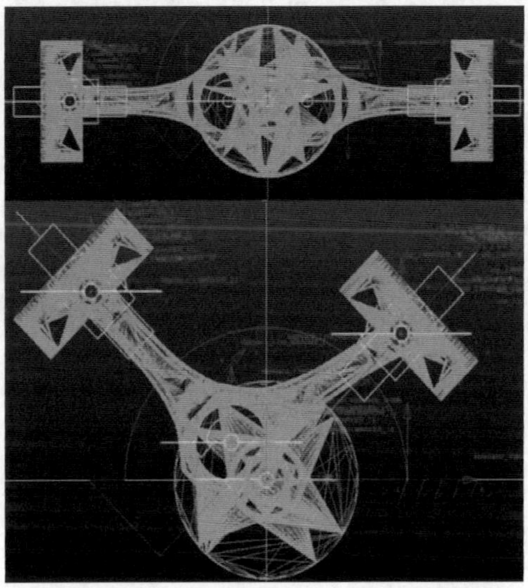

Figure 4. The layouts of dynamic models of boxer-2 and "V-2" engines in "ADAMS View" software.

After running the simulation, the bearing forces are obtained at a specific engine speed. Five more simulations of different duration times are needed at the other engine speeds to obtain the forces desired for the entire engine speed range. The same procedure is repeated for the both engines.'

3. Results and discussion

During the dynamic simulation of a mechanism, ADAMS calculates various kinematic and dynamic quantities. In this particular case, the forces acting on the main bearings of the crankshaft are of interest, so they were extracted from the simulation results. The intensity of forces is best seen if they are shown in the Cartesian coordinate system. On the other side, if the load distribution around the bearing circumference is needed, then the polar coordinate system is the right choice. For these reasons, both views of the main bearing force will be given below.

Figures 5 and 6 show bearing forces acting on the first main bearing of the boxer-2 and V-2 engines, respectively, for the six operating points considered. It can be seen that at low engine speeds, both diagrams have 2 peaks corresponding to the firing interval between the two cylinders. The dominant force at low engine speeds is the gas force. As the engine speed increases, the bearing force also increases in both types of engines but the bearing load is significantly greater in the V-2 engine than in the boxer engine.

IOP Conf. Series: Materials Science and Engineering 1271 (2022) 012006 doi:10.1088/1757-899X/1271/1/012006

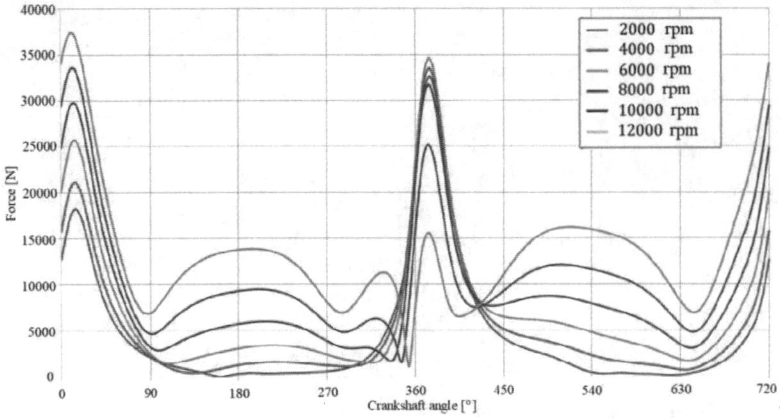

Figure 5. Main bearing 1 force at various engine speeds and full engine load – boxer engine.

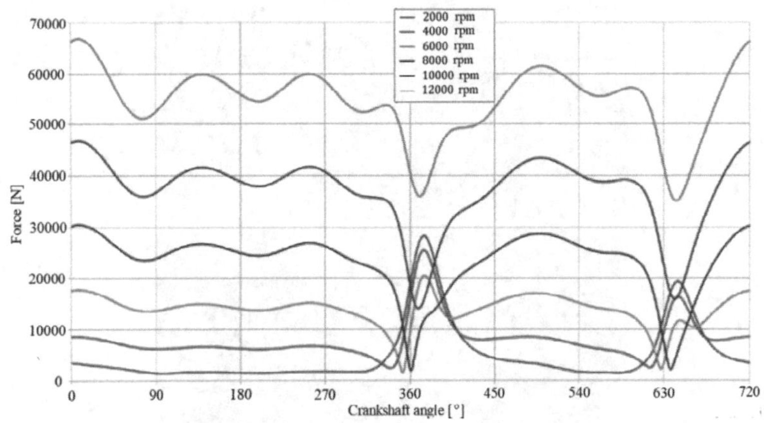

Figure 6. Main bearing 1 force at various engine speeds and full engine load – "V" engine.

The polar diagram is obtained using the horizontal and vertical components of the forces at the relevant joints. Figures 7 and 8 show the polar load diagrams of the first main bearing of the boxer-2 engine and the V-2 engine, respectively, for the six operating points considered. The polar load diagrams of the other bearing are very similar to those in figures 7 and 8, so they will not be shown here due to limited space.

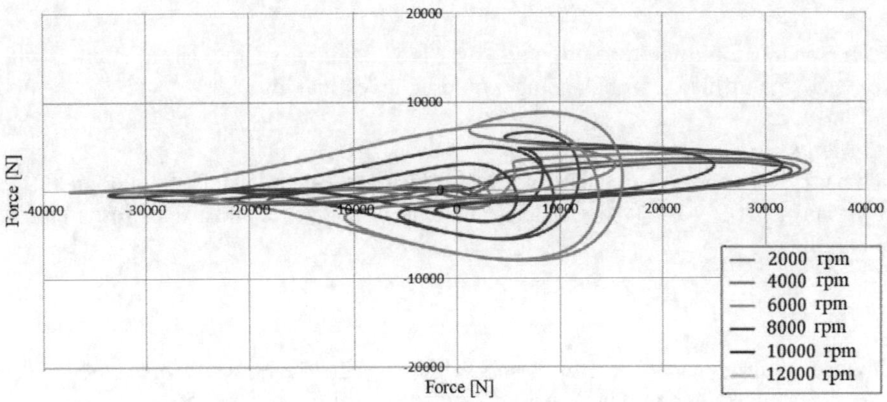

Figure 7. Main bearing 1 force at various engine speeds and full engine load – boxer engine.

Looking at figures 7 and 8, a significant difference can be seen in the shape, but also in the overall intensity of the impact of the main bearing force. Like in figures 5 and 6, here we also notice the increased effect of the gas force at low engine speeds, as well as the increase in inertial forces with the increase of engine speed. On the V-2 engine, this increase has a greater intensity than on the boxer-2 engine. At low engine speeds, it is clearly observed that the force direction approximately follows the directions of the cylinder axes.

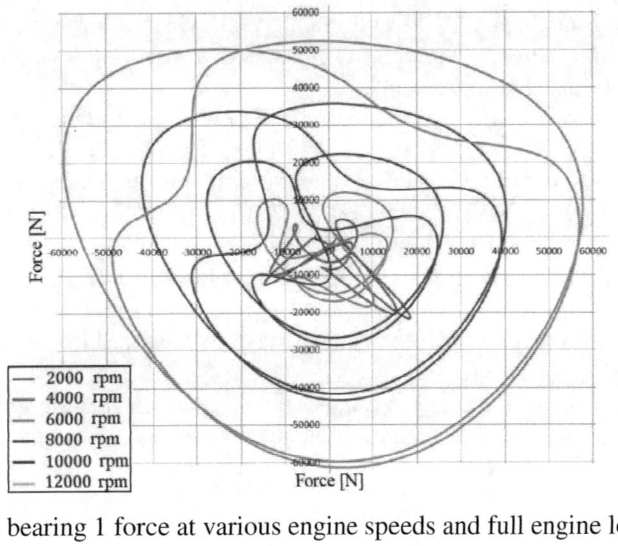

Figure 8. Main bearing 1 force at various engine speeds and full engine load – "V" engine.

4. Conclusion

The aim of the work was to investigate how the cylinder arrangement affects the load of the main crankshaft bearings of two-cylinder motorcycle engines. To that end, boxer and V-type cylinder configurations have been considered. The dynamic models of two such engines have been developed, taking care to ensure the equality of their geometric and mass parameters, so that results would be comparable. By simulating the operation of the crankshaft drives the bearing forces have been determined. Analysis of the obtained results led to the following conclusions:

- Cylinder arrangement has a great impact on the main bearings load,
- The main bearing forces are significantly higher in the V-2 engine compared to the corresponding forces in the boxer-2 engine,
- At low engine speeds, bearing forces are dominantly affected by cylinder pressure, while at higher engine speeds, the inertial forces play dominant role.

Further research on this topic could be directed towards a comparative analysis of engines with balanced crankshafts, where the influence of balancing would be investigated.

Acknowledgments
This research has been supported by the Ministry of Education, Science and Technological Development through project no. 451-03-9/2021-14/ 200156: "Innovative scientific and artistic research from the FTS activity domain".

References

[1] Summers D, Sturgeon A, Atkinson S and Wiggins V 2012 *The Motorbike Book: The Definitive Visual History* (London, UK: Dorling Kindersley Limited) p 322
[2] Coombs M, Haynes J and Shoemark P 2002, *Motorcycle Basics* 2nd ed (Sparkford, UK: Haynes) p 131
[3] Cocco G 2004 *Motorcycle design and technology* (Minneapolis, US: MotorBooks International) p 117
[4] Livesey A 2021 *Motorcycle Engineering* (New York, US: Routledge) p 380
[5] Gantriis P and Von Wartenberg H 2008 *The Art of BMW: 85 Years of Motorcycling Excellence* (Minneapolis, US: MotorBooks International) p 80
[6] Falloon I 2015 The Complete Book of BMW Motorcycles, (Minneapolis, US: MotorBooks International) p 291
[7] Regul'skii, M N, Pogrebnyak A D and Balakovskii O B 2002 Procedure and Results of Investigation into Fatigue Strength Characteristics of Motorcycle Engine Crankshafts. *Strength of Materials* 34, pp 629–635
[8] He C D 2011 Dynamic Analysis of Motorcycle Engine Rod Bending and Fracture Failure *Advanced Materials Research* Vols. 418–420 Trans Tech Publications Ltd pp. 1817–1824
[9] Gao Y S 2008 Analysis of journal bearings under consideration of varying loading and squeezed film (National Taiwan University, Taipei, Taiwan)
[10] Cordon D C, Steciak D J and Beyerlein S 2007 One-dimensional engine modelling and validation using Ricardo Wave (Master's thesis, University of Idaho, Idaho, US)
[11] Gilani R 2011 Engine Simulation Model for a Formula SAE Race Car: Applied Design, Development, Correlation and Optimization (Master's thesis, Lulea University of Technology, Lulea, Sweden)
[12] Crnogorac N 2022 *Comparison of crankshaft drive components load of boxer-2 and V-2 engines* (in Serbian), Master's thesis, Faculty of Technical Sciences, University of Novi Sad, Serbia

Numerical analysis of intake port modification onto fluid flow pattern and flame propagation in particular combustion chamber

Z Masoničić[1], I Grujić[2], A Davinić[2], S Savić[2] and R Pešić[2]

[1] Institute of Nuclear Sciences "Vinča", University of Belgrade, M. Petrovića Alasa 12-14, 11351 Belgrade, Serbia
[2] University of Kragujevac, Faculty of Engineering, Sestre Janjić 6, 34000 Kragujevac, Serbia

E-mail: masonicicz@vinca.rs

Abstract. It is known for a long time that inlet port geometry has a great influence onto fluid flow pattern and flame propagation in spark ignition IC engines. In this paper only the initial results of the very complex study on the effects that proposed inlet port modification has onto the in-cylinder flow processes including the combustion stage, in particular. The study mentioned below incorporates state of the art multidimensional numerical modelling of reactive flow in the inlet port and combustion chamber geometry layout with moving boundaries of spark ignition IC engine with two valves. The main objective of this study is to investigate the possibilities for improvement the IC engine output characteristics through the inlet port modifications. Bearing in mind that flame propagation through unburnt mixture of spark ignition IC engine is primarily controlled by dint of turbulence diffusion, it is of prime importance for engine optimization to correctly identify the influence of certain types of organized flows onto the fluid flow pattern and flame propagation. The efforts for improvement have to be aimed on the type of organized flow or eventually their combination that is of prime importance for further development of the in-cylinder processes and combustion stage.

1. Introduction

In this paper some results concerning the effect of inlet port geometry modification onto fluid flow pattern and flame propagation are elucidated. The results presented in this paper, as a part of much broader research, are intended to clarify the effect of inlet port geometry modification onto fluid flow pattern and flame front shape and its propagation through unburnt mixture of particular combustion chamber. It has long been recognized that geometry configuration inlet port-inlet valve is of great importance for fluid flow pattern and flame propagation [1, 2]. Fluid flow pattern even at the very beginning of induction stroke is very complex and entirely three-dimensional, consisting of numerous vorticies around all three axes. In spite of the aforementioned compelexity of fluid flow pattern there are some types of organized flow that can be positively identified. Also, the results of experimental investigations and numerical simulations suggest that some types of organized flows are of predominant importance for flame front shape and its propagation [3, 4, 5, 6, 7]. The turbulence generated during induction stroke, as the result of intake flow energy convesion, during compression strokes atenuates relatively fast yielding to negative effects especially onto combustion process [1, 2, 8].

Precisely on the basis of these conclusions the possibility for improvement is identified. The main idea relies on assumption that induction of certain types of organized flows yields to suppressing of attenuation as well as the negative effects it brings onto IC engine working process. It is of prime importance to identify the types of organized flows that are of predominant importance. It should also be noted that erroneous identification of organized flows or its excessive intensity can yield to undesirable effects primarily in the form of reduction of charge coefficient as well as reduction of engine output parameters.

The combustion chamber considered in this paper is so called "pent-roof" combustion chamber with flat piston head and four tilted valves. Results were obtained by dint of 3D numerical modeling.

2. Geometric model and computational method

The analysis of fluid flow pattern and flame propagation was based on geometry layout presented in figures 1 to 4.

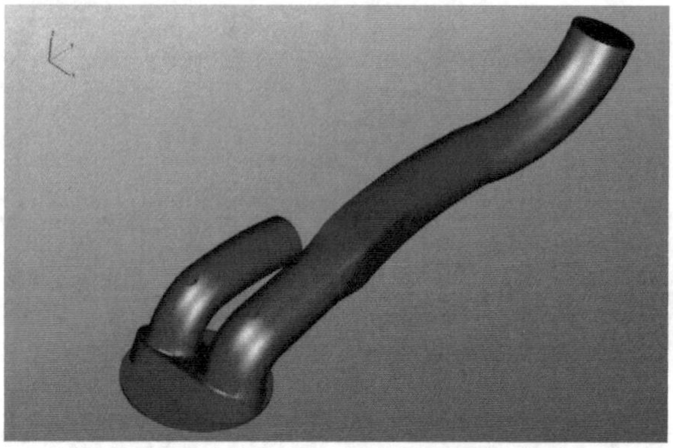

Figure 1. Perspective view of the unmodified combustion chamber geometry layout.

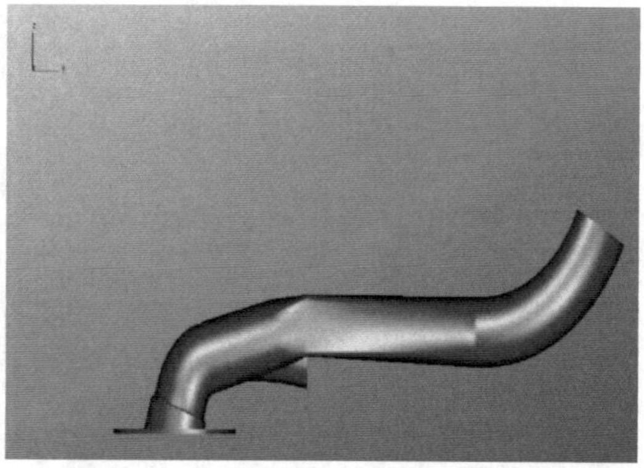

Figure 2. Side view of the unmodified combustion chamber geometry layout.

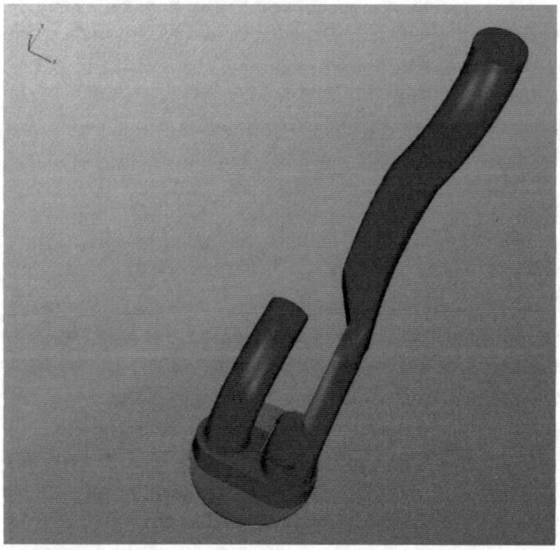

Figure 3. Perspective view of the modified combustion chamber geometry layout.

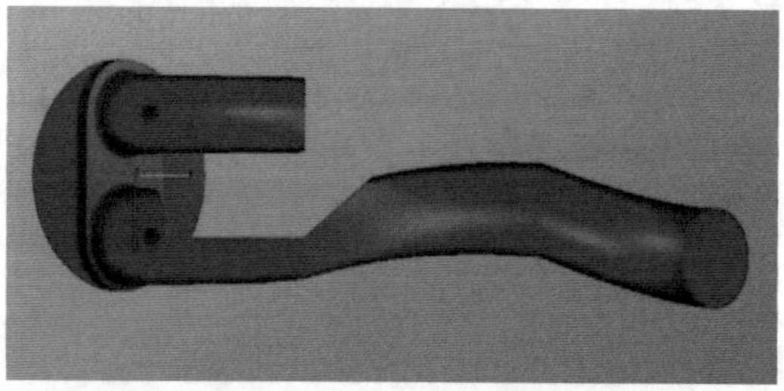

Figure 4. Upper view of the modified combustion chamber geometry layout

From the figures presented, it can be encountered that considered modification of the intake system implies the installation of specially developed deflector in the intake system. Closing of the installed deflector yields to blockage of the one half of the intake channel. The basic assumption that this modification relies on is that blockage of the part of the intake channel will, eventually, yield to improved charge through more intensive tangential flow into cylinder. The improved tangential flow into cylinder enables intensification of turbulence especially at low and medium working regimes when the inlet flow is relatively small. The intensification of turbulence during the intake and compression stroke results in improvements during the combustion process in the form of a reduction in the period of latent combustion and a better quality of the entire process in general, as well as a direct positive impact on the output characteristics and exhaust emissions of the engine. At high working regimes, the flow is significantly better, and then it is necessary to ensure as little resistance as possible in the intake system in order to achieve high values of the charge coefficient as well as the output characteristics of the engine (power, torque, etc.). In this case, the deflector is placed in the open position, which achieves full utilization of the complete flow section of the intake channel. In the following text, the situation related to the case of an open deflector will be denoted by TO, while the case of a closed deflector when one half of the intake channel is "closed" will be denoted by TC.

The combustion chamber considered is so called “pent-roof“ combustion chamber with flat piston head and two tilted valves. The basic block data sheet consists of:

- bore/stroke ratio: 80,5/67,4 mm.
- squish gap: 2,11 mm.
- maximum valves lift (intake/exhaust): 9,2 mm.
- compression ratio: 9,2.

The calculations were carried out at the engine speed of 3300 min^{-1} and mixture quality $\lambda=1$, by dint of AVL FIRE code [9]. The selection of state-of-the art multidimensional techniques is quite logical due to the fact that only these techniques can encompass valve-port geometry in explicit manner. The computational domain is discretized with block-structured grid, consisted of 350 000 ÷ 1 200 000 cells depending of the discrete position of moving boundaries. The numerical solution method is based on a fully conservative finite volume approach where all dependent variables are evaluated at cell center. For the solution of a recast linear system of equations, a conjugate gradient type of solver (CGS) is used. The second-order midpoint rule is used for integral approximation while the diffusion term is incorporated into the surface integral source after employment of special interpolation practice. Bearing in mind that the system of governing equations is open and impossible for resolving, the additional relations are obtained using relatively new k-ξ-f model of turbulence i.e., eddy-viscosity model based on Durbin’s elliptic relaxation concept [10]. This model solves a transport equation for the velocity scale ratio ξ instead of imaginary turbulent normal stress component. In addition, the pertinent hybrid boundary conditions were applied.

The combustion process is modeled using the Extended Coherent Flame Model [9]. This model relies on assumption that, primarily in reciprocating internal combustion engines, chemical time scales are few orders of magnitude smaller than turbulent ones. Therefore, a laminar flamelet concept can be used assuming that mean values of flame front velocity and thickness are dependent only on the pressure, temperature and the richness in fresh gas. The combustion reactions take place within relatively thin layers that separate unburnt and the burn gas. Using this assumption, the mean turbulent reaction rate can be determined using flame surface density and the laminar burning velocity, only.

The results of fluid flow pattern and flame propagation will be presented in three different cut planes, as follows:

- *y-z* plane, *x=const.* passing through intake and exhaust valve.
- *x-z* plane, *y=const.* passing through both intake valves.
- *x-y* plane, *z=const.* passing through the squish zone.

In order to alleviate comparisons of fluid flow patterns particularly in the case of subtle differences colors were employed as well.

3. Results and discussion

On figures 5 to 8 fluid flow pattern at very beginning of induction stroke, for both cases under consideration, is presented.

Figure 5. Fluid flow pattern in y-z plane at 15° ATDC, TO.

Figure 6. Fluid flow pattern in y-z plane at 15° ATDC, TC.

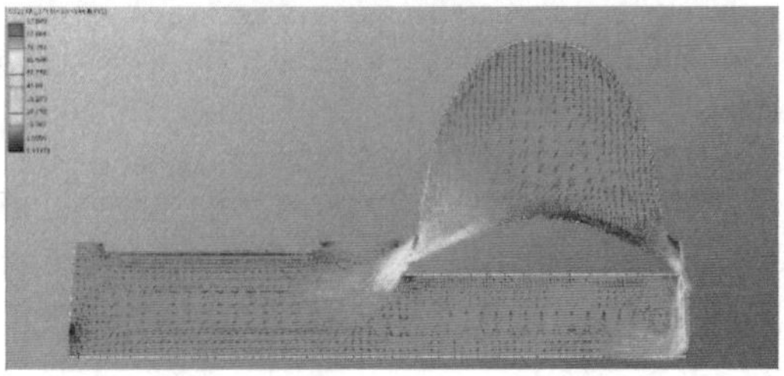

Figure 7. Fluid flow pattern in x-z plane at 15° ATDC, TO.

Figure 8. Fluid flow pattern in x-z plane at 15° ATDC, TC.

As it can be seen, the intake jet hits the piston crown and curls. This is the beginning of formation of vortex flow around y-axis. Also, the formation of another vortex flow located to the right side of the valve is encountered. From the figures 6 and 8 it can, also, be clearly encountered the effect of the modification implemented in the form of more intensive intake jet. Intensification of intake jet through more expressive tangential input, as a consequence of the implemented modification, yields to more intensive squirl flow.

On figures 9 to 16 the evolution of the fluid flow pattern during the rest of the induction stroke is presented.

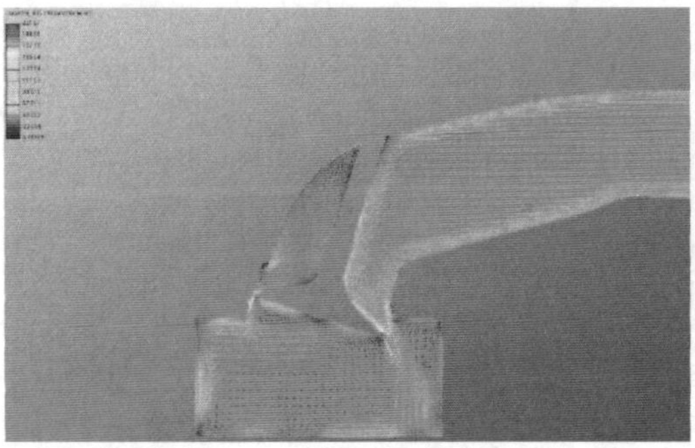

Figure 9. Fluid flow pattern in y-z plane at 75° ATDC, TO.

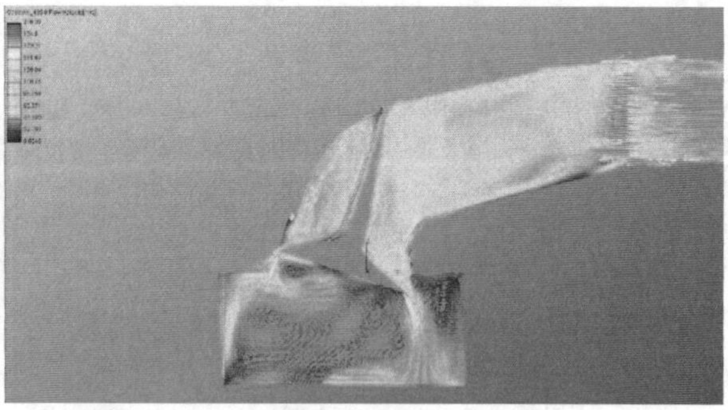

Figure 10. Fluid flow pattern in y-z plane at 75° ATDC, TC.

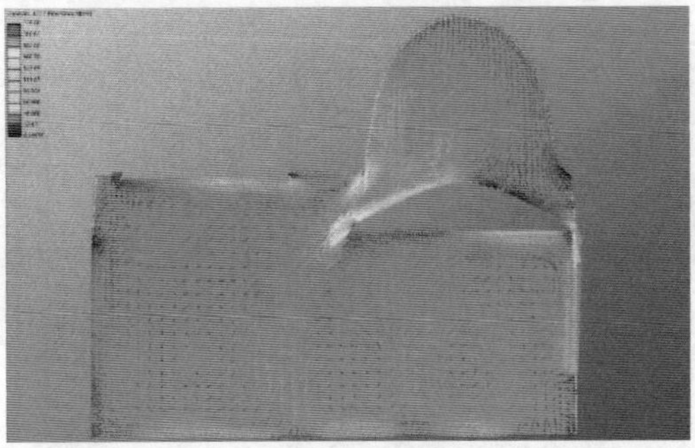

Figure 11. Fluid flow pattern in x-z plane at 75° ATDC, TO.

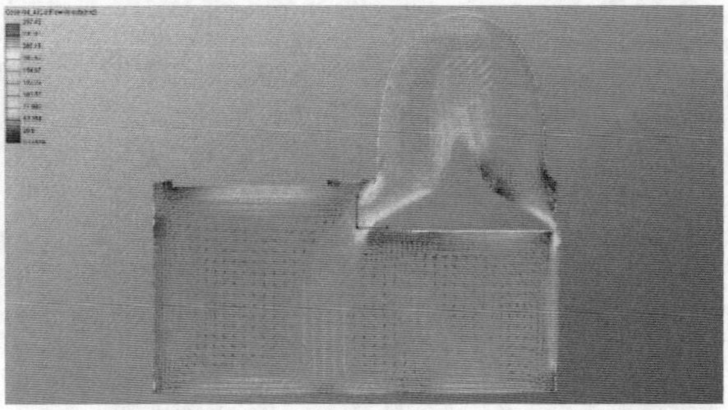

Figure 12. Fluid flow pattern in x-z plane at 75° ATDC, TC.

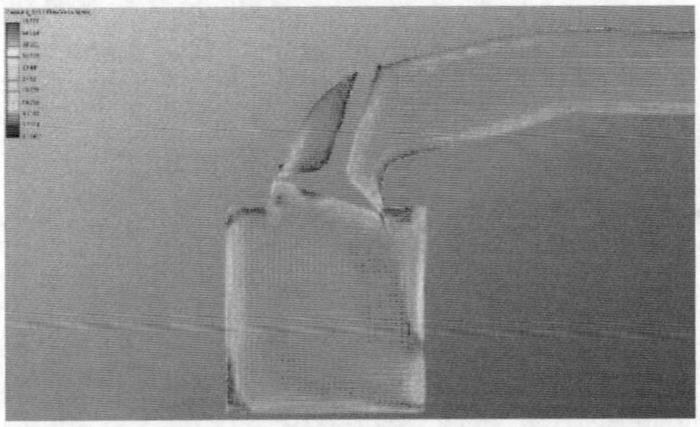

Figure 13. Fluid flow pattern in y-z plane at 180° ATDC, TO.

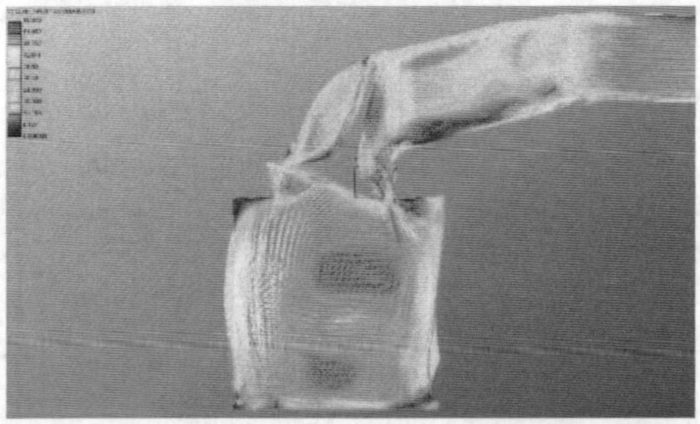

Figure 14. Fluid flow pattern in y-z plane at 180° ATDC, TC.

Figure 15. Fluid flow pattern in x-z plane at 180° ATDC, TO.

Figure 16. Fluid flow pattern in x-z plane at 180° ATDC, TC.

Further movement of piston downward and the incoming intake jet, as the main source of the turbulence motion yield to intensification of the vortex flow around y-axis, so called tumble motion and gradually attenuation of the other vortex flow located on the right side. The fluid flow pattern is dominated by well-formed high intensity tumble motion occupying nearly the entire combustion chamber [1, 2, 4, 5, 8, 11]. The effect of the closed baffle in the intake channel is still visible in the form of more intensive flow around z-axis, squirl motion, figures 17 and 18, but no qualitative differences can be encountered.

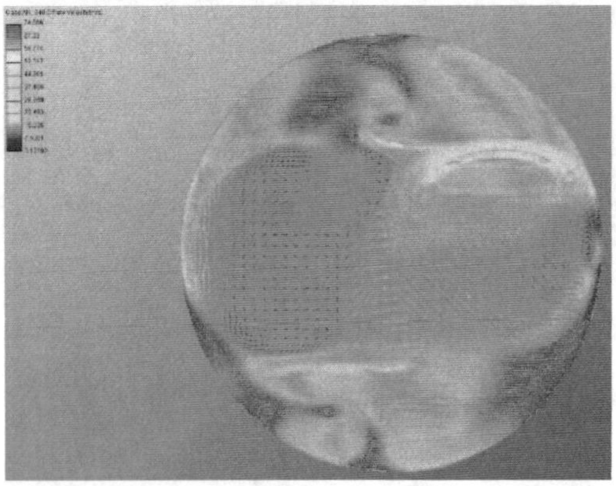

Figure 17. Fluid flow pattern in x-y plane at 180° ATDC, TO.

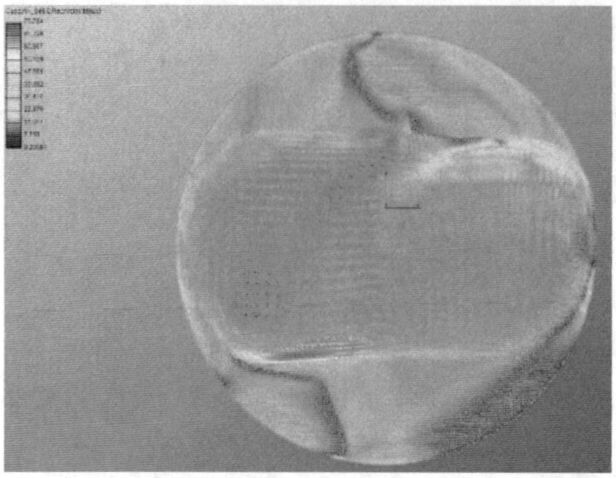

Figure 18. Fluid flow pattern in y-z plane at 180° ATDC, TC.

Change of direction of piston movement and loss of the intake jet during compression, figures 19 to 22, yield to drastic reorganization of fluid flow pattern. Quick decay of vortex motion in the zone beneath intake valve and in the vicinity of the piston crown is encountered. In addition, the restitution of tumble flow with its centre of rotation around x-axis begins. Decrease of space in the combustion chamber yields to stretching of tumble flow and its gradually displacement towards the zone beneath the exhaust valve. This situation is in compliance with "spin-up" theory. Decay of all vortex flows except the tumble flow consequently yields to attenuation of the effects caused by the closed baffle in the intake channel. Only the differences in form of the slightly increased flow velocity still persist.

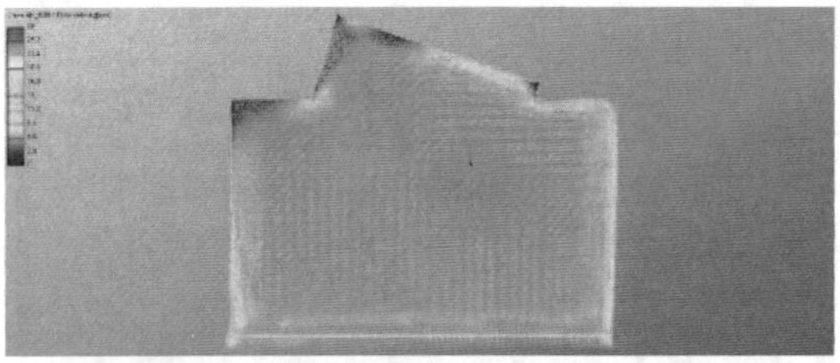

Figure 19. Fluid flow pattern in y-z plane at 270° ATDC, TO.

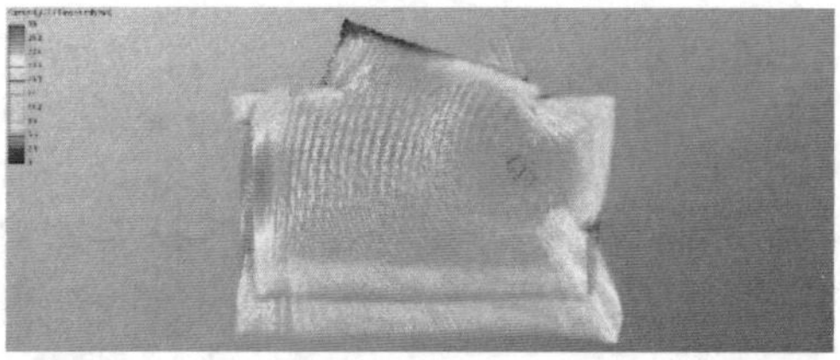

Figure 20. Fluid flow pattern in y-z plane at 270° ATDC, TC.

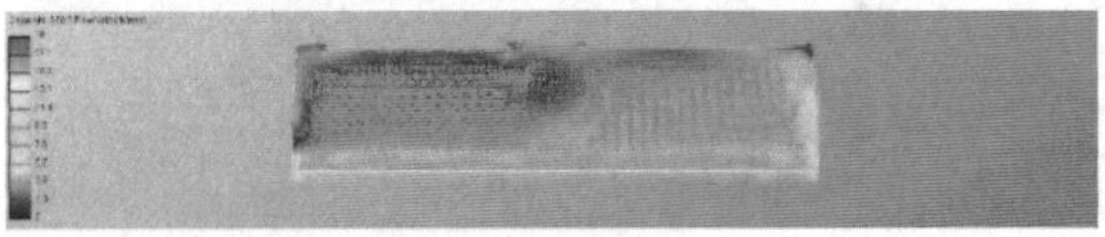

Figure 21. Fluid flow pattern in x-z plane at 330° ATDC, TO.

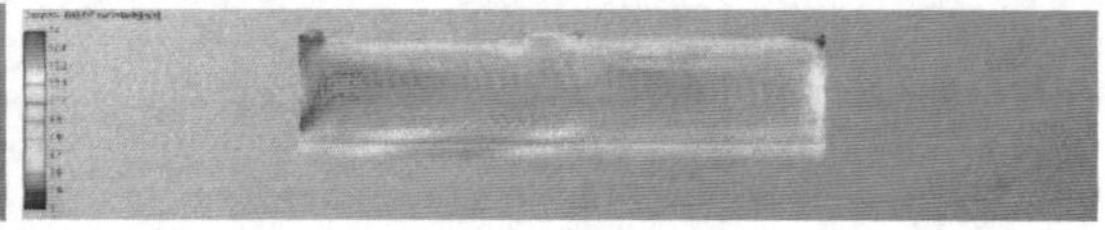

Figure 22. Fluid flow pattern in x-z plane at 330° ATDC, TC.

The flame propagation through unburnt mixture is analysed by dint of spatial distribution of temperature. Flame front is located in the zone with maximum temperature gradient. The flame propagation at the very beginning of combustion for both cases under consideration is shown in figures 23 to 26.

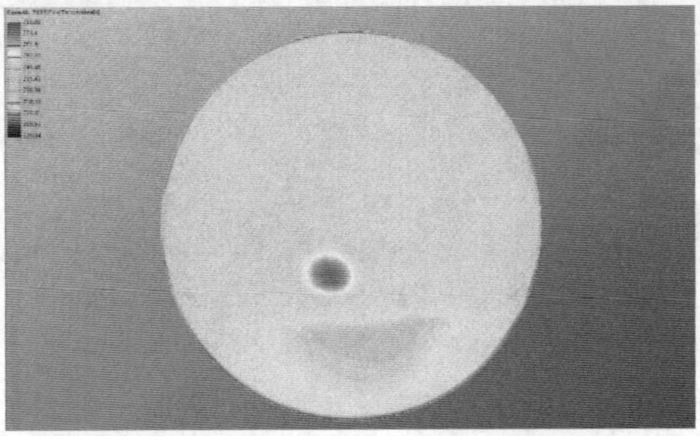

Figure 23. Spatial distribution of temperature in x-y plane at 350° ATDC, TO.

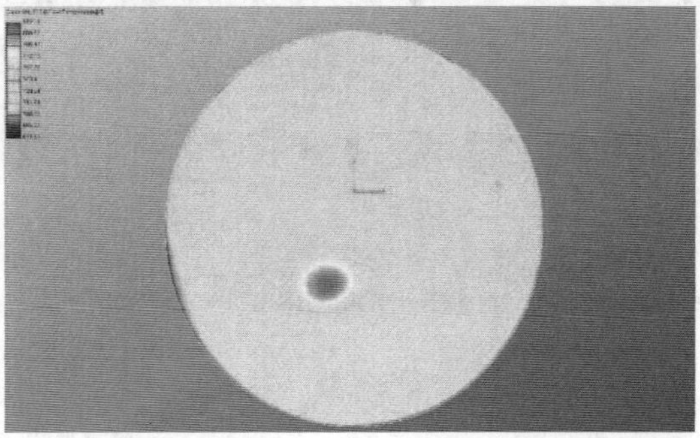

Figure 24. Spatial distribution of temperature in x-y plane at 350° ATDC, TC.

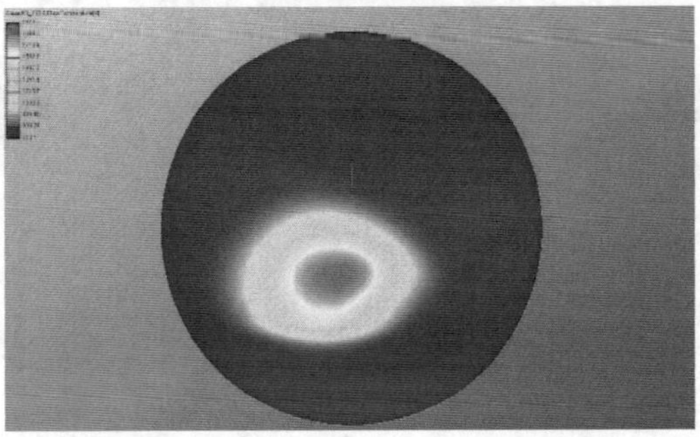

Figure 25. Spatial distribution of temperature in x-y plane at 360° ATDC, TO.

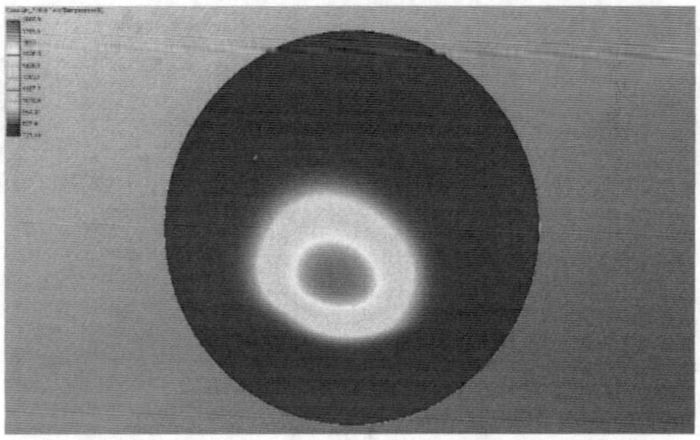

Figure 26. Spatial distribution of temperature in x-y plane at 360° ATDC, TC.

From the figures presented it can be clearly encountered that there are no clear differences in flame front propagation for both cases under consideration. Flame front shape and its displacement are controlled primarily by dint of turbulent diffusion i.e., by high intensity of turbulence and cascade process of tearing or breaking up large vortices into smaller ones and their dissipation into heat. Namely, the well-known "flame dominated fluid flow" is encountered in both cases. Macro flow established before ignition, as the one of six most important mechanisms for determination of flame front shape and propagation [12], through its characteristic time and length scales, defines the turbulent structure. From the figures representing the fluid flow pattern before the ignition one can assume that in case with closed baffle in the intake channel combustion would be faster. This is partially true mainly due to the fact that above mentioned mechanism is only one of six conflicting mechanisms acting in concert that determine flame front shape and its displacement. Slightly higher kinetic energy of turbulence during combustion in case with closed baffle, figures 27 and 28, generates slightly faster flame propagation, as presented in figures 29 and 30.

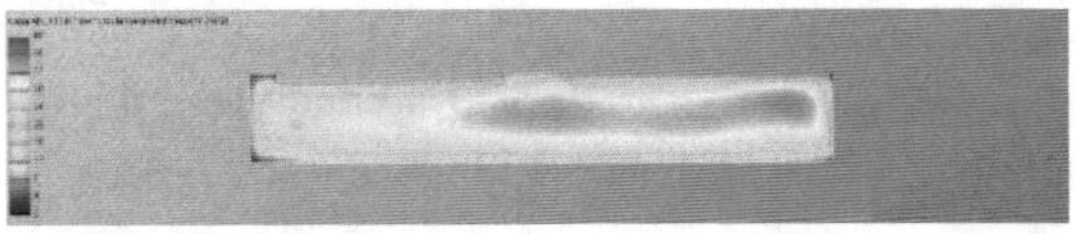

Figure 27. Spatial distribution of kinetic energy of turbulence in x-z plane at 370° ATDC, TO.

Figure 28. Spatial distribution of kinetic energy of turbulence in x-z plane at 370° ATDC, TC.

Figure 29. Spatial distribution of temperature in x-z plane at 370° ATDC, TO.

Figure 30. Spatial distribution of temperature in x-z plane at 370° ATDC, TC.

Nearing the end of the combustion process, figures 31 and 32, differences gradually attenuate confirming the existence of "flame dominated fluid flow" controlled primarily by dint of turbulent diffusion and where the earlier established macro flow is one but not the only mechanism governing the combustion process.

Figure 31. Spatial distribution of temperature in x-z plane at 380° ATDC, TO.

Figure 32. Spatial distribution of temperature in x-z plane at 370° ATDC, TC.

4. Conclusion

It is known for a long time that various types of organized flows in combustion chamber of IC engines are of predominant importance for combustion and output parameters. However, depending on the engine concept and particularly the combustion chamber design, each of these types of organized flows have different impact on the engine under consideration.

The fluid flow pattern in particular combustion chamber is very complex and entirely three-dimensional. Additionally, implementation of the new moving object in computational domain and investigation of its effects using contemporary numerical methods is absolute novelty. Geometry configuration inlet port-inlet valve is of great importance on fluid flow pattern especially during induction. During compression, the influence of these initial conditions gradually decreases and the effect of the combustion chamber geometry layout prevails. With the piston approaching the top dead centre, the drastic reorganization of established fluid flow pattern occurs and only one type of organized flow, i.e., tumble motion becomes dominant. Other types of organized flows, i.e., squirl and squish flows are of secondary importance on overall fluid flow pattern. Consequently, modifications aimed to increase these types of flows are no likely to have any significant effects.

In the innovation process, it is essential to recognize the direction in which to go and where to look for opportunities for improvement. The application of contemporary numerical modeling techniques provides better insight into complexity of IC engine working process and certainly represents one of the indispensable tools both for defining research directions and for analyzing the effects of applied modifications.

5. References

[1] Arcoumanis C and Whitelaw J 1987 Fluid mechanics of internal combustion engines – a review *Proc. Inst. Mech. Eng.* C **201** pp 57–74

[2] Lumley J 1999 *Engines – An introduction* (New York: Cambridge University Press) pp 33–44

[3] Danneman J, Pielhop K, Klaas M and Schröder W 2011 Cycle resolved multi planar flow measurements in a four valve combustion *Exp. Fluids* **50** 4 pp 961–76

[4] Lee K, Bae C and Kang K 2007 The effects of tumble and swirl flows on flame propagation in a four valve SI engine *Appl. Therm. Eng.* **27** pp 2122–30;

[5] Khaligi B 1990 Intake generated swirl and tumble motion in a 4.-valve engine with various intake configurations *SAE Int. J. Engines* **99** pp 354–374

[6] Jovanović Z, Petrović S and Tomić M 2008 The effect of combustion chamber geometry layout on combustion and emission *Therm. Sci.* **12** pp 7–24

[7] Gosman A 1986 *The Thermodynamic and Gas Dynamics of I.C. Engines* vol 2 (New York: Oxford Science Publications) pp 616–772

[8] Hill P and Zang D 1994 The effects of swirl and tumble on combustion in spark-ignited engines *Prog. Energy Combust. Sci.* **20** pp 373–429

[9] *AVL FIRE v2013.2 – CFD Solver v2013.2* 2013 (Graz: AVL List GmbH) pp 1–83

[10] Durbin P 1991 Near wall turbulence closure modeling without damping functions *Theoret. Comput. Fluid Dynamics* **3** pp 1–13

[11] Arcoumanis C, Bae C and Hu Z 1994 Flow and combustion in four-valve spark-ignition optical engine *SAE Int. J. Fuels Lubr.* **103** pp 197–211

[12] Jovanović Z and Zeljković I 1997 Modeliranje strujanja, sagorevanja i emisije kod oto-motora – višedimenzionalni pristup *Tehnologija i opreme smanjenja toksične emisije iz stacionarnih i mobilnih izvora* ed M Radovanović and A Jovović (Beograd: Mašinski Fakultet Univerziteta u Beogradu) pp 281–316 (*Printed in Serbian)*

Novel modular powertrain proposal for e-bicycle

B Stojić[1], A Poznić[1]

[1]University of Novi Sad, Faculty of Technical Sciences, Department for Mechanization and Design Engineering, Trg Dositeja Obradovića 6, 21000 Novi Sad - Serbia

E-mail address: bstojic@uns.ac.rs

Abstract. Bicycles with electric pedal assist systems are becoming increasingly popular, with increasing versatility, a range of intended areas of use, and different design approaches for system realization. This paper deals with the proposal for a novel modular drivetrain concept based on a friction driving wheel, for which a brief design outline is given. Desired performances based on law constraints and average bicycle driving experience analysed, and basic powertrain properties required for their realization assessed.

1. Introduction

Cities around the world are being more and more exposed to problems of urban traffic (jams and congestions, lack of parking etc.) and related environmental issues. Without doubt, one of noteworthy possibilities to mitigate such difficulties is more massive use of bicycles in urban transport. Recent growth of usage of bicycles with electric pedal assist systems (further: e-bikes), as reported e.g. in [1], can probably provide significant contribution to realization of such concept.

E-bikes are becoming increasingly popular, with increasing versatility and range of intended areas of use. They can help certain groups of population that, for whatever reasons, have difficulties in or reluctance to bicycle riding to overcome such issues. Pedals can be turned with virtually no physical resistance; carriage of heavy load is made easier; average cyclist can achieve greater distances and overcome uphill or wind resistances; greater average velocity means shorter stay outside by bad weather conditions, and so on. It can be expected that all these features can make bicycle transport more favourable option to broader public.

Aiming to contribute to further increase of usefulness and flexibility of e-bikes usage, this paper proposes novel modular drivetrain concept based on a friction driving wheel, which can be easily attached to or removed from the bike. Basic design outline of the system will be given. Desired performances based on law constraints and average bicycle driving experience will be defined, and basic design parameters required for their realization calculated.

2. Short review of conventional e-bike concepts

While the term "e-bike", used in broad sense, can assume any bicycle or similar single-track vehicle with electric motor added, it is important to bear in mind what can be understood by this notion in legal sense. National requirements may vary from country to country all over the world so it has to be observed depending on the geographic region. Regarding legal definition that applies on European Union level, EU directive 2002/24/EC defined e-bikes as "cycles with pedal assistance which are equipped with an

IOP Conf. Series: Materials Science and Engineering 1271 (2022) 012008 doi:10.1088/1757-899X/1271/1/012008

auxiliary electric motor having a maximum continuous rated power of 0.25 kW, of which the output is progressively reduced and finally cut off as the vehicle reaches a speed of 25 km/h, or sooner, if the cyclist stops pedalling" [2]. From 2016, named directive was repealed by regulation 168/2013 [3], but the definition remained unchanged.

Most common used layout types of e-bikes are hub drive in the front or rear wheel, or mid drive, Figure 1. In a hub-drive system the motor is installed on the centre of the rear or front wheel and thus the driving torque from the motor is exerted directly on the wheel. A mid-drive motor is built into the frame, between the pedals, so that the motor powers the chainring, from which the torque is being delivered to the rear wheel via the chain in conventional way. Using hub drive is most appropriate way to make an e-bike by converting i.e. upgrading conventional bicycle by attaching necessary components, although there are also many factory – produced e-bikes that use such layout. Mid drive requires dedicated form of the frame at least in the region where the chainset with drive is located and is therefore less suitable for conversion of existing bicycles so that e-bikes of this type are mostly factory - produced.

Figure 1. Rear hub (left) and mid (right) drive [4].

There are many criteria affecting optimal choice of the overall layout and specific individual components, most important being mass, size, desired driving performance and range, impact on centre of gravity location, tyre adhesion with the ground, suitability for implementation and maintenance, vibration characteristics and so on. Beside layout choice, there are other important design parameters, such as above all battery location and size, electronic controls used etc. Overall choice will also depend on designated area of use (urban commuting, open road journey, off-road riding, cargo transport etc.). According to review [5], for contemporary factory produced e-bikes it is most common to encounter bicycle weight in the range from around 15-17 kg for especially lightweight or compact designs to more typical 20 kg or so. Declared driving range varies in broader span, somewhere between 60 and 140 km, and expected battery charging time 2-4 h. Mid and hub drive concepts are more or less evenly represented. Referred list of e-bikes currently assessed as being best on the market by specialized journalists does not include any e-bike friction drive.

3. Background of using friction drive on e-bikes

Friction torque transmission direct to the bicycle tyre is burdened with some inherent drawbacks which prevented it to become more broadly adopted so far, such as above all intensive tyre wear, significant mechanical losses and unsuitability for wet conditions. These circumstances will probably remain unchanged so that this concept should be viewed as niche, but it though features its own potential benefits and merits so it is certainly worthwhile to further deal with development possibilities. Above all, as opposed to hub or mid drive, friction drive offers extended level of modularity. This means that an ordinary bicycle can, by temporarily attaching driving module, be used as e-bike, and be quickly and easily put back to original state whenever desired. A number of cases where such option could be purposeful can be thought of, e.g. as anti-theft measure, to make tyre repair easier, to avoid impact of precipitations, to exploit merits of recreational bicycle driving, to share same device between more bicycle users, to circumvent situations where authority representatives could express doubt in fulfilment of legal requirements for e-bikes and so on. Modularity and occasional use would, furthermore, mitigate the problem with the tyre wear. In addition, costs of such systems are expected to be less in comparison with conventional ones.

Beside listed benefits of modularity from practical viewpoint, it could possibly to some extent have certain societal impact on the attitude towards e-bike usage. Namely, despite listed general features of e-bikes that can make more probable that an average person will decide to use bicycle (especially one with e-assistance, but which can also lead to become fan of "ordinary" bikes through time) as regular means of city transport, it is not rare to receive comments from general cyclist community that are underestimating or negative towards more broad usage of electric assistance on bicycles. Although relatively few and in general not meant to be malicious, such statements could, though slightly, negatively affect general public opinion on this matter. It could be hoped that modular approach, where the bicycle itself remains unchanged and can be used as such whenever needed or wanted, can soften such hard-line attitude and help to make e-bike concept more acceptable to more general cyclists' population.

Lastly, although this is not the topic dealt with by this paper, let it be mentioned that some of the components for e-bike conversion available on the market, final products or individually produced e-bikes, do not fulfil requirements for street-legal use. Although currently in many countries it to some extent remains unclear which way authorities' representatives will check fulfilment of these requirements in individual cases should any doubts arise, in controversial situations it would certainly be beneficial to be able to quickly remove device that could potentially represent legal issue.

4. Powertrain design basics proposal

Currently, several friction drive products are available on the market, e.g. [6]. In general, friction drive uses friction wheel or roller that is attached directly to, or integrated with, the output shaft of electric motor. From this wheel, torque is transmitted to bicycle tyre by means of frictional tangential reaction force between the two. Appropriate means of tension i.e. exerting normal force to realize enough friction between friction wheel and bicycle tyre necessary for friction based torque transmission has to be applied.

In this paper, authors propose especially compact design where the whole device can be placed between the seat tube and seat stays, Figure 1. Such compact design does not interfere with rear bicycle rack as opposed to many of the present design solutions, leaving more options available for bicycle use. Simple design (Figure 2) consists virtually only of electric motor carrier, comprising parts for attachment on the bicycle, and motor with friction wheel itself. Carrier is made from two parts, one of them rigidly attached to the bicycle frame, and the other one, pivoted with respect to first one, which holds the motor with the wheel. Pivoting enables degree of freedom necessary to press friction wheel toward bicycle tyre realizing normal i.e. tensioning force. The latter can be applied by mounting torsional spring directly to the pivot of electric motor carrier, or by using coil spring between the carrier and appropriate mount located on the bicycle frame or seat post. Tensioning system is currently under development.

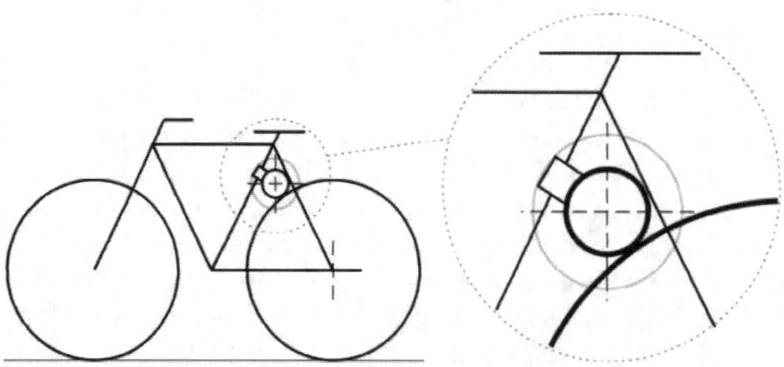

Figure 2. System placement on the bicycle.

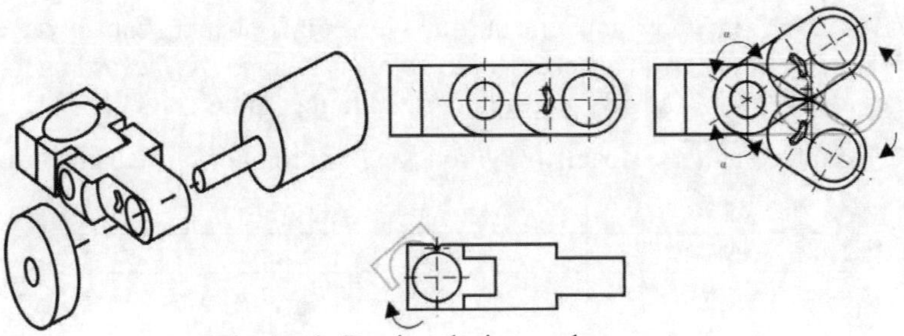

Figure 3. Device design and appearance.

5. System sizing considerations

Key features of driving system for e-bike are its driving performances and the range that can be accessed with fully charged battery. Assuming legal requirement valid in most countries (permanent motor power 250 W, e-assistance cut-off speed 25 km/h) are fulfilled, driving performances will be mostly limited by these factors. Driving range is the feature that can be more affected by designer's choice. However, criteria that should be applied to make the choice are somewhat vague. One of the key parameters, the overall mass of the bicycle, e-drive add-on and driver (possibly with some luggage) can vary in very broad range, strongly affecting energy required to ride certain distance.

Furthermore, it is impossible to predict rider contribution, i.e. share of driving energy supplied by the rider compared to that from the battery, which can be anywhere between 0 and 100%. Different bike riders can exhibit different kinds of behaviour in this view, and the circumstances affecting this behaviour may also vary substantially. This is probably one of the reasons why, to the knowledge of authors, no standard driving cycle has yet been developed for this vehicle category, which could assist range calculations and assessments. Finally, choice of the battery can be also limited by what is economically and technically feasible to be used for this purpose.

From the other hand, it should be also kept in mind that, in order to retain benefits arising out of the system modularity, choice of the battery and way of its attachment to the bike should also be in accordance with overall system concept which means that neither possible nor meaningful to adopt single solution. Dealing with this significant topic is, however, omitted in this paper.

For all the reasons listed, it is only reasonable to give tentative evaluation of range related characteristic. For this, a worst-case scenario will be assumed, meaning that the cyclist exerts no mechanical power on the pedals (except to compensate for mechanical losses, which are very small) i.e. all driving energy comes from the battery. Based on common experiences, it will be adopted that, supplied 250 W mechanical power, average cyclist in average conditions on the level ground will attain velocity of roughly 20 km/h. For 1 h of continuous riding under such conditions ("full-throttle" on the level ground), cyclist will cover 20 km, while total mechanical energy consumed by the motor will be 250 Wh, giving total amount of 250/20 = 12.5 Wh/km. Expanding adopted conditions by introducing variable velocity, and/or road sections with uphill and downhill slopes, would basically not affect this result by itself because of energy conservation. If, however, in such circumstances driver would have to apply friction brakes, which is more realistic scenario, this would introduce additional energy losses and amount of energy required for certain road distance will rise. Besides, efficiency of energy conversion between battery and the motor has also to be taken into consideration. To make final assessment, energy requirement for these general conditions will be adopted by doubling calculated value. Justification for such (largely arbitrary, but experience-influenced) approach can be found in literature, e.g. [7], and this assessment is nevertheless only provisional. So, final assessment for consumption of energy supplied from the battery amounts to approximately 25 Wh/km. Compared to common experience, this consumption assessment may appear too high. It is worthwhile mentioning that real value could be much better, probably 50-100% or even higher in sense of less energy requirements, providing that no full throttle is applied all the time and that the cyclist contributes in a certain level applying his/her own energy to pedals.

To express this result in a way more convenient from viewpoint of battery capacity declaration, which is given in [Ah] by battery producers, it is necessary to obtain current in [A], according to battery voltage. Results for voltages of 24V, 36V and 48V are summarized in the Table 1.

Table 1. Estimated battery capacity required for 1 km road distance using only e-drive.

24 V	1 Ah/km
36 V	0.7 Ah/km
48 V	0.5 Ah/km

6. Conclusions

In this paper new modular friction drive for electric assisted bicycle was proposed. With comparison with existing designs, proposal features noticeable compactness, leaving space for the usage of rear bicycle rack. Introducing modular design, system is used without bicycle conversion, i.e. bicycle itself remains unchanged and can be used anytime as originally built. Idea of modular friction drive is not to replace, but rather to complement existing choice of e-bike layouts. General characteristic of friction drives were analysed and conclusion was made that, despite some inherent drawbacks, this concept can offer a number of advantages making further research and development of this niche approach purposeful. Further research work that should lead to practical prototype realization possibility comprises above all tensioning system development and defining ways the battery can be connected to the system and attached to bike.

Acknowledgment

This research has been supported by the Faculty of technical sciences Novi Sad, Serbia - Department of Mechanization and Design Engineering through project "Testing, design and expertise in the field of mechanization in order to increase the quality of the teaching process and scientific research activities of the Department of Mechanization and Design Engineering".

References

[1] E-bike Market Size, Trends, Growth, Report 2022-2030, Precedence Research, https://www.precedenceresearch.com/table-of-content/1515, accessed August 12th 2022

[2] Directive 2002/24/EC of the European Parliament and of the Council of 18 March 2002 relating to the type-approval of two or three-wheel motor vehicles and repealing Council Directive 92/61/EEC, Eur-Lex, 2013

[3] Regulation (EU) No 168/2013 of the European Parliament and of the Council of 15 January 2013 on the approval and market surveillance of two- or three-wheel vehicles and quadricycles Text with EEA relevance, Eur-Lex, 2020

[4] E-bikes: hub-drive vs mid-drive motors, https://lewisandclarkoutfitters.com/blogs/e-bikes/e-bikes-hub-drive-vs-mid-drive-motors, accessed August 8th 2022

[5] 7 best electric bikes to boost your pedal power and help you arrive feeling fresh, https://www.independent.co.uk/extras/indybest/outdoor-activity/cycling/best-electric-bike-uk-e-b1835667.html, August 8th 2022

[6] Friction-Drive Electric Bike Reviews, https://electricbikereview.com/category/friction-drive/, accessed July 15th 2022

[7] Guzzella L, Sciarretta A: Vehicle Propulsion Systems - Introduction to Modeling and Optimization, Springer, 2013

IOP Conf. Series: Materials Science and Engineering 1271 (2022) 012009 doi:10.1088/1757-899X/1271/1/012009

Development of mathematical model for design of vehicle steering system

Momir Drakulić[1*], Lazar Stojnić[1], Ivan Blagojević[2], Aleksandar Đurić[1], Sreten Perić[1]

[1] Military Academy, University of Defence, Generala Pavla Jurišića Šturma 33, 11000 Belgrade, Serbia
[2] Faculty of Mechanical Engineering, University of Belgrade, Kraljice Marije 16, 11000 Belgrade

[*]drakulic.momir@gmail.com

Abstract. The vehicle steering system is one of the most important systems on the vehicle, both in terms of safety and performance. Based on the geometry of the steering trapezoid, the geometry of the suspension system, the alignment angles of the wheels, as well as the characteristics of the tires, a mathematical model for the design of the formula student vehicle "Road Arrow" steering system was developed. Based on the dynamic state of the vehicle, the mathematical model provides information on the stress of the steering system, the torque on the steering wheel and the slip angles of the steered wheels. The mathematical model was developed using the MATLAB software, and validation of the results was performed by comparing the results with the ones obtained by simulations in MSC ADAMS. Analysing the obtained results, it is possible to conclude that the average value of relative error of the stress is approximately 15%, for slip angles of the steered wheels is approximately 7%, and the error for torque on the steering wheel is approximately 6%.

1. Introduction

The experiences gained in the design of the control system for the formula student vehicle in the "Road Arrow" project, led to the idea of developing a mathematical model for the calculation of the system that would be applied in some software environment and thus increase the quality and speed of the design process. It must be noted that many software packages have been developed for vehicle design purposes, such as: MATLAB, MSC ADAMS, Lotus Shark and others. All of them provide the necessary conditions for quality design process, while requiring a complete simulation model of the vehicle, experience in use and are often complicated and hard to overcome.

The development of the mathematical model is divided into two sections. The first section presents a mathematical model of the steering system, that refers to the influence of the geometry of the system itself (position of the kingpin axis, geometry of the trapezoidal steering linkage, etc.) on its characteristics. In the second section, the influence of tires on the steering system is described based on Pacejka's "magic formula" [3].

The development of a mathematical model also implies the introduction of certain limitations, where the size of the error that may occur, must be taken into account. The reason for introducing restrictions is to simplify mathematical models that can be too complex. The restrictions that have been introduced are:

- the lateral force acts in the direction of the Y axis,
- turning is performed only to the right side of vehicle and returns to the initial position,
- the value of kingpin and castor angles must be positive,
- the change of the angle between the steering arm and the Y-axis during the vertical travel of the wheel is ignored and
- all elements that affect the transfer of forces are massless and absolutely rigid.

2. Mathematical model of the geometry of the steering trapezoid

The mathematical model of the geometry of the steering trapezoid enables the dependence of the left and right wheel rotation to be displayed. In practice, the elements of the steering trapezoid are located in three-dimensional space, but in the mathematical model, their base projections are used, that is, the XY - plane.

2.1. Mathematical model of the rear steering trapezoid

Figure 1 shows two arbitrary positions of the rear steering trapezoid, with initial parameters marked, where:

- b_0 – is front axle width,
- h – is rack length,
- d – is the distance between the rack and the front axle of the vehicle,
- γ – is the angle between the steering arm of the wheel hub and the longitudinal axis of the vehicle,
- p – is the length of the steering arm of the wheel hub and
- l_s, x, c, α, β – derived auxiliary parameters that were used for calculations.

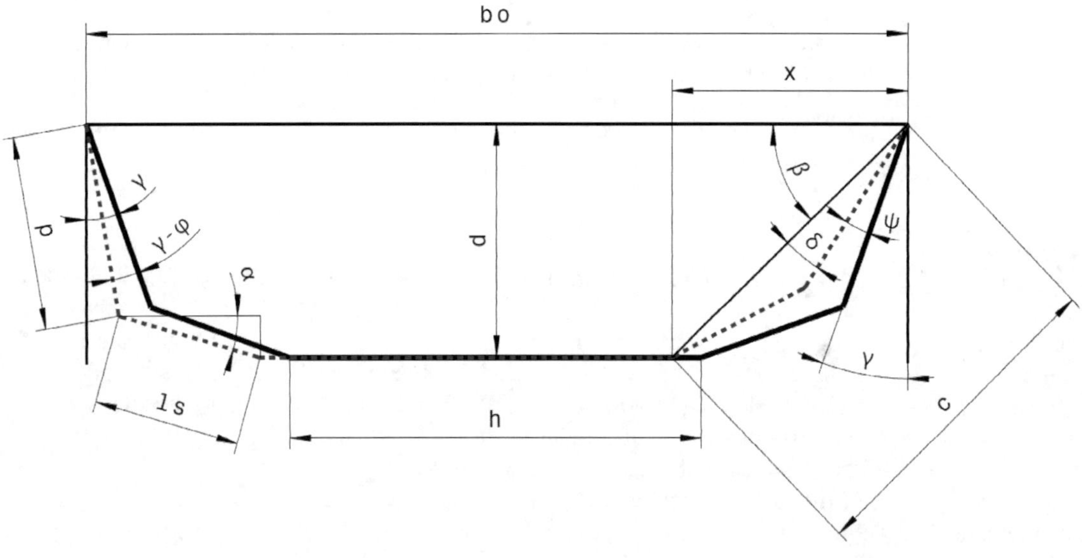

Figure 1. Two arbitrary positions of the rear steering trapezoid.

Using the sine and cosine theorem, the following mathematical equations of the rear steering trapezoid are defined:

$$l_s = \sqrt{\left[\frac{b_0}{2} - \frac{h}{2} - p \cdot \sin(\gamma)\right]^2 + [d - p \cdot cos(\varphi)]^2} \tag{1}$$

$$\alpha = \arcsin\left(\frac{d - p \cdot cos(\gamma - \varphi)}{l_s}\right) \tag{2}$$

$$x = b_0 - p \cdot \sin(\gamma - \varphi) - l_s \cdot cos\alpha - h \tag{3}$$
$$c^2 = x^2 + d^2 \tag{4}$$
$$\beta = arctan\frac{d}{x} \tag{5}$$
$$\delta = \arccos\left(\frac{c^2 + p^2 - l_s{}^2}{2 \cdot c \cdot p}\right) \tag{6}$$

Based on the presented system of equations, the dependence of rotation of the outer and inner wheel can be easily obtained:

$$\psi = 90 - \beta - \delta - \gamma \tag{7}$$

where:

- φ – is turning angle of the outer wheel and
- ψ – is turning angle of the inner wheel.

2.2. *Mathematical model of the front steering trapezoid*

Figure 2 shows two arbitrary positions of the front steering trapezoid, where the parameters marking is identical to that of the rear steering trapezoid. Using the sine and cosine theorem, the following mathematical equations were defined:

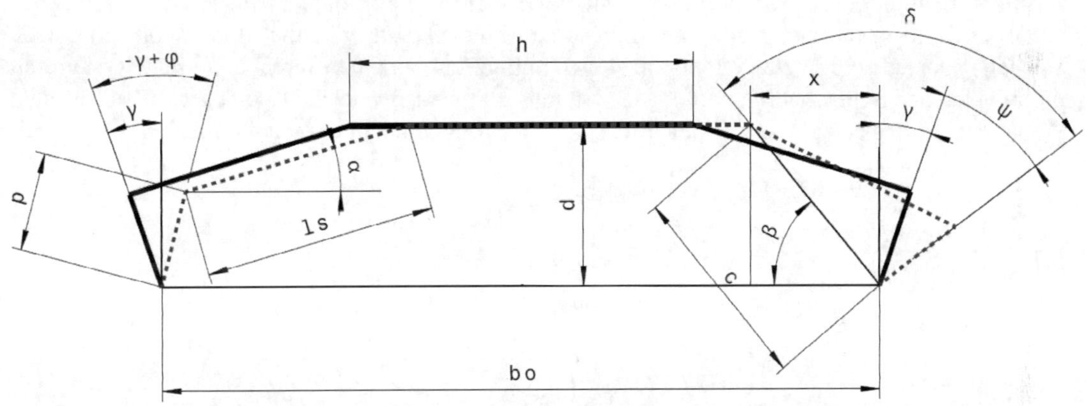

Figure 2. Two arbitrary positions of the front steering trapezoid.

$$l_s = \sqrt{\left[\frac{b_0}{2} - \frac{h}{2} + p \cdot \sin(\gamma)\right]^2 + [d - p \cdot cos(\varphi)]^2} \tag{8}$$
$$\alpha = \arcsin\left(\frac{d - p \cdot cos(-\gamma + \varphi)}{l_s}\right) \tag{9}$$
$$x = b_0 - p \cdot \sin(-\gamma + \varphi) - l_s \cdot cos\alpha - h \tag{10}$$
$$c^2 = x^2 + d^2 \tag{11}$$
$$\beta = arctan\frac{d}{x} \tag{12}$$
$$\delta = \arccos\left(\frac{c^2 + p^2 - l_s{}^2}{2 \cdot c \cdot p}\right) \tag{13}$$

Based on the presented system of equations, the dependence of rotation of the outer and inner wheel is:

$$\psi = \beta + \delta - \gamma - 90 \tag{14}$$

where:

- φ – is turning angle of the outer wheel and
- ψ – is turning angle of the inner wheel.

It should be emphasized that the size b_0 represents the distance between the two penetration points of the kingpin axis and base, that is the XY plane, as shown in Figure 3.

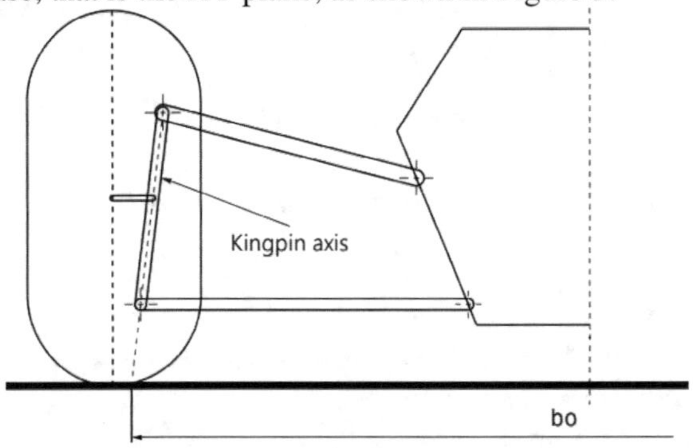

Figure 3. Width of the front axle used in the mathematical model.

3. Mathematical model for calculating the torque about the z axis of the wheels

The analysis of the load transfer from the wheel to the vehicle begins with the adoption of a coordinate system. The adopted coordinate systems are shown in Figure 4, where different coordinate systems are used for the left and right wheel. A coordinate system is set for each wheel in its contact with the ground. The input data for the calculations are the forces that occur in contact between the tire and the ground.

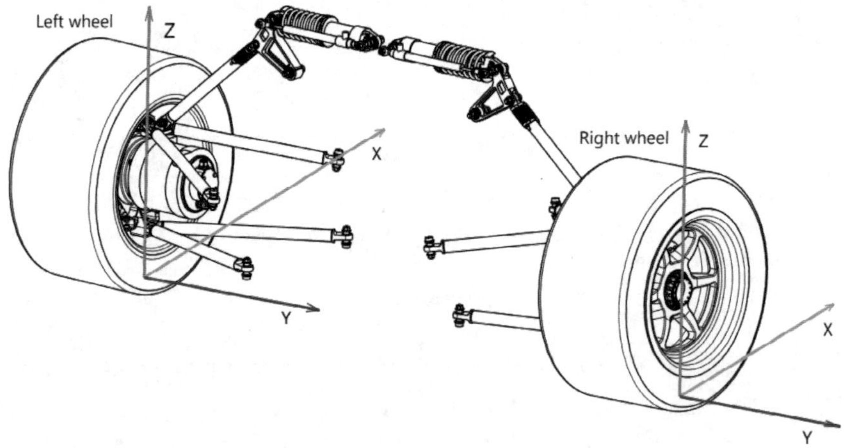

Figure 4. Adopted coordinate systems.

The forces that occur in the contact between the tire and the ground act in the direction of the axes of the shown coordinate systems. The mathematical model was formed so that as a solution it gives the value of the M_z, torque around the vertical axis, which passes through the point of intersection of the kingpin axis and the base (Figure 3.). The specified torque is transmitted to the steering wheel via the elements of the steering trapezoid, and is therefore very important when designing the steering system.

The development of a mathematical model for the calculation of the M_z torque is divided into several parts, so that the influence of the position of the kingpin axis and each force is considered separately. The influence of tire deformation on the M_z torque is described using Pacejka's "magic formula".

IOP Conf. Series: Materials Science and Engineering 1271 (2022) 012009 doi:10.1088/1757-899X/1271/1/012009

3.1. Mathematical model of the influence of the position of the kingpin axis on the M_z torque

The point on which all forces acting on the wheel is determined by the contact between the tire and the ground, while the position of the point around which the M_z torque (the point around which the wheel rotates) is defined by the position of the kingpin axis.

The first step in the development of the mathematical model is to determine the position of the penetration point of the kingpin axis through the base plane (XY - plane). The position of the point is determined by the distance to the coordinate origin l_M, and the angle between the Y-axis and the line that passes through the coordinate system origin and the point of penetration of the kingpin axis - α_M, which is shown in figures from 5 to 12, for both wheels, where:

- r – static wheel radius,
- λ – kingpin inclination angle,
- υ – caster angle,
- d_x – distance between the center of the wheel and the Z axis along the X axis,
- d_y – distance between the center of the wheel and the Z axis along the Y axis and
- d_{pd}, b_{pd}, d_{pl}, b_{pl} – derived auxiliary parameters that were used for calculations

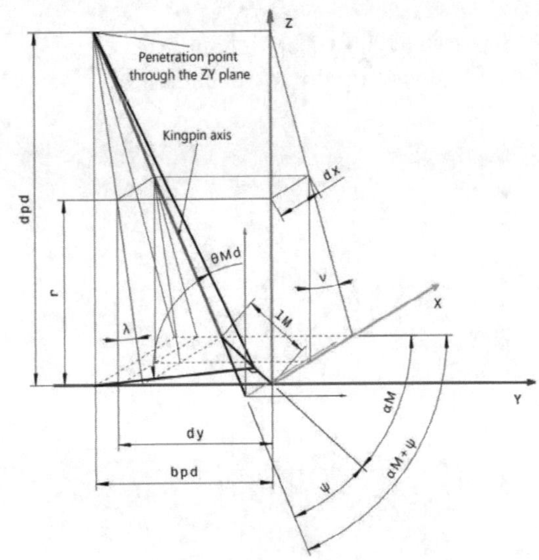

Figure 5. Right wheel kingpin axis position isometric view.

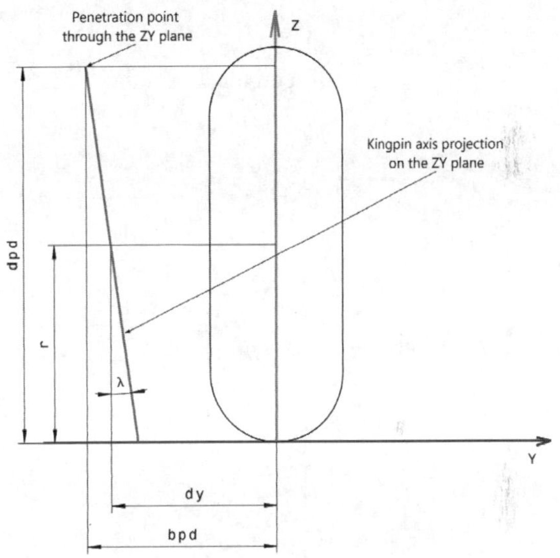

Figure 6. Right wheel kingpin axis projection on the ZY plane.

IOP Conf. Series: Materials Science and Engineering 1271 (2022) 012009 doi:10.1088/1757-899X/1271/1/012009

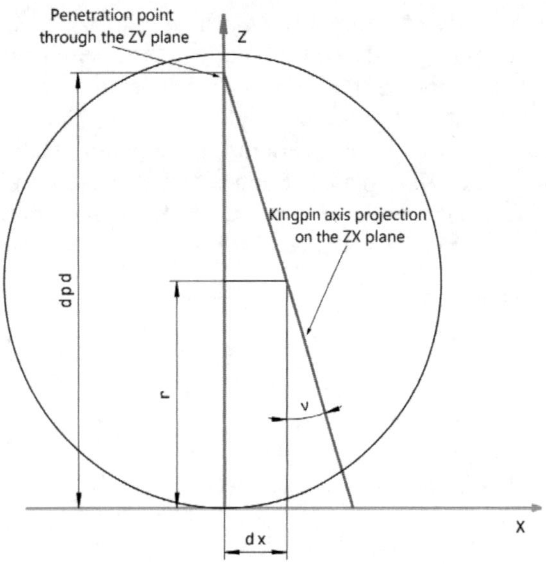

Figure 7. Right wheel kingpin axis projection on the ZX plane.

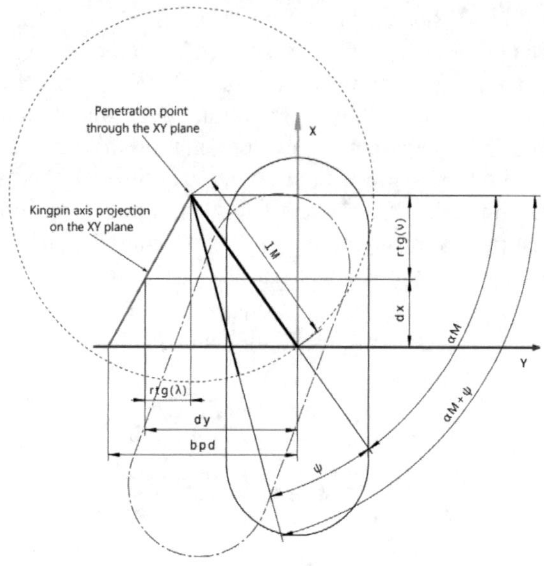

Figure 8. Right wheel kingpin axis projection on the XY plane.

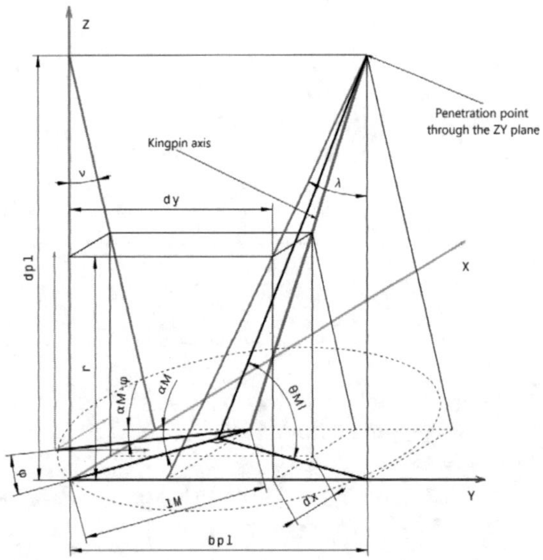

Figure 9. Left wheel kingpin axis position isometric view.

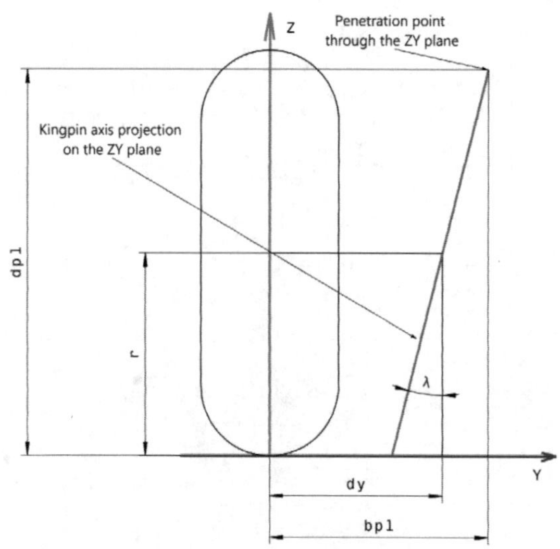

Figure 10. Left wheel kingpin axis projection on the ZY plane.

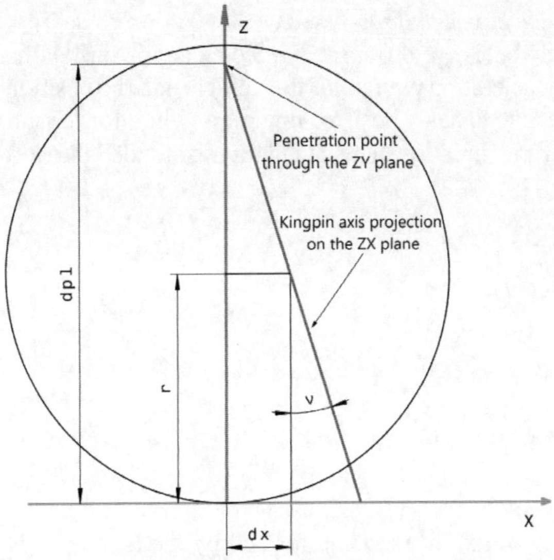

Figure 11. Left wheel kingpin axis projection on the ZX plane.

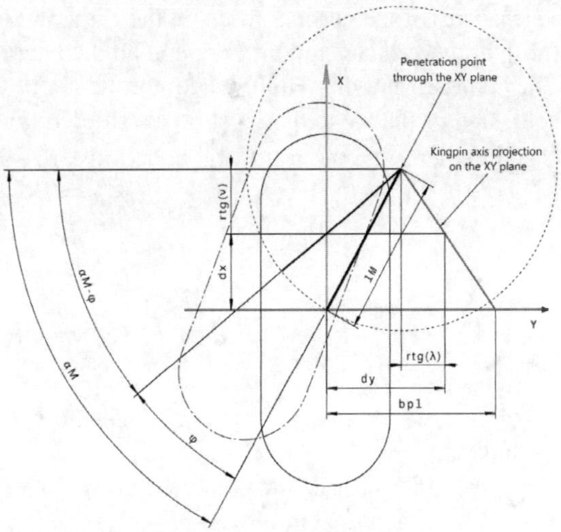

Figure 12. Left wheel kingpin axis projection on the XY plane.

The mathematical model was developed so that all forces are projected normal to the plane formed by the kingpin axis and the l_M arm. In order to project the forces normal to the specified plane, it is necessary to determine the angle between the plane and the Z axis at the point of force action, which can be determined by solving the following equations:

$$\alpha_M = arctg\left[\frac{r\cdot tg(\upsilon)+d_x}{d_y - r\cdot tg(\lambda)}\right] \tag{15}$$

$$l_M = \sqrt{(r\cdot tg(\upsilon)+d_x)^2 + \left(d_y - r\cdot tg(\lambda)\right)^2} \tag{16}$$

$$d_{pd} = \frac{l_M \cdot \sin(\alpha_M + \psi)}{tg(\upsilon)} \tag{17}$$

$$b_{pd} = l_M \cdot cos(\alpha_M + \psi) + d_{pd}\cdot tg(\lambda) \tag{18}$$

$$\theta_{Md} = arctg\left[\frac{d_{pd}}{b_{pd}\cdot sin(\alpha_M + \psi)}\right] \tag{19}$$

$$d_{pl} = \frac{l_M \cdot \sin(\alpha_M - \varphi)}{tg(\upsilon)} \tag{20}$$

$$b_{pl} = l_M \cdot cos(\alpha_M - \varphi) + d_{pl}\cdot tg(\lambda) \tag{21}$$

$$\theta_{Ml} = arctg\left[\frac{d_{pl}}{b_{pl}\cdot sin(\alpha_M - \varphi)}\right] \tag{22}$$

$$\beta_{ZiKP} = arctg\sqrt{tg(\upsilon)^2 + tg(\lambda)^2} \tag{23}$$

where:

- θ_{Md} – is the angle between the Z-axis and the plane formed by the kingpin axis and the l_M arm on the right wheel,
- θ_{Ml} – is the angle between the Z-axis and the plane formed by the kingpin axis and the l_M arm on the left wheel and
- β_{ZiKP} – is the angle between the Z and kingpin axis.

3.2. Mathematical model for M_z torque due to influence of vertical forces

The intensity of the vertical force on the right and left wheels are different and are changing during the turning of the vehicle, and they depend on the lateral acceleration, weight of the vehicle and the position of the center of gravity. Right wheel torque about the Z axis has a positive direction, while for the left wheel, sign of the value b_{pl} must be considered for torque direction. The calculation is divided into left and right wheel, and the following equations are formed:

$$F_{Znd} = F_{Zd} \cdot \cos(\theta_{Md}) \quad (24)$$

$$M_{VKPd} = F_{Znd} \cdot l_M \quad (25)$$

$$M_{Vd} = M_{VKPd} \cdot \cos(\beta_{ZiKP}) \quad (26)$$

$$\begin{matrix} F_{Znl} = -F_{Zl} \cdot \cos(\theta_{Ml}), b_{pl} > 0 \\ F_{Znl} = F_{Zl} \cdot \cos(\theta_{Ml}), \ b_{pl} < 0 \end{matrix} \quad (27)$$

$$M_{VKPl} = F_{Znl} \cdot l_M \quad (28)$$

$$M_{Vl} = M_{VKPl} \cdot \cos(\beta_{ZiKP}) \quad (29)$$

where:

- F_{Zd} – is vertical force on the right wheel,
- F_{Znd} – is projection of the F_{Zd} vertical force, normal to the plane formed by the kingpin axis and the l_M arm on the right wheel,
- M_{VKPd} – is torque about the kingpin axis of the right wheel, formed by the vertical force,
- M_{Vd} – is torque about the Z-axis on the right wheel, formed by the vertical force,
- F_{Zl} – is vertical force on the left wheel,
- F_{Znl} – is projection of the F_{Zl} vertical force, normal to the plane formed by the kingpin axis and the l_M arm on the left wheel,
- M_{VKPl} – is torque about the kingpin axis of the left wheel, formed by the vertical force and
- M_{Vl} – is torque about the Z-axis on the left wheel, formed by the vertical force.

3.3. Mathematical model for M_z torque due to influence of traction/braking forces

Traction and braking forces on the right and left wheels should be of similar intensity, and this depends directly on the design of the drivetrain and braking system.

In motorsport world, front-wheel drive racing vehicles are very rare. Most often, they have rear-wheel or all-wheel drive.

The input data for this part of the calculation are the traction and braking force values. In the case of traction force, F_t (longitudinal force) sign is positive, while otherwise it is negative. In the case of traction forces, the torque about the Z-axis for the right wheel has a positive direction, while for the left wheel it has a negative direction and vice versa. Taking above mentioned into account, the following equations were formed:

$$F_{tnXYd} = F_{td} \cdot \cos(\alpha_M + \psi) \quad (30)$$

$$F_{tnd} = F_{tnXYd} \cdot \cos(90 - \theta_{Md}) \quad (31)$$

$$M_{tKPd} = F_{tnd} \cdot l_M \quad (32)$$

$$M_{td} = M_{tKPd} \cdot \cos(\beta_{ZiKP}) \quad (33)$$

$$F_{tnXYl} = F_{tl} \cdot \cos(\alpha_M - \varphi) \quad (34)$$

$$F_{tnl} = -F_{tnXYl} \cdot \cos(90 - \theta_{Ml}) \quad (35)$$

$$M_{tKPl} = F_{tnl} \cdot l_M \quad (36)$$

$$M_{tl} = M_{tKPl} \cdot \cos(\beta_{ZiKP}) \quad (37)$$

where:

- F_{td} – is longitudinal force (traction/braking) on the right wheel,
- F_{tnXYd} – is projection of the F_{td} longitudinal force (in the XY plane), normal to the plane formed by the kingpin axis and the l_M arm on the right wheel,

- F_{tnd} – is projection of the F_{tnXYd} longitudinal force, normal to the plane formed by the kingpin axis and the l_M arm on the right wheel,
- M_{tKPd} – is torque about the kingpin axis of the right wheel, formed by the longitudinal force,
- M_{td} – is torque about the Z-axis on the right wheel, formed by the longitudinal force,
- F_{tl} - is longitudinal force (traction/braking) on the left wheel,
- F_{tnXYl} – is projection of the F_{tl} longitudinal force (in the XY plane), normal to the plane formed by the kingpin axis and the l_M arm on the left wheel,
- F_{tnl} – is projection of the F_{tnXYl} longitudinal force, normal to the plane formed by the kingpin axis and the l_M arm on the left wheel,
- M_{tKPl} – is torque about the kingpin axis of the left wheel, formed by the longitudinal force and
- M_{tl} – is torque about the Z-axis on the left wheel, formed by the longitudinal force.

3.4. Mathematical model for M_z torque due to influence of lateral forces

Lateral force on the right and left wheel is not of the same intensity, and it depends directly on the vertical force, lateral acceleration and the characteristics of the tires (adhesion coefficient). The input data for this part of the calculation is lateral force values. Torque about the Z-axis for both wheels has the same direction, and it has the greatest influence on the turning resistance of the vehicle. Taking above mentioned into account, the following equations were formed:

$$F_{lnXYd} = F_{ld} \cdot \sin(\alpha_M + \psi) \tag{38}$$
$$F_{lnd} = F_{lnXYd} \cdot \cos(90 - \theta_{Md}) \tag{39}$$
$$M_{lKPd} = F_{lnd} \cdot l_M \tag{40}$$
$$M_{ld} = M_{lKPd} \cdot \cos(\beta_{ZiKP}) \tag{41}$$
$$F_{lnXYl} = F_{ll} \cdot sin(\alpha_M - \varphi) \tag{42}$$
$$F_{lnl} = F_{lnXYl} \cdot \cos(90 - \theta_{Ml}) \tag{43}$$
$$M_{lKPl} = F_{lnl} \cdot l_M \tag{44}$$
$$M_{ll} = M_{lKPl} \cdot \cos(\beta_{ZiKP}) \tag{45}$$

where:

- F_{ld} - is lateral force on the right wheel,
- F_{lnXYd} - is projection of the F_{ld} lateral force (in the XY plane), normal to the plane formed by the kingpin axis and the l_M arm on the right wheel,
- F_{lnd} – is projection of the F_{lnXYd} lateral force, normal to the plane formed by the kingpin axis and the l_M arm on the right wheel,
- M_{lKPd} – is torque about the kingpin axis of the right wheel, formed by the lateral force,
- M_{ld} – is torque about the Z-axis on the right wheel, formed by the lateral force,
- F_{ll} - is lateral force on the left wheel,
- F_{lnXYl} – is projection of the F_{ll} lateral force (in the XY plane), normal to the plane formed by the kingpin axis and the l_M arm on the left wheel,
- F_{lnl} – is projection of the F_{lnXYl} lateral force, normal to the plane formed by the kingpin axis and the l_M arm on the left wheel,
- M_{lKPl} – is torque about the kingpin axis of the left wheel, formed by the lateral force and
- M_{ll} – is torque about the Z-axis on the left wheel, formed by the lateral force.

3.5. Mathematical model for M_z torque due to influence of tire deformations

A mathematical model that represents the influence of tires on the forces transfer, was obtained using the "magic formula", developed by *Hans Bastiaan Pacejka*. The paper presents a general "magic formula", but it should be emphasized that in the mathematical model it is adjusted independently for left and right tire. The deviation exists due to the difference in vertical and lateral forces on the left and right wheels, as well as the lateral camber of the wheel. For the calculation, part of the "magic formula" was used, which describes the characteristic of the tire in pure lateral sliding to simplify the model. The model gives data on the maximum lateral force that the tire can transmit depending on the vertical force. This type of model is not realistic in all situations, but in this case it is acceptable considering that it is a racing vehicle that tends to use the maximum performance of the tires.

$$F'_{z0} = F_{z0} \cdot \lambda_{F_{zo}} \tag{46}$$

$$df_z = \frac{F_z - F'_{z0}}{F'_{z0}} \tag{47}$$

$$\gamma_y = \gamma \cdot \lambda_{\gamma_y}$$

$$S_{Hy} = \left(P_{Hy1} + P_{Hy2} \cdot df_z\right) \cdot \lambda_{Hy} + P_{Hy3} \cdot \gamma_y \tag{48}$$

$$S_{Vy} = F_z \cdot \left\{\left(P_{Vy1} + P_{Vy2} \cdot df_z\right) \cdot \lambda_{Vy} + \left(P_{Vy3} + P_{Vy4} \cdot df_z\right) \cdot \gamma_y\right\} \cdot \lambda_{Vy} \tag{49}$$

$$\alpha_y = \alpha + S_{Hy} \tag{50}$$

$$C_y = P_{Cy1} \cdot \lambda_{Cy} \tag{51}$$

$$K_y = P_{Ky1} \cdot F_{z0} \cdot sin\left[2 \cdot arctg\left\{\frac{F_z}{P_{Ky2} \cdot F_{z0} \cdot \lambda_{F_{zo}}}\right\}\right] \cdot \left(1 - P_{Ky3} \cdot |\gamma_y|\right) \cdot \lambda_{F_{zo}} \cdot \lambda_{Ky} \tag{52}$$

$$\mu_y = \left(P_{Dy1} + P_{Dy2} \cdot df_z\right) \cdot \left(1 - P_{Dy3} \cdot {\gamma_y}^2\right) \cdot \lambda_{\mu_y} \tag{53}$$

$$D_y = \mu_y \cdot F_z \tag{54}$$

$$B_y = \frac{K_y}{C_y \cdot D_y} \tag{55}$$

$$E_y = \left(P_{Ey1} + P_{Ey2} \cdot df_z\right) \cdot \left\{1 - \left(P_{Ey3} + P_{Ey4} \cdot \gamma_y\right) \cdot sgn(\alpha_y)\right\} \cdot \lambda_{Ey} \tag{56}$$

$$F_{y0} = D_y \cdot sin\left[C_y \cdot arctg\left\{B_y \cdot \alpha_y - E_y \cdot \left(B_y \cdot \alpha_y - arctg\left(B_y \cdot \alpha_y\right)\right)\right\}\right] + S_{Vy} \tag{57}$$

$$\gamma_z = \gamma \cdot \lambda_{\gamma_z} \tag{58}$$

$$C_t = q_{Cz1} \tag{59}$$

$$S_{Ht} = q_{Hz1} + q_{Hz2} \cdot df_z + \left(q_{Hz3} + q_{Hz4} \cdot df_z\right) \cdot \gamma_z \tag{60}$$

$$S_{Hr} = S_{Hy} + \frac{S_{Vy}}{K_y} \tag{61}$$

$$\alpha_r = \alpha + S_{Hr} \tag{62}$$

$$\alpha_t = \alpha + S_{Ht} \tag{63}$$

$$B_t = \left(q_{Bz1} + q_{Bz2} \cdot df_z + q_{Bz3} \cdot {df_z}^2\right) \cdot \left(1 + q_{Bz4} \cdot \gamma_z + q_{Bz5} \cdot |\gamma_z|\right) \cdot \frac{\lambda_{K_y}}{\lambda_{\mu_y}} \tag{64}$$

$$B_r = q_{Bz9} \cdot \frac{\lambda_{K_y}}{\lambda_{\mu_y}} + q_{Bz10} \cdot B_y \cdot C_y \tag{65}$$

$$D_t = F_z \cdot \left(q_{Dz1} + q_{Dz2} \cdot df_z\right) \cdot \left(1 + q_{Dz3} \cdot \gamma_z + q_{Dz4} \cdot {\gamma_z}^2\right) \cdot \left(\frac{r}{F_{z0}}\right) \cdot \lambda_t \tag{66}$$

$$D_r = F_z \cdot \left(\left(q_{Dz6} + q_{Dz7} \cdot df_z\right) \cdot \lambda_r + \left(q_{Dz8} + q_{Dz9} \cdot df_z\right) \cdot \gamma_z\right) \cdot r \cdot \lambda_{\mu_y} \tag{67}$$

$$E_t = \left(q_{Ez1} + q_{Ez2} \cdot df_z + q_{Ez3} \cdot {df_z}^2\right) \cdot \left\{1 + \left(q_{Ez4} + q_{Ez5} \cdot \gamma_z\right) \cdot \frac{2}{\pi} \cdot arctg\left(B_t \cdot C_t \cdot \alpha_t\right)\right\} \tag{68}$$

$$M_{zr}(\alpha_r) = D_r \cdot cos[arctg(B_r \cdot \alpha_r)] \cdot \cos(\alpha) \tag{69}$$

$$t(\alpha_t) = D_t \cdot cos\left[C_t \cdot arctg\left\{B_t \cdot \alpha_t - E_t \cdot \left(B_t \cdot \alpha_t - arctg(B_t \cdot \alpha_t)\right)\right\}\right] \cdot \cos(\alpha) \tag{70}$$

$$M_{z0} = -t \cdot F_{y0} + M_{zr} \tag{71}$$

where:

- F_{z0} – is nominal value of vertical force,
- F_z – is value of the vertical force,
- r – is radius of the unloaded tire,
- α– is sideslip angle,
- γ– is camber angle,
- F_{y0}– the value of the lateral force and

IOP Conf. Series: Materials Science and Engineering 1271 (2022) 012009 doi:10.1088/1757-899X/1271/1/012009

- M_z– value of the aligning torque (returning the wheels to the initial position).

Other parameter values are given in tabular form, and are obtained by testing tires. [12]

3.6. Mathematical model for calculation of the torque on the steering wheel

The input data for the calculation of the torque on the steering wheel are the aforementioned torques around the Z axis, for both wheels, which are:

$$M_{zd} = M_{Vd} + M_{td} + M_{vd} + M_{z0d} \quad (72)$$

$$M_{zl} = M_{Vl} + M_{tl} + M_{vl} + M_{z0l} \quad (73)$$

where:

- M_{zd} – is total torque about the Z-axis on the right wheel,
- M_{zl} – is total torque about the Z-axis on the left wheel,
- M_{z0d} – total torque around the Z-axis on the right wheel, due to tire deformation and
- M_{z0l} – total torque around the Z-axis on the left wheel, due to tire deformation.

As the total torque on the steering wheel is transmitted from the wheels via the steering trapezoid, it is concluded that the load transfer depends significantly on the geometry of the steering trapezoid. For the rear steering trapezoid, the torque on the steering wheel is:

$$\alpha_{sd} = \arcsin\left(\frac{(d - p \cdot cos(\gamma + \psi))}{l_s}\right) \quad (74)$$

$$F_{sd} = \frac{M_{zd}}{p \cdot (\cos(\alpha_{sd}) \cdot \cos(\gamma + \psi) + \sin(\alpha_{sd}) \cdot \sin(\gamma + \psi))} \quad (75)$$

$$F_{zld} = F_{sd} \cdot \cos(\alpha_{sd}) \quad (76)$$

$$\alpha_{sl} = \arcsin\left(\frac{(d - p \cdot cos(\gamma - \varphi))}{l_s}\right) \quad (77)$$

$$F_{sl} = \frac{M_{zl}}{p \cdot (\cos(\alpha_{sl}) \cdot \cos(\gamma - \varphi) + \sin(\alpha_{sl}) \cdot \sin(\gamma - \varphi))} \quad (78)$$

$$F_{zll} = F_{sl} \cdot \cos(\alpha_{sl}) \quad (79)$$

$$F_{zl} = F_{zld} + F_{zll} \quad (80)$$

$$M_{tu} = F_{zl} \cdot \frac{Dzzl}{2} \quad (81)$$

where:

- α_{sd} – is angle between right steering arm and Y-axis,
- F_{sd} – is force in the right steering arm,
- F_{zld} – force in the rack, coming from the right wheel,
- α_{sl} – is angle between left steering arm and Y-axis,
- F_{sl} – is force in the left steering arm,
- F_{zll} – is force in the rack, coming from the left wheel
- F_{zl} – is total force in the rack,
- M_{tu} – steering wheel torque and
- $Dzzl$ – pinion gear diameter.

For the front steering trapezoid, the mathematical model for calculating the torque on the steering wheel is described by the following equations:

$$\alpha_{sd} = \arcsin\left(\frac{(d - p \cdot cos(\gamma + \psi))}{l_s}\right) \quad (82)$$

$$F_{sd} = \frac{M_{zd}}{p \cdot (\cos(\alpha_{sd}) \cdot \cos(\gamma + \psi) + \sin(\alpha_{sd}) \cdot \sin(\gamma + \psi))} \quad (83)$$

$$F_{zld} = F_{sd} \cdot \cos(\alpha_{sd}) \quad (84)$$

IOP Conf. Series: Materials Science and Engineering 1271 (2022) 012009 doi:10.1088/1757-899X/1271/1/012009

$$\alpha_{sl} = \arcsin\left(\frac{(d - p \cdot cos(-\gamma + \varphi))}{l_s}\right) \tag{85}$$

$$F_{sl} = \frac{M_{zl}}{p \cdot (\cos(\alpha_{sl}) \cdot \cos(-\gamma + \varphi) + \sin(\alpha_{sl}) \cdot \sin(-\gamma + \varphi))} \tag{86}$$

$$F_{zll} = F_{sl} \cdot \cos(\alpha_{sl}) \tag{87}$$

$$F_{zl} = F_{zld} + F_{zll} \tag{88}$$

$$M_{tu} = F_{zl} \cdot \frac{Dzzl}{2} \tag{89}$$

where:

- α_{sd} – is angle between right steering arm and Y-axis,
- F_{sd} – is force in the right steering arm,
- F_{zld} – force in the rack, coming from the right wheel,
- α_{sl} – is angle between left steering arm and Y-axis,
- F_{sl} – is force in the left steering arm,
- F_{zll} – is force in the rack, coming from the left wheel
- F_{zl} – is total force in the rack,
- M_{tu} – steering wheel torque and
- $Dzzl$ – pinion gear diameter.

4. Mathematical model validation

The last portion in the development of a mathematical model is validation. The best form of mathematical validation would be an experimental load test of the steering system. Due to the limiting circumstances, no experimental measurements were performed on the vehicle, but the obtained results of the mathematical model were compared with the results obtained in the ADAMS Car software package.

A vehicle motion simulation was performed in the ADAMS Car software package, where the vehicle moves in a straight line at a speed of 50 $^{km}/_{h}$, when the steering wheel starts turning to the right up to 70°, after which it returns to the initial position. Figure 13 shows the change in the angle of rotation of the steering wheel during the simulation.

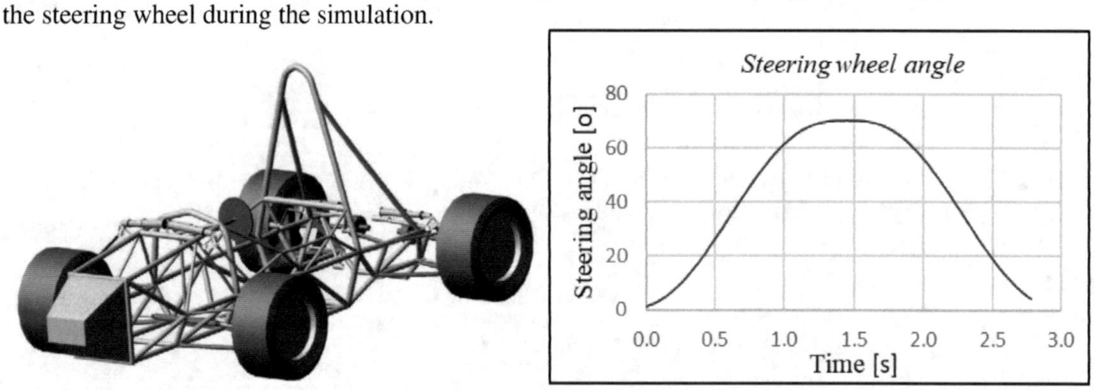

Figure 13. Steering wheel angle during the simulation in the ADAMS Car software package.

The first part of the analysis refers to the mathematical model of the tire, that is the "Magic Formula.". Figure 14 shows a comparison of the tire characteristics obtained in the ADAMS Car software package and the mathematical model for the calculation of the steering system for different values of the vertical force. Figure is showing the change in aligning torque that occurs as a result of tire deformation. This analysis has two goals, the first goal is to check whether the same tire is used (due to the large influence on results) in the simulation and the mathematical model, while the second goal is to check how much the results obtained by the mathematical model deviate from the results obtained in the ADAMS Car software package.

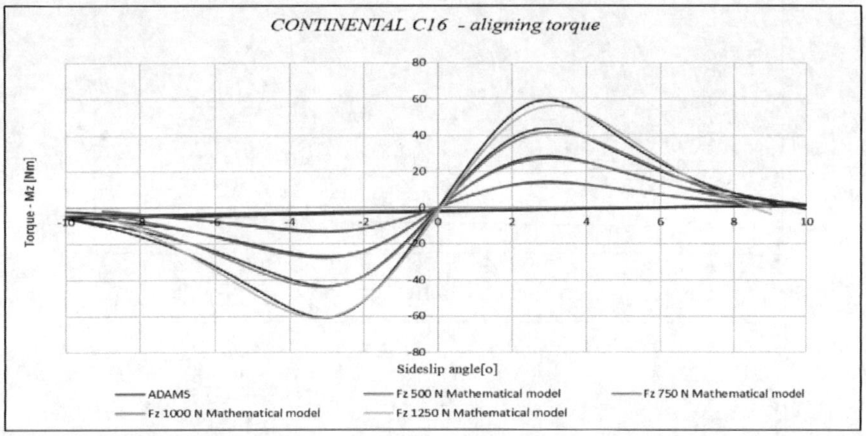

Figure 14. Change in the intensity of the aligning torque due to tire deformation.

Further analysis, also represents a check of the mathematical model of the tire, and is related to the sideslip angle. The results shown in Figures 15 and 16 represent a comparative view of the change in the sideslip angle.

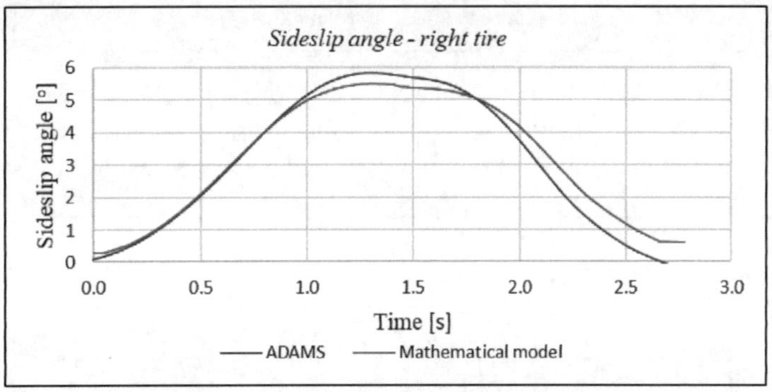

Figure 15. Right tire sideslip angle.

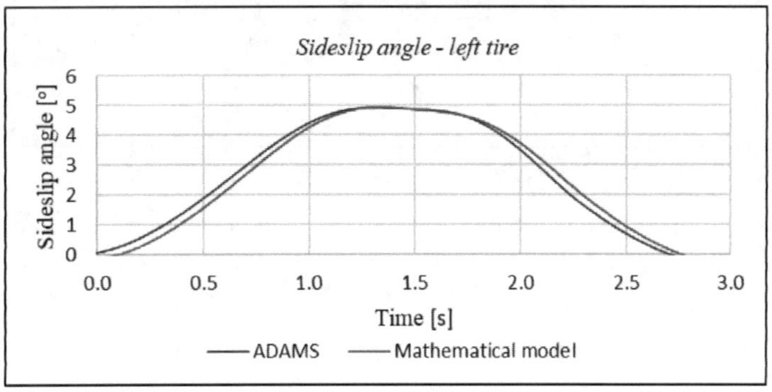

Figure 16. Left tire sideslip angle.

The second part of the analysis refers to the load transfer from the wheel to the rack and further to the steering wheel. Figures 17 and 18 show the change in force intensity in the rack from both the right and left wheel.

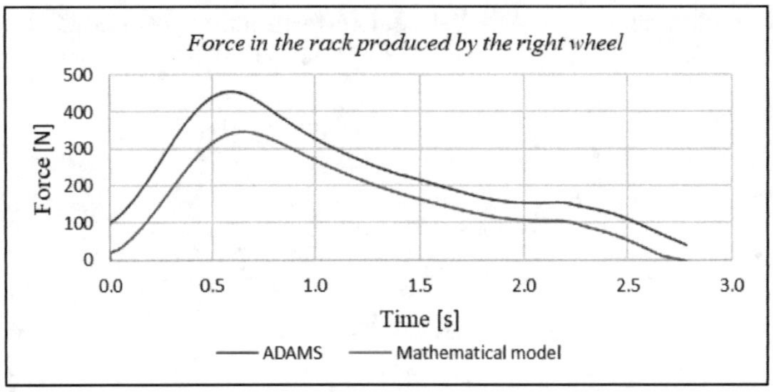

Figure 17. Change in force intensity in the rack produced by the right wheel.

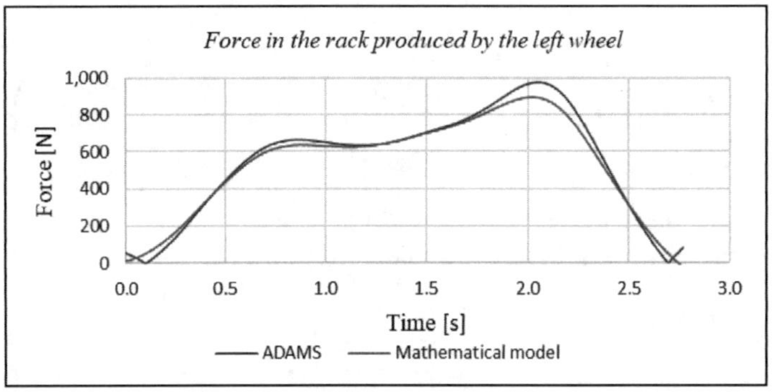

Figure 18. Change in force intensity in the rack produced by the left wheel.

The last part of the analysis relates to the torque on the steering wheel. This is the most important analysis, because the value of this torque is extremely important when designing a steering system.

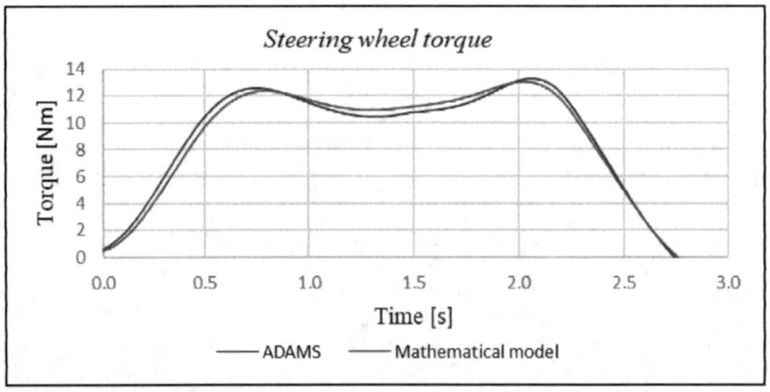

Figure 19. Change in the intensity of the torque on the steering wheel.

The analysis of the obtained results leads to the conclusion that the average value of the relative load error of the elements of the steering system is 15%, the error of the steering wheel angles is 7%, and the torque error on the steering wheel is 6%, which is shown in Figures 20 to 24.

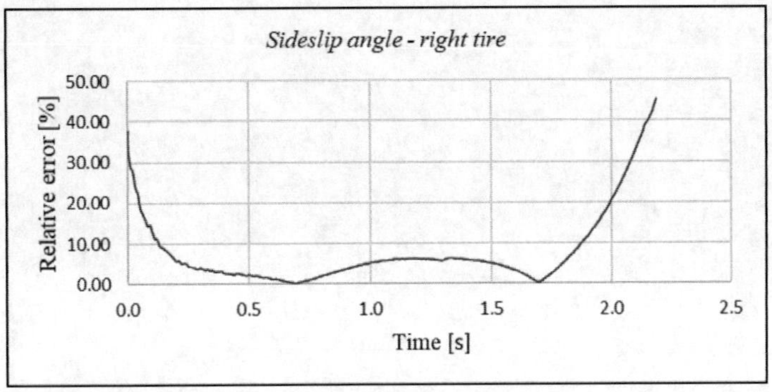

Figure 20. Relative tire sideslip angle error – right tire.

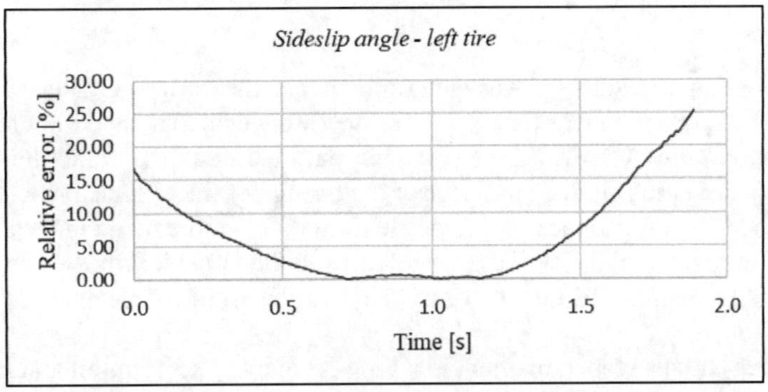

Figure 21. Relative tire sideslip angle error – left tire.

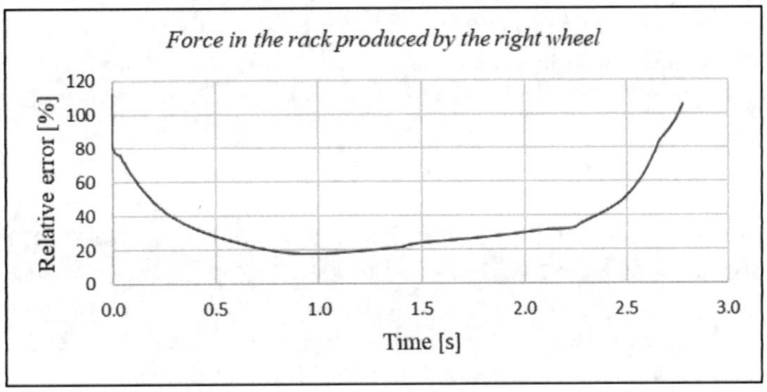

Figure 22. Relative load error of steering system elements – right wheel.

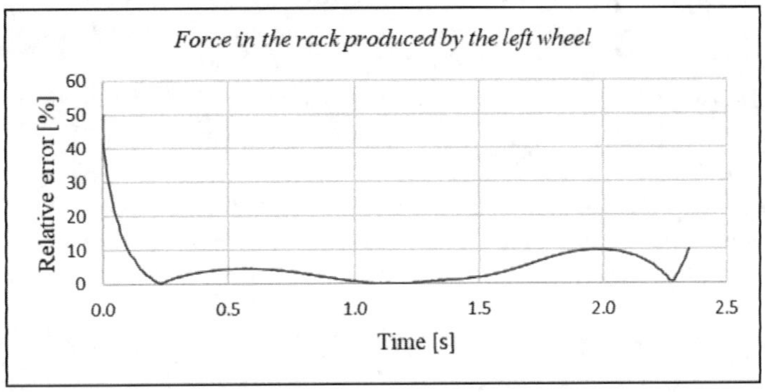

Figure 23. Relative load error of steering system elements – left wheel.

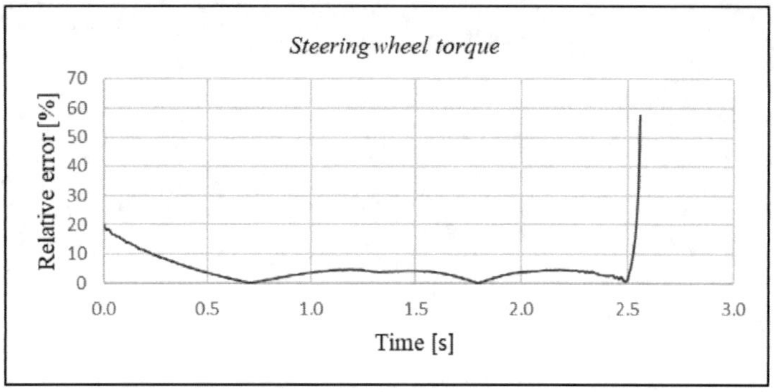

Figure 24. Relative steering wheel torque error.

5. Conclusion

By comparing the obtained results, it was determined that the results obtained using the software package are within 10% relative error for torque on steering wheel. There are deviations in relation to the results obtained in the ADAMS Car software package, but they are the result of adopted simplifications. Also, deviations in the character of the change of the obtained results in a time interval of about 1.8 seconds are a consequence of one of the limitations of the mathematical model. According to Figure 13, in that time interval the steering wheel turns to the left side, while the mathematical model allows calculation only for turning the steering wheel to the right side, which is why the deviation occurred.

It is concluded that the developed mathematical model met the set requirements, and as such it can be used to evaluate the operation of the steering system, that is, it can be used for the calculation and design of the steering system for the formula student vehicle. The mathematical model can be expanded so that calculations can also be made in the case of turning the vehicle to the left. Also, by using the MATLAB software package, a sophisticated application can be created with the help of which calculations and optimization of the formula student vehicle steering system would be done relatively quickly and in a simple way.

References

[1] Milliken W and Milliken D 2003 *Race Car Vehicle Dynamics* (Warrendale: SAE International)
[2] Smith C 1978 *Tune to Win* (Fallbrock: Aero Publishers)
[3] Pacejka H 2012 *Tire and Vehicle Dynamics* (Oxford: Butterworth-Heinemann)
[4] Trzesniowski M 2008 *Rennwagentechnik* (Graz: Springer Vieweg Wiesbaden)
[5] Rajamani R 2005 *Vehicle Dynamics and Control* (Minnesota: Springer)
[6] Gillespie T 1992 *Fundamentals of Vehicle Dynamics* (Warrendale: SAE International).
[7] Jazar R 2008 *Vehicle Dynamics, Theory and Application* (New York: Springer)
[8] Popovic Z 2007 *Teorija kretanja motornih vozila* (Beograd: Vojna akademija)
[9] MATLAB 2018 *9.7.0.1190202 (R2019b)* (Natick: MathWorks Inc.)
[10] Jonson A and Olsson E 2016 *A Metodology for indentification od Magic Formula Tire Model Parameters from In-Vehicle Measurements* (Gothenburg: Chalmers University of Technology)
[11] Harsh D 2018 Full Vehicle Model of a formula student car (Delft: Delft University of Technilogy)
[12] Continental 2016 *Competition Tire 2016 (C16) – Documentation* (Hanover: Continental)
[13] Ojbouboh I 2012 *Data algorithm for vehicle dynamics and tire modeling* (Boulder: University of Colorado)

Lightweight Materials for Automobiles

Nina Busarac[1], Dragan Adamovic[1], Nenad Grujovic[1] and Fatima Zivic[1]

[1] University of Kragujevac, Faculty of Engineering, Sestre Janjic 6, 34000 Kragujevac, Serbia

busaracn@gmail.com

Abstract. This paper deals with vehicle lightweighting, as a strategy to help attain sustainability in the automotive industry by facilitating improved fuel economy. We reviewed innovative materials appropriate for the manufacturing of low-carbon vehicles (LCVs), such as advanced high-strength steel (AHSS), aluminum alloys, magnesium alloys, as well as novel composite materials commonly used for lightweight construction applications. Research shows that vehicle curb weight greatly affects fuel consumption. Primary weight reduction refers to body-in-white (BIW), which can subsequently lead to secondary weight reductions in terms of engine and powertrain size. This review takes into account the environmental aspect of the car body material and the possibility of closed-loop recycling, especially for aluminum and magnesium alloys.

1. Introduction

Non-governmental organizations (NGOs) and governments worldwide are putting pressure on original equipment manufacturers (OEMs) in the automotive sector to increase fuel efficiency and sustainability. European Union set the fleet average CO_2 emission per kilometer value at 95 g/km for the year 2021 [1], and Corporate Average Fuel Economy (CAFE) requirements in the US are set at 55.8 miles per gallon (mpg) for 2025 [2]. UK government set out The Net Zero Strategy with a goal to reach net-zero emissions by 2050, implying a future emphasis on electric vehicles [3]. With the ever-increasing environmental protection requirements, manufacturers have turned to lightweighting, as vehicle curb weight greatly affects fuel consumption and therefore greenhouse gas (GHG) emissions. For internal combustion engine (ICE) vehicles, a 10% reduction in vehicle weight can potentially result in a 6 to 8% savings in fuel consumption [4–9]. On the other hand, battery electric vehicles (BEVs) need to employ the lightweighting strategy to offset the heavy battery weight, which amounts to 70% of the BEV weight by some estimates [4], and also to increase the vehicle range.

It is important to mention that official policies are not the only driving forces of change toward low-carbon vehicles (LCVs). A rise in consumer awareness, as well as the threat of energy insecurity, play a major role too.

As we said, battery weight is a big constraint for electric and hybrid vehicles, but batteries are not the only components making today's cars heavier. Integrating more safety functions and vehicle connectivity equipment (vehicle-to-vehicle or V2V wireless networks etc) does so too [4]. Additional mass brought about by these systems needs to be offset through weight reductions.

Primary weight reduction refers to body-in-white (BIW), which can subsequently lead to lesser power needs in terms of acceleration and braking. A lighter car body would require a potentially smaller engine and powertrain, which would represent a secondary weight reduction or a phenomenon known as mass decompounding [10]. It was and still is common to produce BIW from steel pressings that are then welded together. Such construction makes up about 20% of vehicle weight [11]. If we substituted steel for aluminum alloys, BIW weight reduction would be around 23% [11]. Overall vehicle weight reduction has an impact on fuel economy in the range of 0.3 to 0.4 liter per 100 km per 100 kg of removed mass [2].

Looking at the bigger picture, novel light materials could require lighter tooling, and potentially facilitate higher production rates. Another indirect cost saving can be found in easier transportation of lighter vehicles to dealerships [12].

Lightweighting encompasses many aspects, one of them being material selection, where the multi-material concept is regarded as the future [13]. In this concept, the best materials are chosen for each application [6]. This means moving away from traditional mild steel and towards advanced high-strength steel (AHSS), aluminum, magnesium, and composite materials, according to part requirements. Car parts made of different optimally chosen materials need to be joined together. This is where difficulties in joining dissimilar materials must be overcome. Aluminum alloys and AHSS are hard to weld and the risk of galvanic corrosion and undesirable intermetallic compounds should be considered. Aluminum superplastic forming is a possible strategy to minimize the number of welds, by creating continuous complex parts [6]. To minimize welding manufacturers could design unibodies or monocoques. A drawback of this strategy is that in case of a crash the whole construction needs to be changed since it is essentially one piece [8]. Further, joints of polymer-based fiber composites and metals are problematic due to internal stresses induced by temperature changes, because of significantly different thermal expansion coefficients of these materials [6].

There are more aspects to consider when switching to advanced lightweight materials. Traditional low-carbon steel cannot just be substituted with contemporary counterparts while keeping the old sheet metal designs, and characterization and validation methods. New materials bring new manufacturing challenges to tackle, such as understanding failure mechanisms, establishing new material testing methods, and then design guidelines. Carbon Fiber Reinforced Composites (CFRC) followed this path prior to application in series production, which will be further discussed in a later section.

The recent pandemic caused supply chain disturbances, halts in production, imposed restrictions, etc, that weakened the performance of automakers and impacted the overall economy. This could hinder the growth of the lightweight material market. Despite this, some estimates suggest that the market will reach $126.76 billion (USD) by 2030 globally and $50.45 billion (USD) in Europe, from $71.65 billion and $26.15 billion estimated in 2020, respectively [14]. Metals are expected to hold the largest share in the market for lightweight materials for electric vehicles [8].

Beyond fuel economy, other benefits that make lightweight materials attractive are the possibility of achieving higher levels of performance [12], in terms of acceleration, cornering, top speeds, stopping distances, etc; and increased safety (lesser mass factoring into kinetic energy).

Lastly, improving sustainability in the automotive industry also means finding end-of-life vehicle (ELV) solutions. European Directive on end-of-life vehicles requires the reuse and recovery be a minimum of 95% by an average weight per vehicle and year [15]. Researchers are working on tighter loop and closed-loop recycling for aluminum alloys [16] and magnesium alloys [17], which would decrease primary production-related emissions.

2. Advanced High-Strength Steel

Steels have long been used in automaking applications due to their high modulus of elasticity, strength, formability, weldability, crashworthiness, etc, and low cost. Automotive steel can be plain carbon steel, high-strength steel (HSS), or advanced high-strength steel (AHSS). Advanced high-strength steels are characterized by having tensile strength higher than 550 MPa [18], allowing for smaller cross-sections, as well as better load-carrying capacity and crashworthiness. First-generation AHSS encompasses dual phase (DP) steels, transformation-induced plasticity (TRIP) steels, complex phase (CP) steels, and martensitic (MS) steels, while the second-generation refers to twinning-induced plasticity (TWIP) steels, lightweight steels with induced plasticity (L-IP), and austenitic stainless steel (AUST.SS) [18]. Figure 1 illustrates the comparison between properties of conventional steel, first- and second-generation AHSS, as well as a possible third generation.

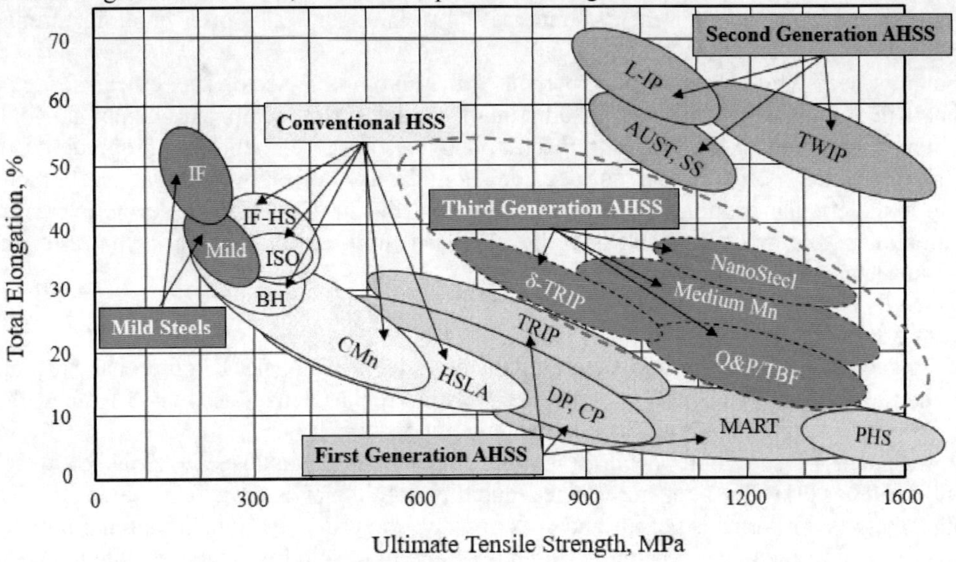

Figure 1. Comparison between properties of different steels.

Switching from low-carbon steel to AHSS can reduce component weight by up to 25% [6,7]. Other lightweighting materials such as light alloys and carbon fiber composites could offer even greater weight savings but AHSSs have advantages in terms of cost, although these steel grades deteriorate stamping molds faster than lesser grades [7]. Another possible disadvantage is the metastable nature of the AHSS microstructure, which can be disrupted during welding [6] due to heat input.

3. Aluminum Alloys

Material density is an important parameter to consider for lightweighting applications. The density of aluminum is about one-third that of steel, meanwhile, the strength-to-density ratio of aluminum alloys is comparable to first-generation AHSS. Other advantages of aluminum alloys are lower modulus of elasticity (in terms of collision energy absorption), good machinability, formability, castability, etc, plus switching to aluminum in many cases would not require significant reengineering [19].

Aluminum is a fast-growing lightweight material. The average aluminum content per vehicle for European manufacturers is estimated to reach 198.8 kg by 2025, while global premium brands have long surpassed that value, e. g. Jaguar and Land Rover with Al content around 500 kg [19]. An example of aluminum-intensive BEV could be Tesla Model S, with its body and chassis made almost entirely of aluminum and total Al content of 190 kg [20]. Interestingly, using aluminum for automobile construction is not a new idea, e. g. Ford's Model T Touring Car had used aluminum body panels prior to 1918 when the material of choice was switched to steel because of lower cost [7]. Nowadays, that course is changing, due to new circumstances.

The use of aluminum alloys can allow for 30% to 60% vehicle weight reduction [7]. Following the mixed material design approach, Al and AHSS can be used in conjunction to replace mild steel, such as the example of the Ford F-150 which achieved a weight reduction of more than 300 kg by combining an AHSS frame and Al body [6].

Combining dissimilar metals such as aluminum and steel brings up welding difficulties, because of the difference in properties such as thermal and electrical conductivity, melting temperature, etc, the formation of brittle intermetallic compounds, and possible galvanic corrosion. Intermetallic compounds are mainly problematic for arc and laser welding, while sold-state friction welding offers higher quality joints, and friction stir welding is especially convenient for long linear joints. Friction stir spot welding was also developed to substitute resistance spot welding, for dissimilar joint applications. Despite its issues, welding is still preferred to riveting or adhesive bonding, because both of these other options require overlap, therefore increasing weight. They also offer less joint strength and have longer cycle times for assembly [21].

As aluminum is becoming increasingly important in the automotive sector it is crucial to define strategies that will minimize the amount of aluminum ending up as scrap. Aluminum could be considered a permanent material [22], that is a material that does not lose its intrinsic properties in the recycling loop, or in other words, it does not degrade significantly. For that reason use of aluminum can contribute to the circular economy. One of the ideas for closing the production cycle is creating lightweight aluminum foam panels or extruded profiles from non-recyclable aluminum waste, that could be used in vehicle production [23].

4. Magnesium Alloys

Magnesium is a good candidate for a lightweighting material due to being 75% lighter than steel and 33% lighter than aluminum [6]. The 2017 Chrysler Pacifica replaced its steel rear liftgate with a magnesium one, which resulted in a significantly lighter component [7].

As the lowest density structural metal [24], Mg offers clear advantages in terms of weight reduction, but it still has drawbacks regarding price, ductility, joining, and corrosion behavior [6]. The oxide layer that magnesium forms is porous and does not have the protective function of aluminum or steel oxide layers. Due to melting temperature differences and immiscibility, fusion welding between magnesium and steel is not realistic. Other methods of joining with different materials can lead to galvanic corrosion, which does not accord with multi-material concepts. Formability is also an issue – drawing depths are limited and elevated processing temperatures are a necessity. These and other challenges explain why magnesium accounts for, on average, less than 1% of vehicle weight [24]. Overcoming them could allow automakers to exploit Mg to its full potential.

If magnesium were to be implemented in automotive applications on a larger scale, primary production related emissions and energy consumption (which are higher than for steel or aluminum) could become a sustainability issue [25]. In accord with the circular economy model, researchers working on an EU-funded project found a way to recycle magnesium from industrial waste and produce low-cost high-performance alloys, by combining it with secondary rare earth elements, and designed a prototype tailgate [26].

5. Polymers

Polymers are already widely used in automotive applications and lightweighting is only intensifying it. Plastics are not only lightweight but also have high absorption properties, favorable from the point of safety standards. ICE vehicles require high-temperature resistant special engineering plastics, which could decrease in importance if the trend of vehicle electrification continues. Instead, BEV components will be made of polyamides (battery housing), polycarbonate (sensors and LEDs), polypropylene (car interior and exterior), etc [27]. Polymers are also gaining importance in form of polymer matrix composites, also called reinforced plastics.

6. Composite Materials

When strict fuel savings are required, non-metallic materials such as polymer matrix composites are the answer. Adequate composites can be up to 35% lighter than aluminum and 60% lighter than steel [5]. While being much lighter, composites can also be up to 10 times stronger than steel and 8 times stronger than aluminum [9].

Automakers started experimenting with composite materials long ago. In the 1950s, Chevy Corvette cars were made with fiber-reinforced plastic (FRP) body panels, which marked the first widespread use of FRPs in the automotive industry. Sheet molding compounds or SMCs (resin systems reinforced by fabric) started being used in the 1970s [7]. Today, carbon fiber reinforced polymers (CFRP), glass fiber reinforced polymers (GFRP) and even natural fiber alternatives are some of the relevant advanced composite materials that show great promise.

Carbon fiber composites are characterized by good stiffness and strength, can be made into complex shapes, and show potential vehicle weight reduction of 50% to 70% [7]. One of the biggest issues for CFRPs, besides joining to metals, is the high price. The price of automotive-grade CFRP can be up to 3900% that of plain steel. Glass fiber reinforced polymers are less expensive, with the price around 400% that of steel [8,14]. Carbon fiber is a more expensive choice, but it has better properties. It is 30% lighter than glass fiber while being twice as strong. Although carbon fiber production is an energy-intensive process, its use in lightweight vehicles shows energy net savings in overall lifecycle studies [5].

CFRP started as a material for aerospace and other high-tech applications, due to initial cost, but it is now expanding towards the mainstream. Currently, it is used in sports cars, luxury cars, and some BEVs [2]. When manufacturers are switching from old material, like sheet metal, to a new material like composite panel, new material-specific problems arise, such as anisotropy, ply drop-off [28], edge effect [29], etc. Before carbon fiber composites could be commercially used, scientists and engineers had to investigate and understand fiber/matrix failure mechanisms in order to establish simulation and validation approaches, testing methods, and eventually design car parts safely using this new material.

Another option for non-structural and semi-structural parts is natural fiber reinforcement. There are already some cars in series production that feature natural-fiber composites, such as Porsche 718 Cayman GT4 Clubsport [4]. Some other manufacturers are also implementing natural-fiber composites, like Ford, who is using cellulose fibers in the armrests of the Lincoln MKX [9]. Amongst other natural fibers being explored are bamboo and kenaf [6]. These fibers are additionally interesting from a sustainability standpoint as they are derived from renewable sources.

In addition to polymer matrix composites discussed so far, metal matrix composites are gaining importance with the development of advanced aluminum matrix composites or AMCs, reinforced with ceramic particles like SiC, Al_2O_3, etc [30]. Advanced AMCs show promise for application in BEVs, e. g. components of an axial flux motor [31]. AMCs could also be suitable for other applications, due to their stiffness which can increase structural integrity and decrease vibrations.

On the other hand, titanium monoboride reinforced titanium matrix composites (Ti/TiB) found application in ICE vehicles as a lighter alternative material for engine valves. Ti/TiB composites are a powder metallurgy solution that exploits already good characteristics of Ti and improves formability during hot working. TiB particles suppress grain growth at high temperatures and forging cracks, making Ti/TiB composites suitable for high-temperature net-shape forming. Necessary wear resistance can be obtained at a reasonable cost, by oxidation treatments [32].

7. Other materials

Like the aforementioned Ti matrix composites imply, titanium is another potential lightweighting material. Ti alloys are known for high strength, durability, and corrosion resistance [33], but their high cost and difficulties in casting, forming, and machining prevent them from being widely used in the automotive sector. Lower heat and wear resistance, relative to steel (figure 2), also represent a drawback [32]. Titanium-aluminum alloys are most common in automotive applications, e. g. Ti-6Al-4V alloy used for manufacturing intake valves. To achieve the wear resistance required for the application this alloy needs expensive coatings such as molybdenum thermal spray coating [33]. High costs related to production and surface treatment greatly limit the scope of application of Ti alloys to racing cars and the like.

Figure 2. Comparison of relevant characteristics of TiB reinforced composites, Ti alloy, and steel.

Another important material that has recently shown lightweighting progress is stainless steel. A new family of low-density stainless steel has been developed. The new dispersion-strengthened Fe-Mn-Al-C-Cr system alloys are characterized by 17% lower density than conventional stainless steel, as well as ultra-high strength (>1 GPa) and high ductility [34].

Lightweight porous structures or foams are one more way to decrease vehicle weight. Aluminum foams have been shown as a suitable material for automotive applications due to their strength, stiffness, vibration damping, and energy absorbing capacity. Mechanical properties can be further improved by ceramic, metallic, or other reinforcements [35]. Polymer foams such as modified polypropylene copolymer (co-PP) foam are used for automotive trim parts, due to their flexural properties, impact properties, and sound insulation. Researchers are developing nanocomposites based on co-PP foams reinforced with graphene nanoplatelets, which would improve mechanical properties by regulating cell size and distribution [36].

8. Conclusion

High-performance and safe lightweight vehicles are a necessity for achieving sustainability in the automotive sector. Lightweighting has transcended mere steel substitutions and has for some time now been progressing towards mixed material concepts, accompanied by innovative design and multi-objective optimization. As greater and greater emphasis is being put on battery electric vehicles in the interest of environmentalism, lighter structural materials will be the key to offsetting battery weight and extending vehicle range. Even though we are witnessing an increase in fossil fuel vehicle phase-out tendencies, internal combustion vehicles are still being put on the market, though with tightened fuel economy regulations. Improving fuel economy is, again heavily dependent on weight reductions.

References

[1] Constantin C, Băţăus M, Renţea C and Frăţilă G 2022 Effectiveness of the measures for CO_2 emission reduction in real world *IOP Conf. Ser. Mater. Sci. Eng.* **1220** 012015
[2] Müller N and Ellringmann T 2016 It’s time to reassess materials for automotive lightweighting *Automot. World*
[3] Baker K 2022 Climate change: UK’s net zero target is under threat because there’s no plan to pay for it *Conversat. UK*
[4] Shell C 2020 Vehicle Fuel Economy – Innovative Materials and Processes That Lighten the Load *Cleantech*
[5] Pervaiz M, Panthapulakkal S, Kc B, Sain M and Tjong J 2016 Emerging Trends in Automotive

Lightweighting through Novel Composite Materials *Mater. Sci. Appl.* **07** 26–38
[6] Weber A 2018 Lightweighting Is Top Priority for Automotive Industry *Assem. Mag.*
[7] Mueller M 2016 Timeline: A Path to Lightweight Materials in Cars and Trucks
[8] MarketsandMarkets Research Private Ltd. Automotive Lightweight Material Market by Material (Metal, Composite, Plastic, Elastomer), Application & Component (Frame, Engine, Exhaust, Transmission, Closure, Interior), Vehicle Type (ICE, Electric & Hybrid), Region - Global Forecast to 2025
[9] Ravichandran K 2020 Automotive lightweight materials: The age of the composite *PreScoute*
[10] Alonso E, Lee T M, Bjelkengren C, Roth R and Kirchain R E 2012 Evaluating the Potential for Secondary Mass Savings in Vehicle Lightweighting *Environ. Sci. Technol.* **46** 2893–901
[11] Leggett D 2020 Lightweighting and sustainability *Just Auto*
[12] Czerwinski F 2021 Current Trends in Automotive Lightweighting Strategies and Materials *Materials* **14** 6631
[13] Koenig B 2019 Lightweighting's New Phase *SME*
[14] Bachal O and Mutreja S 2021 Automotive Lightweight Materials Market by Vehicle Type (IC Engine Powered, Electric Powered, and Others), Material Type (Metals, Composites, Plastics, and Elastomer), and Component (Frame, Powertrain, Interior Systems & Components, and Exterior Systems & Components): Global Opportunity Analysis and Industry Forecast, 2021–2030
[15] European Parliament, Council of the European Union 2000 Directive 2000/53/EC of the European Parliament and of the Council of 18 September 2000 on end-of life vehicles
[16] Paraskevas D, Kellens K, Renaldi, Dewulf W and Duflou J R 2013 Closed and Open Loop Recycling of Aluminium: A Life Cycle Assessment perspective *Proceedings of the 11th Global Conference on Sustainable Manufacturing - Innovative Solutions* (Berlin) pp 305–10
[17] D'Errico F, Tauber M and Just M 2022 Magnesium Alloys for Sustainable Weight-Saving Approach: A Brief Market Overview, New Trends, and Perspectives *Current Trends in Magnesium (Mg) Research* ed S S. Sunkari (IntechOpen)
[18] Mallick P K 2012 Advanced materials for automotive applications: An overview *Advanced materials in automotive engineering* ed J Rowe (Cambridge, United Kingdom: Woodhead Publishing) pp 5–27
[19] Adamović D, Vujinović T, Živić F, Živković J and Topalović M 2021 Application of Aluminum and ITS Alloys in the Automotive Industry With Special Emphasis on Wheel Rims *Traffic Transp. Theory Pract.* **6** 87–95
[20] Lutwyche G 2021 Material Innovations That Are Making Vehicles Lighter *Grainger Worrall*
[21] Weber A 2017 New Techniques for Joining Steel and Aluminum *Assem. Mag.*
[22] Moulin J-M 2022 How designers using permanent materials like aluminium contribute to the circular economy *Shapes*
[23] CORDIS 2019 Aluminium recycling produces metal foam for industrial products
[24] Kleinbaum S 2019 Magnesium for Automotive Lightweighting: Status and Challenges *Magnesium Technology 2019* The Minerals, Metals & Materials Series ed V V Joshi, J B Jordon, D Orlov and N R Neelameggham (Cham: Springer International Publishing) pp 13–4
[25] Golroudbary S R, Makarava I, Repo E, Kraslawski A and Luukka P 2022 Magnesium Life Cycle in Automotive Industry *Procedia CIRP* **105** 589–94
[26] CORDIS 2019 New recovery processes for high-performance and low-cost rare earth-magnesium alloys
[27] Khemka P 2019 Plastics in the Automotive Industry – Which Materials Will Be the Winners and Losers? Special Report *GPCA*
[28] Minakuchi S and Takeda N 2020 Ply Curving Termination to Suppress Delamination in Composite Ply Drop-Off *ICAF 2019 – Structural Integrity in the Age of Additive Manufacturing* Lecture Notes in Mechanical Engineering ed A Niepokolczycki and J Komorowski (Cham: Springer International Publishing) pp 124–32

[29] Fan H, Jin F and Fang D 2009 Characterization of edge effects of composite lattice structures *Compos. Sci. Technol.* **69** 1896–903
[30] Koli D K, Agnihotri G and Purohit R 2015 Advanced Aluminium Matrix Composites: The Critical Need of Automotive and Aerospace Engineering Fields *Mater. Today Proc.* **2** 3032–41
[31] Wehring O 2018 Metal matrix composites deliver claimed 40% weight saving for electric motor rotors *Just Auto*
[32] Saito T 2020 Affordable High-performance Ti/TiB Metal Matrix Composites Tailored for Automobile Engine Components: Materials Design, Manufacturing Process, and Remaining Issues *RD Rev. Toyota CRDL* **51** 1–30
[33] Salihu S A, Suleiman Y I, Eyinavi A I and Usman A 2019 Classification, Properties and Applications of titanium and its alloys used in aerospace, automotive, biomedical and marine industry- A Review *Int. J. Precious Eng. Res. Appl.* **4** 23–36
[34] Moon J, Ha H-Y, Kim K-W, Park S-J, Lee T-H, Kim S-D, Jang J H, Jo H-H, Hong H-U, Lee B H, Lee Y-J, Lee C, Suh D-W, Han H N, Raabe D and Lee C-H 2020 A new class of lightweight, stainless steels with ultra-high strength and large ductility *Sci. Rep.* **10** 12140
[35] Parveez B, Jamal N A, Maleque A, Yusof F, Jamadon N H and Adzila S 2021 Review on advances in porous Al composites and the possible way forward *J. Mater. Res. Technol.* **14** 2017–38
[36] Girginer Ozunlu B and Guner F S 2022 An Industrial Case for Polypropylene Nanocomposite Foams: Lightweight, Soundproof Exterior Automotive Parts *Polymers* **14** 1192

A review of hypereutectic aluminum piston materials

Slavica Miladinovic[1], Blaža Stojanović[1], Sandra Gajević[1], Lozica Ivanović[1], Aleksandar Skulić[1]

[1]University of Kragujevac, Faculty of Engineering, Department for mechanical constructions and mechanization, Sestre Janjic 6, 34000 Kragujevac, Serbia

E-mail: slavicam@kg.ac.rs

Abstract. Demands of today's automotive industry are increasing; there is a constant tendency of weight reduction and higher reliability of constructions. Pistons and piston liners for vehicles are mostly manufactured from aluminum-silicon alloys. With different manufacturing methods and/or the addition of some chemical elements, base alloy properties can be modified. Methods for hypereutectic aluminum alloys and composites manufacturing and their influence on material properties will be given in this paper. The influence of Si percentage in base alloy and how an addition of some chemical elements affects the properties of these materials will be observed as well.

1. Introduction

Expansion of automobile industry and industry in general, has created a need to use light weight materials, especially metals. Mostly used light weight metals in vehicles are steel, aluminum, magnesium, plastics and different types of composites. Use of light weight materials in vehicles has a great importance due to their influence on environment and economy. The most widely used lightweight materials in vehicles are aluminum and its alloys. Some useful properties of aluminum are good heat and electrical conductivity, light weight (only one third of the density of steel), corrosion and wear resistance, high recyclability, good castability and weldability [1].

Usage of aluminum and its alloys is increasing every year. In 2006 there was estimation, in [2], of how much the consumption of aluminum is going to increase globally. In figure 1 is represented aluminum percentual usage for every sector in year 2020, where can be observed that aluminum was mostly used for construction, followed by automotive and transportation, foil and packaging, etc.

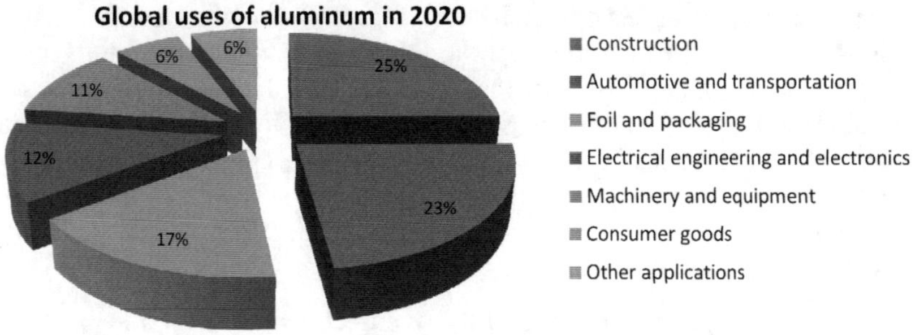

Figure 1. Major global use for different sectors in year 2020. [3]

Aluminum and its alloy in vehicles are used for: power train, chassis and suspension and car body. In power trains aluminum is used for engine parts (pistons, engine blocks, cylinder heads), fuel system and liquid lines; in chassis aluminum is used for cradle, axis wheels and suspension components; and in car body is used for hoods, bumper, interiors, doors, etc. [4, 5]. On figure 2 are illustrated car parts made of aluminum and its alloys.

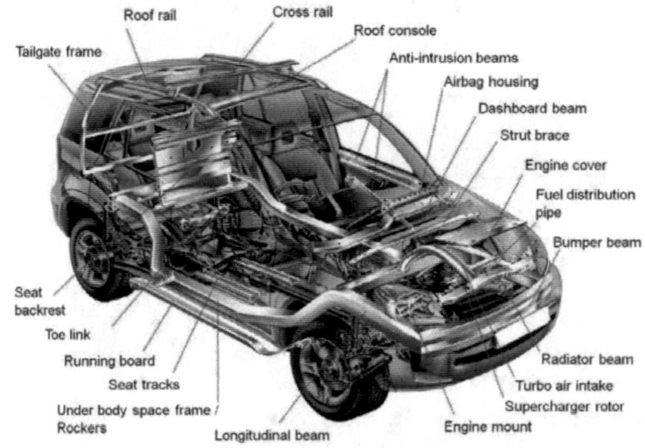

Figure 2. Car body parts made of aluminum and its alloys [6]

One of aluminum alloys that is used mostly for engine parts, especially pistons, due to their high wear and corrosion resistance, good thermal conductivity and low thermal expansion and small specific gravity [5, 7, 8] are hypereutectic aluminum alloys. The mentioned properties depend on the microstructure of hypereutectic aluminum alloys, dendrite cell size, morphology and distribution of primary and eutectic silicon particles. Primary silicon morphology depends of conditions during solidification, alloying elements and chemical composition, and it can induce lower wear resistance, and also negatively affect the machinability and mechanical properties of these alloys. Some of the techniques by which this can be mitigated i.e. which can refine primary and eutectic silicon particles in hypereutectic aluminum alloys are given in [9]. The application of electric current on the cast microstructure of Al-13wt%Si (Al13Si) and Al-20wt%Si (Al20Si) was investigated. It was observed that after electric current application during solidification of Al13Si and Al20Si, for all cases except for bottom of Al20Si, there was reduction of the average dendrite cell size. Size of largest primary silicon particles wasn't changed but there was an increase in amount of primary silicon particles of a smaller size in the Al20Si. The electric current didn't cause any changes in Al13Si. Properties of hypereutectic aluminum alloys can be changed by addition of modifiers like Phosphorus, Erbium, Cobalt, Magnesium, Strontium, etc, and/or refiners like Al-1Ti-3B. Modifiers are added to the alloy in purpose of refinement of Si particles (primary and eutectic) as well as better and more homogenous distribution of particles. In case that there is iron present in the alloy it acts like impurity reducing wear rate, but with addition of Manganese it is possible to null the effect of iron on the alloy and increase wear rate [10-14]. As a modifier in hypereutectic alloys was even used Calcium in [15]. Addition of Calcium resulted in refinement of primary silicon particles and enhancement of mechanical properties especially elongation. In further enhancement of hypereutectic aluminum alloys, it is possible to add reinforcement to them and by it creating composite. As reinforcements for composites with aluminum base are most commonly used alumina, silicon carbide, basalt, glass fibre, carbon fibre and even biomaterials such as bamboo fibres [16]. There aren't a large number of investigations of hypereutectic aluminum composites, because these materials are new, but in recent years there have been some investigations.

In this paper the manufacturing methods and mechanical characteristics of hypereutectic aluminum alloys and composites will be reviewed.

2. Methods for manufacturing of hypereutectic aluminum alloys

Aluminum alloys and composites, in general, can be manufactured by liquid state processing, semi-solid processing and solid state processes. Liquid state processing are different types of casting such as gravity casting, stir casting, squeeze casting, ultrasonic assisted casting and others. Semi-solid process is newer manufacturing method which includes rheocasting and tixocasting, while solid process includes powder metallurgy processing, friction stir processing, diffusion bonding and sintering [17, 18].

Hypereutectic alloys and composites are, usually, manufactured by gravity casting. Nowadays most of these materials are manufactured by centrifugal casting. Next to the gravity casting and centrifugal casting manufacturing methods for hypereutectic aluminum alloys are: stir casting, pressure die casting, tixocasting, rheocasting and some very new methods like selective laser melting, electron beam melting and spray-deposition.

2.1 Gravity casting
Gravity casting or just conventional casting, is one of the oldest processes which includes melting of the material and if needed (depending of the material which is being manufactured) addition of refinement/modifier/reinforcement, and pouring the melt into the mould. The name is given because the pouring into the mould is done with the influence of gravity without any pressure. In [15] was investigated influence of calcium addition on B390 alloy with the chemical composition Al–18.7Si-4.0Cu–0.56Mg–0.68Zn–0.13Mn–1.0Fe (wt%) produced by gravity casting. The alloy was melt and held at 750°C and treated with Titanium chloride (Ti2Cl6) for modification, and after 30 min the melt was poured into mould. Al4.5wt%Ca and AlCuP were added to the base alloy as well, and after observing mechanical properties, it can be seen that with decrease in Calcium content, the tensile strength increases as well as elongation, while the size of primary silicon increases. Minimum size of primary Si was 20.3μm. In [11] Al–30Si alloys were melted and degassed with C_2Cl_6 for 15 min, and the modification treatment was carried out with the addition of Al–3P and Al–10Sr master alloys at different temperatures (770 °C, 1000°C and 1300°C). It was concluded that with addition of P and Sr there was refinement of Si particles and tensile strength was increased. In [19] the mechanical properties of hypereutectic Al17.5Si (Al–17.5Si–4.5Cu–1Zn–0.7Mg–0.5Ni) alloy were investigated. Specimens of hypereutectic alloy were hot deformed and aged for 4–16h at temperatures of 120, 150 and 180°C. For specimen aged at 120°C there was no significant difference in ultimate tensile strength when aging time increased from 4 to 16h. When aging time at temperature at 180°C increased the ultimate tensile strength decreased. The best results gave the specimen aged at 150°C, where with the increase of aging time ultimate tensile strength increased and was highest for aging time of 16h (396 MPa).

2.2 Stir casting
Stir casting is a process where melt is stirred continuously and then pouring into the mould. In the production of composites there is also degassing and the reinforcement addition during the stirring of the melt. Degassing is applied due to oxidation during the stirring which adversely affects properties of base material [20]. In [21] Al17Si hypereutectic aluminum alloy was reinforced with 2wt% graphite and/or 2 wt% carbon fibers and produced as-cast (only matrix alloy) and stir casted by vortex method. [22] Vortex method includes stirring of the melt at high speed to form vortex and then adding reinforcement at the side of the vortex. Highest ultimate tensile strength and hardness are observed for the hybrid composite [21]. Another hypereutectic AlSi composite, with base of Al22Si alloy, reinforced with graphene (0.25 and 1%), is manufactured by stir casting by vortex method in [23]. Hardness of specimens was measured at edge and center of the casted cylinders. The highest hardness was for composite with 0.25% of graphene nanosheets, followed by composite with 1% of graphene nanosheets and stir casted base alloy. Characteristics of stir casted A390 alloy (Al17Si) with the addition of Mg were compared before and after T6 treatment in [24]. Amount of Mg in the alloy was 0.5-10%. Hardness of untreated samples increased with the addition of Mg gradually, but was lower than hardness of samples that were T6 treated. For T6 treated samples highest hardness was for alloy with 0.5%Mg, and the lowest for alloy with 5%Mg, which means that even if hardness increases with Mg for untreated samples after T6 that is not the case.

2.3 Centrifugal casting
Centrifugal casting of materials is casting where the mould cavities are filled with the influence of centrifugal force that pushes molten metal to the walls of the mould. This type of casting is usually used for rings, hollow cylinders and other hollow parts [25]. Centrifugally casted hypereutectic alloys Al18Si reinforced at the piston head with SiC were investigated in [26]. During the material manufacturing two different slurry and mould temperatures (850, 800°C and 600, 500°C respectively)

were applied at the constant rotational speed of the mould of 800rpm. All the results were compared to Mahle piston (17-19wt%Si). The highest hardness value had the piston obtained with combination of slurry/mould temperatures 850/600°C respectively at the piston head. A variation of centrifugal casting is centrifuge casting which means that there is rotating arm centrally mounted on motor shaft, and on one side of rotating arm is mould and on the other is counterweight. Manufacturing of Al18Si alloy by centrifuge casting was done in [27], and during the manufacturing pouring temperatures were varied (750 and 850°C) while arm rotational speed was constantly 200 rpm. With increase of pouring temperature the microstructure is better and due to centrifugal force there is higher primary silicon content at top region of specimen, providing better mechanical characteristics.

2.4 Other manufacturing processes

Squeeze casting is a form of liquid state processing that includes application of pressure during solidification of casted material in the mould [28]. Squeeze casted Al 15-22Si alloys were compared to the gravity casted ones and the squeezed casted samples had higher hardness and ultimate tensile strength, and even better elongation. The casted alloy Al22Si had the highest hardness (≈76HV) and Al17.5Si had the highest ultimate tensile strength (153MPa) [29].

As already said, next to the liquid state processes there are semi solid state processes. First ever semi solid process that was commercialized was tixocasting. Tixocasting is a type of process where billets that are pre casted are reheated to the semi-solid casting temperature, and then the material is casted again [18]. Advantage of tixocasting is better quality of the product due to uniform globular microstructure, while some disadvantages are higher cost and the fact that scraps from production cannot be recycled [30]. Tixocasted and T6 treated aluminum alloys with 15.3%Si and 15.7%Si were investigated in [31], and results were compared to casted alloy Al15.9Si. Hardness before T6 treatment was the highest for Al15.7Si (77HRB≈145.41HV), while after T6 treatment the casted alloy Al15.9Si had the highest hardness of 91HRB which was ≈197.5HV. T6 treatment didn't have a great impact on wear of the alloys. Same three alloys were compared with Al16.11Si produced by lost foam technology, Al16.2Si and Al19.85Si produced by squeeze casting and T6 treated [32]. For all alloys except casted Al15.9Si Phosphorus was used as primary Si refiner. Before T6 treatment, like in [31], hardness was the highest for Al15.7Si alloy, but after T6 treatment, hardness was the highest for alloy Al16.11Si (93HRB≈207.5HV). Observation of hardness and microstructure of A390 (Al16.83Si) alloy fabricated by cooling slope (CS) casting and tixocasting was done in [33]. CS casting is type of casting where ingot of material is melted and then cooled to the temperature slightly above liquid temperature, but in [33] this temperature was slightly under liquidus temperature. Melted alloy was then poured on the inclined steel plate leading to the mould. These ingots were used for tixocasting, and some tixocasted specimens were treated with T5 and T6 heat treatment. Even without heat treatments, tixocasted specimens had higher hardness than specimens fabricated by CS casting.

Electric pulse modification (EPM) can be applied in order to control solidification of alloys. In [34] this technology is applied during the solidification of Al13Si alloy rich with Ni. EPM was applied to the melted alloy for 30s at pulse frequency of 3Hz at different pulse voltages (0-700V). Samples without EPM had coarse primary Si and large eutectic Si unevenly distributed and a high amount of α-Al dendrites around which Ni-rich phases formed. With application of EPM to the melted Al13Si alloy with the increase in voltage the hardness increased.

Another modification of hypereutectic AlSi alloys is possible by electron beam melting of their surface. In [35] EBM is applied on A390 hypereutectic (16-18%Si) AlSi alloy. More about this process can be found in the mentioned literature. It's basically use of plasma and magnetic field to help form an electron beam that was used for surface remelting. Voltages used for EBM are from 0-36kV. With the increase in voltage the hardness increased and was the highest for AlSi alloy treated with EBM at 36kV. Solid state process-powder metalurgy fabricated Al17Si alloy and forged was investigated in [36] in the as-forged condition (AF) and after direct quenching (DQ). After DQ in water hardness of hypereutectic AlSi alloy was slightly increased.

3. Microstructure

Microstructure of manufactured alloys and composites influences their mechanical and tribological characteristics.

Gravity casted materials. Microstructure of gravity casted B390 alloy was noted in [15] and it consisted of: primary Si particles, eutectic Si, α dendrite, eutectic α, Al_8Fe_2Si compound and Al_5FeSi intermetallic compound. With the addition of Calcium of 16ppm primary Si particles were refined and well distributed in the alloy. But higher amount of Calcium leads to increase of size of primary Si particles and it affects their distribution. Gravity casted alloy Al30Si in its microstructure had coarse irregularly shaped primary Si particles (platelets and polygon of average size 203.8μm), while eutectic Si was needle shaped. With the addition of Sr the primary Si particles size decreased and eutectic Si had granular structure. With the addition of P primary Si can be refined even to 32.8μm, and Sr addition mostly modifies eutectic Si [11]. In [19] microstructure of as-cast (Al17.5Si) alloy shows fine primary Si particles, α-aluminum dendrites, eutectic Si and complex intermetallic phases, and the microstructure was much finer than that of the commercial Al-Si alloy. After T6 treatment and aging at 150°C for 16h the α-aluminum dendritic structure was eliminated. In samples aged for 16h at 120°C no precipitates were observed, only some spheroidal particles were dissipated in the matrix. In samples aged at 150°C for 16h two major precipitates were observed: the rod-shaped precipitates (of size of 20 and 3nm) and lath-shaped precipitates (rectangular or circular cross section ≈2-3nm) dispersed in the matrix. With increasement of the aging time in samples aged at 180°C can be seen that with increase of aging time density of the plate-shaped precipitate decreases while their size increases.

Stir casted materials. In [22] SEM metalography showed uniform distribution of reinforcements (graphite and/or carbon fibers) in hypereutectic Al-Si base. Metalographic examination of samples obtained in [24] for as-cast material contains mainly of primary Si, eutectic Si and eutectic α. Stir casted samples with or without graphene show that stirring can affect Si particle size, shape and distribution. In samples obtained by stirring of base material the Si particles are globular and finer but not very well distributed. The better distribution and refinement of Si particles was achieved with addition of low graphene content (0.25-1wt%) which resulted in better mechanical characteristics.

Centrifugal casted materials. Microstructure of centrifugaly casted piston with hypereutetic AlSi base depends of the combination slurry/mould temperatures. In [26] in microstructure of piston obtained with slurry/mould temperature combination 800/500°C there were no zones observed. When temperatures increase there are three zones that can be distinguished: zone with a number of pores and oxides which is called the defects gathering zone, zone that doesn't contain SiC particles and impurities (mostly contains α-Al dendrite and eutectic Si) and is called the metal matrix zone and the zone where SiC particles are segregated (zone where SiC particles are well distributed in the base material and embedded in primary Si can be found SiC particles). In the last zone, the SiC segregation zone, is observed that with decrease in slurry and mould temperature there is decrease in number of SiC particles and some pores appear. Higher pouring temperature and centrifugal force influence that the distribution of Si particles is better and primary Si content is higher at the top region resulting in higher ultimate tensile strength and hardness at the top of specimen [27].

Squeeze casted materials. When compared to gravity casted samples in [29] squeeze casted samples had much finer structure, eutectic Si phase was fine, size and amount of primary Si particles was reduced and there were α-Al dendrites present in the microstructure. With increase of Si content in the alloy concentration of primary Si particles and their size were increasing while the concentration of primary α-Al dendrite decreased.

Tixocasted materials. When compared microstructure of casted alloy Al15.9Si and thixocasted alloys Al15.3Si and Al15.7Si; casted alloy Al15.9Si in its microstructure had a eutectic Al-Si phase which was very fine and some clusters of voluminous primary Si particles. Different from that the thixocasted alloys Al15.3Si and Al15.7Si had a uniform microstructure with globules of α aluminum, high amount of primary Si particles and small fraction of eutectic Si [31, 32]. In [32] alloy Al15.9Si produced by ingot metallurgy had rough primary Si particles and not homogenous distribution in its microstructure. Al16.11Si produced by lost foam technology had a high amount of primary Si particles in its microstructure, and squeeze casted alloys Al16.2Si and Al19.85Si had fine microstructure. Influence of T6 treatment on observed materials resulted in some changes in their microstructure; eutectic Si was globular and tended to agglomerate, while the most intermetalic phases dissolved. Effect of microstructure like disposition of particles and their size on hardness is high. In this case hardness is affected by amount of primary Si particles and their size, as well as volumetric amount of Al2Cu particles in the alloys. Lower pouring temperature of CS casted specimens in [33] resulted in formation of primary Si particles of a small size, well dispersed and surrounded by α-Al rosettes. In the microstructure of CS casted specimens were some very small intermetalic particles present, and because of their small size they weren't identifiable. Tixoformed specimens had α-Al globules in its microstructure and between those globules were Si particles. After reheating CS casted specimens for tixocasting aluminum rosettes have started spheroidisation and coarsening. Primary Si particles have coarsened as well. Tixocasted sample had no porosity, while die casted sample had finer microstructure but porosity was present.

Casted material with applied EPM. Samples without EPM had coarse primary Si and large eutectic Si unevenly distributed and a high amount of α-Al dendrites around which Ni-rich phases formed. With the application of EPM the following changes to the microstructure of the alloy Al13Si occur: with increase of pulse voltage size of primary Si particles decreases, α-Al dendrite crystals disappear, and Ni-rich phases around α-Al are refined [34].

4. Mechanical properties of hypereutectic AlSi alloys and composites

There are many influential factors on mechanical properties of hypereutectic AlSi alloys and composites. In this paper was investigated how manufacturing process and composition of hypereutectic AlSi alloys and composites affects their mechanical properties and is shown in table 1.

Table 1. Review of mechanical properties of hypereutectic AlSi alloys and composites

Reference	**[11]**	**[19]**	**[15]**	**[37]**	**[21]**
Manufacturing process	gravity casting	gravity casting, T6 and aged	gravity casting	gravity casting	stir casting
Base material (wt%Si)	Al30Si	Al17.5Si	B390 (Al18.7Si)	Al16.69Si	Al17Si
Type of reinforcement/ modificator	P and Sr	-	Ti_2Cl_6, Al –4.5 wt-%Ca followed by P addition of 30 ppm of the weight of the melt, using an AlCuP	-	Gr and short carbon fibers
Size of reinforcement	-	-	-	-	Gr: 10-30μm carbon fibers: 1:1000 at 2wt%
Amount of reinforcement [wt%]	-	-	0.6	-	2
Hardness [HV]	-	-	-	112.65	120-139
Tensile strength/Ultimate tensile strength [MPa]	50-135	282-396	125-162	250	144-169

Table 1. Review of mechanical properties of hypereutectic AlSi alloys and composites-continued

Reference	**[23]**	**[24]**	**[26]**	**[27]**	**[31]**
Manufacturing process	stir casting	stir cast and T6 heat treatment	centrifugal casting	centrifuge casting	tixoforming and T6
Base material (wt%Si)	Al22Si	A390 (Al–17Si–4.5Cu–0.5Mg)	Al18Si	Al-18Si	Al15Si
Type of reinforcement/ modificator	graphene nanosheets	Mg	SiC particles	-	-
Size of reinforcement		-	15 and 30 μm	-	-
Amount of reinforcement	0.25-1%	0.5, 6 and 10%	17.2vol%	-	-
Hardness [HV]	124-137	as-cast 116-131 after T6 139-163HV	Al18Si 129-227 Mahle piston 115-160	750°C 81 850°C 34.5-38 83	tixocasted 116.89-187.57 after T6 120.67- 197.5
Tensile strength/Ultimate tensile strength	-	-	-	-	-

IOP Conf. Series: Materials Science and Engineering 1271 (2022) 012011 doi:10.1088/1757-899X/1271/1/012011

Table 1. Review of mechanical properties of hypereutectic AlSi alloys and composites-continued

Reference	[32]	[33]	[30]	[35]	[35]
Manufacturing process	ingot metallurgy-casting lost-foam technology, tixoforming squeeze casting and T6	coling slope casting and tixoforming, and T5 and T6 treated tixoformed material	gravity casting squeeze casting	EBM	Casting with electric pulse solidification method
Base material (wt%Si)	Al15.3-19.85Si	Al16.83Si	Al 15-22Si	Al16-18Si	Al13Si
Type of reinforcement/ modificator/refiner	P	-	-	-	-
Size of reinforcement	-	-	-	-	-
Amount of reinforcement	-	-	-	-	-
Hardness [HV]	as-cast 85-142 after T6 140-199	die cast 68-92 tixocasted 86-147	gravity casting 52-62 squeeze casting 60-76	as-cast 69.1-93.9 EBM 89.7-188.8	132.12-144.56
Tensile strength/Ultimate tensile strength	-	-	gravity casting 92.33-115.67 squeeze casting 113.33-153	-	60-70

A lot of hypereutectic AlSi alloys and composites with different Si content are observed in table 1. In some papers only tensile strength/ultimate tensile strength was observed and from those papers we can conclude that with aging time tensile strength increases. Higher Si content doesn't mean that tensile strength is higher, but it is hard to compare due to lack of investigations (not enough number of papers to make a proper comparison). When compared to gravity casting all other manufacturing methods give higher hardness. As for microstructure maybe the best ones are obtained for centrifugal, stir casted and tixocasted materials. After T6 treatment hardness increases in some cases significantely.

5. Conclusions

There are not many researches about neither mechanical characteristics of hypereutectic aluminum composites nor of their tribological characteristics. That is the reason this research was observing hypereutectic aluminum alloys as well. The conclusions that can be drawn from this research are following:

- There are many fabrication methods, and a lot more will be developed in future;
- All fabrication methods for hypereutectic AlSi alloys and composites have their advantages and disadvantages, the biggest disadvantage is cost of production and complexity of the process;
- Fabrication methods, heat treatments and Si content affect microstructure and mechanical properties of hypereutectic AlSi alloys;
- Due to cost and simplicity of the process, maybe, the best processes for fabrication of hypereutectic AlSi alloys and composites are centrifugal and stir casting.

6. References

[1] Lumley R N 2011 *Fundamentals of Aluminium Metallurgy: Production, processing and applications* ed R Lumley (Cambridge: Woodhead Publishing) pp 1-19

[2] Menzie W D, Barry JJ, Bleiwas D I, Bray E L, Goonan T G and Matos G 2010 The global flow of aluminum from 2006 through 2025: U.S. Geological Survey Open-File Report 2010–1256, 73 p., available at https://pubs.usgs.gov/ofr/2010/1256/.

[3] Aluminum facts. https://www.nrcan.gc.ca/our-natural-resources/minerals-mining/minerals-metals-facts/aluminum-facts/20510 accessed 23.07.2022.

[4] Hirsch J 2014 Recent development in aluminium for automotive applications *Trans. Nonferrous Met. Soc. China,* **24**, 1995–2002.

[5] Gong C, Tu H, Wu C, Wang, J and Su X 2018 Study on Microstructure and Mechanical Properties of Hypereutectic Al–18Si Alloy Modified with Al–3B *Materials* **11** 456. https://doi.org/10.3390/ma11030456

[6] Barberis Pinlung S, 2015, 7xxx Aluminum Sheets for Automotive Applications, *Electronic Theses and Dissertations* 5504.

[7] Li J, Elmadagli M, Gertsmana, V Y, Lo J and Alpas A T 2006 FIB and TEM characterization of subsurfaces of an Al–Si alloy (A390) subjected to sliding wear *Mater. Sci. Eng.* A **421** 317–327.

[8] Elmadagli M, Perry T and Alpas A T 2007 A parametric study of the relationship between microstructure and wear resistance of Al–Si alloys *Wear* **262** 79–92.

[9] Plotkowski A J 2012 Refinement of the Cast Microstructure of Hypereutectic Aluminum-Silicon Alloys with an Applied Electric Potential *Masters Theses* 15.

[10] Xing P, Gao B, Zhuang Y and Liu K 2010 On the modification of hypereutectic Al-Si alloys using rare earth Er *Acta Metall. Sin. (Engl. Lett.)* **23** 327-333.

[11] Zuo M, Zhao D, Teng X, Geng H and Zhang, Z 2013 Effect of P and Sr complex modification on Si phase in hypereutectic Al–30Si alloys *Mater. Des.* **47** 857-864.

[12] Rao P V C S, Devi A S and Kumar K G B 2012 Influence of melt treatments on dry sliding wear behavior of hypereutectic Al-15Si-4Cu cast alloys, *Jordan j. mech. ind. eng.* **6** 55-61.

[13] Sha M, Wu S, Wang X, Wan L and An P 2012 Effects of cobalt content on microstructure and mechanical properties of hypereutectic Al–Si alloys *Mater. Sci. Eng. A* **535** 258-263.

[14] Abouei V, Shabestari S G and Saghafian H 2010 Dry sliding wear behaviour of hypereutectic Al-Si piston alloys containing iron-rich intermetallics *Mater. Charact.* **61** 1089-1096.

[15] Kim H-J 2003 Effect of calcium on primary silicon particle size in hypereutectic Al–Si alloys *Mater. Sci. Technol.* **19** 915-918.

[16] Jordan S P, Christian G and Jeffs S P 2021 An Overview on Aluminum Matrix Composites: Liquid State Processing *World Academy of Science, Engineering and Technology, World Acad Sci Eng Technol International Journal of Materials and Metallurgical Engineering* **15** 43-49.

[17] Kawther W A Al-Helal 2013 New Approaches to Casting Hypereutectic Al-Si Alloys to Achieve Simultaneous Refinement of Primary Silicon and Modification of Eutectic Silicon, PhD Thesis, Brunel Centre for Advanced Solidification Technology (BCAST), Brunel University.

[18] Midson S P and Jackson A 2006 A Comparison of Thixocasting and Rheocasting *Proceedings of 67th World foundry congress* vol **1** *(June 5 – 7, 2006, Harrogate, UK)* pp 1081-1090

[19] He K, Yu F-X, Zhao D-Z and Zuo L 2012 Characterization of precipitates in a hot-deformed hypereutectic Al–Si alloy *J. Alloys Compd.* **539** 74–81

[20] Annigeri Veeresh Kumar U. K. G. B. 2017 Method of stir casting of Aluminum metal matrix composites: a review *Mater. Today Proc.* **4** 1140–1146

[21] Ramesh C S and Prasad T B 2009 Friction and Wear Behavior of Graphite-Carbon Short Fiber Reinforced Al–17%Si Alloy Hybrid composites *J. Tribol.* **131** 014501-1-014501-5.

[22] Pal A, Srivastava A, Singh AK, Singh V, Bajpai A, Kumar A and Kandpal B C 2015 A Review on Technological Advancements in Aluminium Based Metal Matrix Composites *Int. j.*

comput. sci. math. sci. **4** 4-9

[23] Alghamdi A S, Ramadan M, Abdel Halim K S and Fathy N 2018 Microscopical characterization of cast hypereutectic Al-Si alloys reinforced with graphene nanosheets *Eng. Technol. Appl. Sci. Res.* **8** 2514–2519.

[24] Hekmat-Ardakan A, Liu X, Ajersch F and Chen X-G 2010 Wear behaviour of hypereutectic Al–Si–Cu–Mg casting alloys with variable Mg contents , *Wear* **269** (2010) 684–692

[25] Campbell J 2015 *Complete casting handbook: metal casting processes, metallurgy, techniques and design* (Oxford: Butterworth-Heinemann Elsevier Ltd) chapter 16 pp 821-882

[26] Huang X, Liu C, Lv X, Liu G and Li F 2011 Aluminum alloy pistons reinforced with SiC fabricated by centrifugal casting *J. Mater. Process. Technol.* **211** 1540–1546

[27] Mallikarjun, Sukruth M, Chethan K S and Dr. Kiran Aithal S 2017 Study of Functionally Graded Al-Si Alloy Fabricated by Centrifuge Casting Technique *International Journal of Advances in Scientific Research and Engineering (IJASRE)* **3** 70-75

[28] Wyatt L M 1994 *Mechanical Engineer's Reference Book* 12 ed E H Smith (Oxford: Butterworth-Heinemann Elsevier Ltd) chapter 16.5 pp 998-1109

[29] Li R, Liu L, Zhang L, Sun J, Shi Y and Yu, B 2017 Effect of Squeeze Casting on Microstructure and Mechanical Properties of Hypereutectic Al-xSi Alloys *J. Mater. Sci. Technol.* **33** 404–410.

[30] Husain N H, Ahmad A H and Rashidi M M 2017 An overview of thixoforming process 4th International Conference on Mechanical Engineering Research vol 257 (1 – 2 August 2017, Kuantan, Malasya) IOP Conf. Ser.: Mater. Sci. Eng. pp 012053

[31] Lasa L and Rodriguez-Ibabe J M 2002 Effect of composition and processing route on the wear behaviour of Al–Si alloys *Scripta Materialia* **46** 477-481

[32.] Lasa L and Rodriguez-Ibabe J M 2003 Wear behaviour of eutectic and hypereutectic Al–Si–Cu–Mg casting alloys tested against a composite brake pad *Mater. Sci. Eng.:A* **363** 193-202

[33] Birol Y 2008 Cooling slope casting and thixoforming of hypereutectic A390 alloy *J. Mater. Process. Technol.* **207** 200-203

[34] Wang B, Wang J, Tang L, Qi J and Wang J 2016 Effect of electric pulse modification on microstructure and properties of Ni-rich Al-Si piston alloy *China Foundry* **13** 139-142.

[35] Walker J C, Murray J, Narania S and Clare A T 2012 Dry Sliding Friction and Wear Behaviour of an Electron Beam Melted Hypereutectic Al–Si Alloy *Tribol. Lett.* **45** 49–58.

[36] Timmermans G and Froyen L 1999 Fretting wear behaviour of hypereutectic P/M Al–Si in oil environment *Wear* **230** 105-117

[37] Al-Samarai R A, Haftirman Ahmad K R and Al-Douri Y 2012 The influence of roughness on the wear and friction coefficient under dry and lubricated sliding *Int. j. sci. eng. res.* **3** 1–6.

Acknowledgments

This paper presents the results obtained during research within the framework of the project TR 35021, supported by the Ministry of Education, Science and Technological Development of the Republic of Serbia.

IOP Conf. Series: Materials Science and Engineering 1271 (2022) 012012 doi:10.1088/1757-899X/1271/1/012012

A review of hypereutectic aluminum piston materials

S Miladinovic[1], S Gajević[1], L Ivanović[1], A Skulić[1], B Stojanović[1]

[1]University of Kragujevac, Faculty of Engineering, Department for mechanical constructions and mechanization, SestreJanjic 6, 34000 Kragujevac, Serbia

E-mail: slavicam@kg.ac.rs

Abstract. Demands of today's automotive industry are increasing; there is a constant tendency of weight reduction and higher reliability of constructions. Pistons and piston liners for vehicles are mostly manufactured from aluminum-silicon alloys. With different manufacturing methods and/or the addition of some chemical elements, base alloy properties can be modified. Methods for hypereutectic aluminum alloys and composites manufacturing and their influence on material properties will be given in this paper. The influence of Si percentage in base alloy and how an addition of some chemical elements affects the properties of these materials will be observed as well.

1. Introduction

Expansion of automobile industry and industry in general, has created a need to use light weight materials, especially metals. Mostly used light weight metals in vehicles are steel, aluminum, magnesium, plastics and different types of composites. Use of light weight materials in vehicles has a great importance due to their influence on environment and economy. The most widely used lightweight materials in vehicles are aluminum and its alloys. Some useful properties of aluminum are good heat and electrical conductivity, light weight (only one third of the density of steel), corrosion and wear resistance, high recyclability, good castability and weldability [1].

Usage of aluminum and its alloys is increasing every year. In 2006 there was estimation, in [2], of how much the consumption of aluminum is going to increase globally. In figure 1 is represented aluminium percentual usage for every sector in year 2020, where can be observed that aluminum was mostly used for construction, followed by automotive and transportation, foil and packaging, etc.

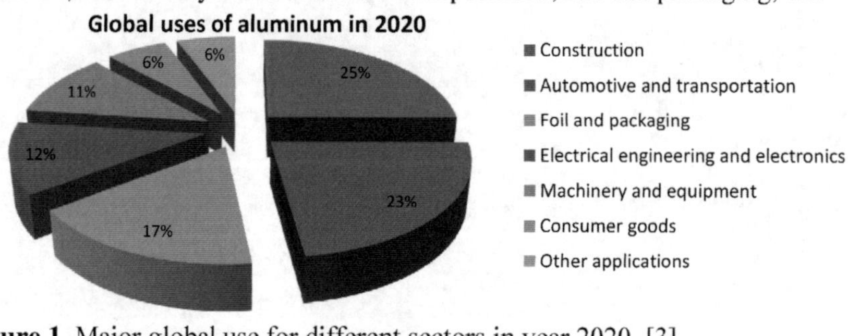

Figure 1. Major global use for different sectors in year 2020. [3]

Aluminum and its alloy in vehicles are used for: powertrain, chassis and suspension and car body. In power trains aluminum is used for engine parts (pistons, engine blocks, cylinder heads), fuel system and liquid lines; in chassis aluminum is used for cradle, axis wheels and suspension components; and in car body is used for hoods, bumper, interiors, doors, etc. [4, 5]. On figure 2 are illustrated car parts made of aluminum and its alloys.

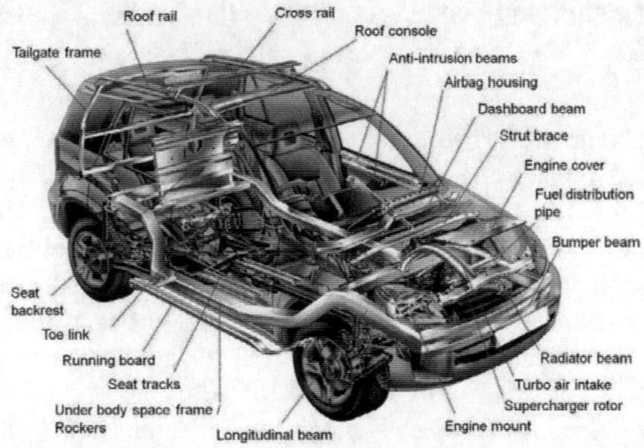

Figure 2. Car body parts made of aluminum and its alloys [6]

One of aluminum alloys that is used mostly for engine parts, especially pistons, due to their high wear and corrosion resistance, good thermal conductivity and low thermal expansion and small specific gravity [5,7, 8] are hypereutectic aluminum alloys. The mentioned properties depend on the microstructure of hypereutectic aluminum alloys, dendrite cell size, morphology and distribution of primary and eutectic silicon particles. Primary silicon morphology depends of conditions during solidification, alloying elements and chemical composition, and it can induce lower wear resistance, and also negatively affect the machinability and mechanical properties of these alloys. Some of the techniques by which this can be mitigated i.e. which can refine primary and eutectic silicon particles in hypereutectic aluminum alloys are given in [9]. The application of electric current on the cast microstructure of Al-13wt%Si (Al13Si) and Al-20wt%Si (Al20Si) was investigated. It was observed that after electric current application during solidification of Al13Si and Al20Si, for all cases except for bottom of Al20Si, there was reduction of the average dendrite cell size. Size of largest primary silicon particles wasn't changed but there was an increase in amount of primary silicon particles of a smaller size in the Al20Si. The electric current didn't cause any changes in Al13Si. Properties of hypereutectic aluminum alloys can be changed by addition of modifiers like Phosphorus, Erbium, Cobalt, Magnesium, Strontium, etc, and/or refiners like Al-1Ti-3B. Modifiers are added to the alloy in purpose of refinement of Si particles (primary and eutectic) as well as better and more homogenous distribution of particles. In case that there is iron present in the alloy it acts like impurity reducing wear rate, but with addition of Manganese it is possible to null the effect of iron on the alloy and increase wear rate [10-14]. As a modifier in hypereutectic alloys was even used Calcium in [15]. Addition of Calcium resulted in refinement of primary silicon particles and enhancement of mechanical properties especially elongation. In further enhancement of hypereutectic aluminum alloys, it is possible to add reinforcement to them and by it creating composite. As reinforcements for composites with aluminum base are most commonly used alumina, silicon carbide, basalt, glass fibre, carbon fibre and even biomaterials such as bamboo fibres [16]. There aren't a large number of investigations of hypereutectic aluminum composites, because these materials are new, but in recent years there have been some investigations.

In this paper the manufacturing methods and mechanical characteristics of hypereutectic aluminum alloys and composites will be reviewed.

2. Methods for manufacturing of hypereutectic aluminum alloys

Aluminum alloys and composites, in general, can be manufactured by liquid state processing, semi-solid processing and solid state processes. Liquid state processing are different types of casting such as gravity casting, stir casting, squeeze casting, ultrasonic assisted casting and others. Semi-solid process is newer manufacturing method which includes rheocasting and tixocasting, while solid process includes powder metallurgy processing, friction stir processing, diffusion bonding and sintering [17,18].

Hypereutectic alloys and composites are, usually, manufactured by gravity casting. Nowadays most of these materials are manufactured by centrifugal casting. Next to the gravity casting and centrifugal casting manufacturing methods for hypereutectic aluminum alloys are: stir casting, pressure die casting, tixocasting, rheocasting and some very new methods like selective laser melting, electron beam melting and spray-deposition.

2.1Gravity casting

Gravity casting or just conventional casting, is one of the oldest processes which includes melting of the material and if needed (depending of the material which is being manufactured) addition of refinement/modifier/reinforcement, and pouring the melt into the mould. The name is given because the pouring into the mould is done with the influence of gravity without any pressure. In [15] was investigated influence of calcium addition on B390 alloy with the chemical composition Al–18.7Si-4.0Cu–0.56Mg–0.68Zn–0.13Mn–1.0Fe (wt%) produced by gravity casting. The alloy was melt and held at 750°C and treated with Titanium chloride (Ti2Cl6) for modification, and after 30 min the melt was poured into mould. Al4.5wt%Ca and AlCuP were added to the base alloy as well, and after observing mechanical properties, it can be seen that with decrease in Calcium content, the tensile strength increases as well as elongation, while the size of primary silicon increases. Minimum size of primary Si was 20.3μm. In [11] Al–30Si alloys were melted and degassed with C_2Cl_6 for 15 min, and the modification treatment was carried out with the addition of Al–3P and Al–10Sr master alloys at different temperatures (770 °C, 1000°C and 1300°C). It was concluded that with addition of P and Sr there was refinement of Si particles and tensile strength was increased. In [19] the mechanical properties of hypereutectic Al17.5Si (Al–17.5Si–4.5Cu–1Zn–0.7Mg–0.5Ni) alloy were investigated. Specimens of hypereutectic alloy were hot deformed and aged for 4–16h at temperatures of 120, 150 and 180°C. For specimen aged at 120°C there was no significant difference in ultimate tensile strength when aging time increased from 4 to 16h. When aging time at temperature at 180°C increased the ultimate tensile strength decreased. The best results gave the specimen aged at 150°C, where with the increase of aging time ultimate tensile strength increased and was highest for aging time of 16h (396 MPa).

2.2Stir casting

Stir casting is a process where melt is stirred continuously and then pouring into the mould. In the production of composites there is also degassing and the reinforcement addition during the stirring of the melt. Degassing is applied due to oxidation during the stirring which adversely affects properties of base material [20]. In [21] Al17Si hypereutectic aluminum alloy was reinforced with 2wt% graphite and/or 2 wt% carbon fibers and produced as-cast (only matrix alloy) and stir casted by vortex method. [22] Vortex method includes stirring of the melt at high speed to form vortex and then adding reinforcement at the side of the vortex. Highest ultimate tensile strength and hardness are observed for the hybrid composite [21]. Another hypereutectic AlSi composite, with base of Al22Si alloy, reinforced with graphene (0.25 and 1%), is manufactured by stir casting by vortex method in [23]. Hardness of specimens was measured at edge and center of the casted cylinders. The highest hardness was for composite with 0.25% of grapheme nanosheets, followed by composite with 1% of grapheme nanosheets and stir casted base alloy. Characteristics of stir casted A390 alloy (Al17Si) with the addition of Mg were compared before and after T6 treatment in [24]. Amount of Mg in the alloy was 0.5-10%. Hardness of untreated samples increased with the addition of Mg gradually, but was lower than hardness of samples that were T6 treated. For T6 treated samples highest hardness was for alloy with 0.5%Mg, and the lowest for alloy with 5%Mg, which means that even if hardness increases with Mg for untreated samples after T6 that is not the case.

2.3Centrifugal casting

Centrifugal casting of materials is casting where the mould cavities are filled with the influence of centrifugal force that pushes molten metal to the walls of the mould. This type of casting is usually used for rings, hollow cylinders and other hollow parts [25]. Centrifugally casted hypereutectic alloys Al18Si reinforced at the piston head with SiC were investigated in [26]. During the material manufacturing two different slurry and mould temperatures (850, 800°C and 600, 500°C respectively) were applied at the constant rotational speed of the mould of 800rpm. All the results were compared to Mahle piston (17-19wt%Si). The highest hardness value had the piston obtained with combination of slurry/mould temperatures 850/600°C respectively at the piston head. A variation of centrifugal casting is centrifuge casting which means that there is rotating arm centrally mounted on motor shaft, and on one side of rotating arm is mould and on the other is counterweight. Manufacturing of Al18Si alloy by centrifuge casting was done in [27], and during the manufacturing pouring temperatures were varied (750 and 850°C) while arm rotational speed was constantly 200 rpm. With increase of pouring temperature the microstructure is better and due to centrifugal force there is higher primary silicon content at top region of specimen, providing better mechanical characteristics.

2.4Other manufacturing processes

Squeeze casting is a form of liquid state processing that includes application of pressure during solidification of casted material in the mould [28]. Squeeze casted Al 15-22Si alloys were compared to the gravity casted ones and the squeezed casted samples had higher hardness and ultimate tensile strength, and even better elongation. The casted alloy Al22Si had the highest hardness (≈76HV) and Al17.5Si had the highest ultimate tensile strength (153MPa) [29].

As already said, next to the liquid state processes there are semi solid state processes. First ever semi solid process that was commercialized was tixocasting. Tixocasting is a type of process where billets that are pre casted are reheated to the semi-solid casting temperature, and then the material is casted again [18]. Advantage of tixocasting is better quality of the product due to uniform globular microstructure, while some disadvantages are higher cost and the fact that scraps from production cannot be recycled [30].Tixocasted and T6 treated aluminum alloys with 15.3%Si and 15.7%Si were investigated in [31], and results were compared to casted alloy Al15.9Si. Hardness before T6 treatment was the highest for Al15.7Si (77HRB≈145.41HV), while after T6 treatment the casted alloy Al15.9Si had the highest hardness of 91HRB which was ≈197.5HV. T6 treatment didn't have a great impact on wear of the alloys. Same three alloys were compared with Al16.11Si produced by lost foam technology, Al16.2Si and Al19.85Si produced by squeeze casting and T6 treated [32]. For all alloys except casted Al15.9Si Phosphorus was used as primary Si refiner. Before T6 treatment, like in [31], hardness was the highest for Al15.7Si alloy, but after T6 treatment, hardness was the highest for alloy Al16.11Si (93HRB≈207.5HV). Observation of hardness and microstructure of A390 (Al16.83Si) alloy fabricated by cooling slope (CS) casting and tixocasting was done in [33]. CS casting is type of casting where ingot of material is melted and then cooled to the temperature slightly above liquid temperature, but in [33] this temperature was slightly under liquidus temperature. Melted alloy was then poured on the inclined steel plate leading to the mould. These ingots were used for tixocasting, and some tixocasted specimens were treated with T5 and T6 heat treatment. Even without heat treatments, tixocasted specimens had higher hardness than specimens fabricated by CS casting.

Electric pulse modification (EPM) can be applied in order to control solidification of alloys. In [34] this technology is applied during the solidification of Al13Si alloy rich with Ni. EPM was applied to the melted alloy for 30s at pulse frequency of 3Hz at different pulse voltages (0-700V). Samples without EPM had coarse primary Si and large eutectic Si unevenly distributed and a high amount of α-Al dendrites around which Ni-rich phases formed. With application of EPM to the melted Al13Si alloy with the increase in voltage the hardness increased.

Another modification of hypereutectic AlSi alloys is possible by electron beam melting of their surface. In [35] EBM is applied on A390 hypereutectic (16-18%Si) AlSi alloy. More about this process can be found in the mentioned literature. It's basically use of plasma and magnetic field to help form an electron beam that was used for surface remelting. Voltages used for EBM are from 0-36kV. With the increase in voltage the hardness increased and was the highest for AlSi alloy treated with EBM at 36kV. Solid state process-powder metalurgy fabricated Al17Si alloy and forged was investigated in [36] in the as-forged condition (AF) and after direct quenching (DQ). After DQ in water hardness of hypereutectic AlSi alloy was slightly increased.

3. Microstructure

Microstructure of manufactured alloys and composites influences their mechanical and tribological characteristics.

Gravity casted materials.Microstructure of gravity casted B390 alloy was noted in [15] and it consisted of: primary Si particles, eutectic Si, α dendrite, eutectic α, Al_8Fe_2Si compound and Al_5FeSi intermetallic compound. With the addition of Calcium of 16ppm primary Si particles were refined and well distributed in the alloy. But higher amount of Calcium leads to increase of size of primary Si particles and it affects their distribution. Gravity casted alloy Al30Si in its microstructure had coarse irregularly shaped primary Si particles (platelets and polygon of average size 203.8μm), while eutectic Si was needle shaped. With the addition of Sr the primary Si particles size decreased and eutectic Si had granular structure. With the addition of P primary Si can be refined even to 32.8μm, and Sr addition mostly modifies eutectic Si [11].In [19] microstructure of as-cast (Al17.5Si)alloy shows fine primary Si particles, α-aluminum dendrites, eutectic Si and complex intermetallic phases, and the microstructure was much finer than that of the commercial Al-Si alloy. After T6 treatment and aging at 150°C for 16h the α-aluminum dendritic structure was eliminated. In samples aged for 16h at 120°C no precipitates were observed, only some spheroidal particles were dissipated in the matrix. In samples aged at 150°C for 16h two major precipitates were observed: the rod-shaped precipitates (of size of 20 and 3nm) and lath-shaped precipitates (rectangular or circular cross section ≈2-3nm) dispersed in the matrix. With increasement of the aging time in samples aged at 180°C can be seen that with increase of aging time density of the plate-shaped precipitate decreases while their size increases.

Stir casted materials. In [22] SEM metalography showed uniform distribution of reinforcements (graphite and/or carbon fibers) in hypereutectic Al-Si base. Metalographic examination of samples obtained in [24] for as-cast material contains mainly of primary Si, eutectic Si and eutectic α. Stir casted samples with or without graphene show that stirring can affect Si particle size, shape and distribution. In samples obtained by stirring of base material the Si particles are globular and finer but not very well distributed. The better distribution and refinement of Si particles was achieved with addition of low graphene content (0.25-1wt%) which resulted in better mechanical characteristics.

Centrifugal casted materials. Microstructure of centrifugaly casted piston with hypereutectic AlSi base depends of the combination slurry/mould temperatures. In [26] in microstructure of piston obtained with slurry/mould temperature combination 800/500°C there were no zones observed. When temperatures increase there are three zones that can be distinguished: zone with a number of pores and oxides which is called the defects gathering zone, zone that doesn't contain SiC particles and impurities (mostly contains α-Al dendrite and eutectic Si) and is called the metal matrix zone and the zone where SiC particles are segregated (zone where SiC particles are well distributed in the base material and embedded in primary Si can be found SiC particles). In the last zone, the SiC segregation zone, is observed that with decrease in slurry and mould temperature there is decrease in number of SiC particles and some pores appear. Higher pouring temperature and centrifugal force influence that the distribution of Si particles is better and primary Si content is higher at the top region resulting in higher ultimate tensile strength and hardness at the top of specimen [27].

Squeeze casted materials. When compared to gravity casted samples in [29] squeeze casted samples had much finer structure, eutectic Si phase was fine, size and amount of primary Si particles was reduced and there were α-Al dendrites present in the microstructure. With increase of Si content in the alloy concentration of primary Si particles and their size were increasing while the concentration of primary α-Al dendrite decreased.

Tixocasted materials. When compared microstructure of casted alloy Al15.9Si and thixocasted alloys Al15.3Si and Al15.7Si; casted alloy Al15.9Si in its microstructure had a eutectic Al-Si phase which was very fine and some clusters of voluminous primary Si particles. Different from that the thixocasted alloys Al15.3Si and Al15.7Si had a uniform microstructure with globules of α aluminum, high amount of primary Si particles and small fraction of eutectic Si [31,32]. In [32] alloy Al15.9Si produced by ingot metallurgy had rough primary Si particles and not homogenous distribution in its microstructure. Al16.11Si produced by lost foam technology had a high amount of primary Si particles in its microstructure, and squeeze casted alloys Al16.2Si and Al19.85Si had fine microstructure. Influence of T6 treatment on observed materials resulted in some changes in their microstructure; eutectic Si was globular and tended to agglomerate, while the most intermetalic phases dissolved. Effect of microstructure like disposition of particles and their size on hardness is high. In this case hardness is affected by amount of primary Si particles and their size, as well as volumetric amount of Al_2Cu particles in the alloys. Lower pouring temperature of CS casted specimens in [33] resulted in formation of primary Si particles of a small size, well dispersed and surrounded by α-Al rosettes. In the microstructure of CS casted specimens were some very small intermetalic particles present, and because of their small size they weren't identifiable. Tixoformed specimens had α-Al globules in its microstructure and between those globules were Si particles. After reheating CS casted specimens for tixocasting aluminum rosettes have started spheroidisation and coarsening. Primary Si particles have coarsened as well. Tixocasted sample had no porosity, while die casted sample had finer microstructure but porosity was present.

Casted material with applied EPM. Samples without EPM had coarse primary Si and large eutectic Si unevenly distributed and a high amount of α-Al dendrites around which Ni-rich phases formed. With the application of EPM the following changes to the microstructure of the alloy Al13Si occur: with increase of pulse voltage size of primary Si particles decreases, α-Al dendrite crystals disappear, and Ni-rich phases around α-Al are refined [34].

4. Mechanical properties of hypereutectic AlSi alloys and composites

There are many influential factors on mechanical properties of hypereutectic AlSi alloys and composites. In this paper was investigated how manufacturing process and composition of hypereutectic AlSi alloys and composites affects their mechanical properties and is shown in table 1.

Table 1. Review of mechanical properties of hypereutectic AlSi alloys and composites

Reference	**[11]**	**[19]**	**[15]**	**[37]**	**[21]**
Manufacturing process	gravity casting	gravity casting, T6 and aged	gravity casting	gravity casting	stir casting
Base material (wt%Si)	Al30Si	Al17.5Si	B390 (Al18.7Si)	Al16.69Si	Al17Si
Type of reinforcement/ modificator	P and Sr	-	Ti_2Cl_6, Al –4.5 wt-%Ca followed by P addition of 30 ppm of the weight of the melt, using an AlCuP	-	Gr and short carbon fibers
Size of reinforcement	-	-	-	-	Gr: 10-30μm carbon fibers: 1:1000 at 2wt%
Amount of reinforcement [wt%]	-	-	0.6	-	2
Hardness [HV]	-	-	-	112.65	120-139
Tensile strength/Ultimate tensile strength [MPa]	50-135	282-396	125-162	250	144-169

Table 1. Review of mechanical properties of hypereutectic AlSi alloys and composites-continued

Reference	**[23]**	**[24]**	**[26]**	**[27]**	**[31]**
Manufacturing process	stir casting	stir cast and T6 heat treatment	centrifugal casting	centrifuge casting	tixoforming and T6
Base material (wt%Si)	Al22Si	A390 (Al–17Si–4.5Cu–0.5Mg)	Al18Si	Al-18Si	Al15Si
Type of reinforcement/ modificator	Grapheme nanosheets	Mg	SiC particles	-	-
Size of reinforcement		-	15 and 30 μm	-	-
Amount of reinforcement	0.25-1%	0.5, 6 and 10%	17.2vol%	-	-
Hardness [HV]	124-137	as-cast 116-131 after T6 139-163HV	Al18Si 129-227 Mahle piston 115-160	750°C 81 850°C 34.5-38 83	tixocasted 116.89-187.57 after T6 120.67- 197.5
Tensile strength/Ultimate tensile strength	-	-	-	-	-

Table 1. Review of mechanical properties of hypereutectic AlSi alloys and composites-continued

Reference	[32]	[33]	[30]	[35]	[34]
Manufacturing process	ingot metallurgy-casting lost-foam technology, tixoforming squeeze casting and T6	coling slope casting and tixoforming, and T5 and T6 treated tixoformed material	gravity casting squeeze casting	EBM	casting with electric pulse solidification method
Base material (wt%Si)	Al15.3-19.85Si	Al16.83Si	Al 15-22Si	Al16-18Si	Al13Si
Type of reinforcement/ modificator/refiner	P	-	-	-	-
Size of reinforcement	-	-	-	-	-
Amount of reinforcement	-	-	-	-	-
Hardness [HV]	as-cast 85-142 after T6 140-199	die cast 68-92 tixocasted 86-147	gravity casting 52-62 squeeze casting 60-76	as-cast 69.1-93.9 EBM 89.7-188.8	132.12-144.56
Tensile strength/Ultimate tensile strength	-	-	gravity casting 92.33-115.67 squeeze casting 113.33-153	-	60-70

A lot of hypereutectic AlSi alloys and composites with different Si content are observed in table 1. In some papers only tensile strength/ultimate tensile strength was observed and from those papers we can conclude that with aging time tensile strength increases. Higher Si content doesn't mean that tensile strength is higher, but it is hard to compare due to lack of investigations (not enough number of papers to make a proper comparison). When compared to gravity casting all other manufacturing methods give higher hardness. As for microstructure maybe the best ones are obtained for centrifugal, stir casted and tixocasted materials. After T6 treatment hardness increases in some cases significantely.

5. Conclusions

There are not many researches about neither mechanical characteristics of hypereutectic aluminum composites nor of their tribological characteristics. That is the reason this research was observing hypereutectic aluminum alloys as well. The conclusions that can be drawn from this research are following:

- There are many fabrication methods, and a lot more will be developed in future;
- All fabrication methods for hypereutectic AlSi alloys and composites have their advantages and disadvantages, the biggest disadvantage is cost of production and complexity of the process;
- Fabrication methods, heat treatments and Si content affect microstructure and mechanical properties of hypereutectic AlSi alloys;
- Due to cost and simplicity of the process, maybe, the best processes for fabrication of hypereutectic AlSi alloys and composites are centrifugal and stir casting.

6. References

[1] Lumley R N2011 *Fundamentals of Aluminium Metallurgy: Production, processing and applications* ed R Lumley(Cambridge: Woodhead Publishing)pp1-19

[2] Menzie W D, Barry JJ, Bleiwas D I, Bray E L, Goonan T G and Matos G2010 The global flow of aluminum from 2006 through 2025: U.S. Geological Survey Open-File Report 2010–1256, 73 p., available at https://pubs.usgs.gov/ofr/2010/1256/.

[3] Aluminum facts. https://www.nrcan.gc.ca/our-natural-resources/minerals-mining/minerals-metals-facts/aluminum-facts/20510 accessed 23.07.2022.

[4] Hirsch J 2014 Recent development in aluminium for automotive applications*Trans. Nonferrous Met. Soc. China,* **24**, 1995–2002.

[5] Gong C, Tu H, Wu C, Wang, J and Su X 2018 Study on Microstructure and Mechanical Properties of Hypereutectic Al–18Si Alloy Modified with Al–3B *Materials* **11** 456. https://doi.org/10.3390/ma11030456

[6] Barberis Pinlung S, 2015, 7xxx Aluminum Sheets for Automotive Applications, *Electronic Theses and Dissertations* 5504.

[7] Li J, Elmadagli M, Gertsmana V Y, Lo J and Alpas A T 2006 FIB and TEM characterization of subsurfaces of an Al–Si alloy (A390) subjected to sliding wear*Mater. Sci. Eng*. A **421** 317–327.

[8] Elmadagli M, Perry T and Alpas A T2007 A parametric study of the relationship between microstructure and wear resistance of Al–Si alloys *Wear* **262** 79–92.

[9] Plotkowski A J 2012 Refinement of the Cast Microstructure of Hypereutectic Aluminum-Silicon Alloys with an Applied Electric Potential*Masters Theses* 15.

[10] Xing P, Gao B, Zhuang Y and Liu K 2010 On the modification of hypereutectic Al-Si alloys using rare earth Er *Acta Metall. Sin. (Engl. Lett.)***23**327-333.

[11] Zuo M, Zhao D, Teng X, Geng H and Zhang, Z 2013 Effect of P and Sr complex modification on Si phase in hypereutectic Al–30Si alloys*Mater. Des.***47**857-864.

[12] Rao P V C S, Devi A S and Kumar K G B 2012 Influence of melt treatments on dry sliding wear behavior of hypereutectic Al-15Si-4Cu cast alloys, *Jordan j. mech. ind. eng.***6** 55-61.

[13] Sha M, Wu S, Wang X, Wan L and An P 2012 Effects of cobalt content on microstructure and mechanical properties of hypereutectic Al–Si alloys*Mater. Sci. Eng. A***535** 258-263.

[14] Abouei V, Shabestari S G and Saghafian H 2010 Dry sliding wear behaviour of hypereutectic Al-Si piston alloys containing iron-rich intermetallics *Mater. Charact.***61** 1089-1096.

[15] Kim H-J 2003 Effect of calcium on primary silicon particle size in hypereutectic Al–Si alloys*Mater. Sci. Technol.***19** 915-918.

[16] Jordan S P, Christian G and Jeffs S P 2021 An Overview on Aluminum Matrix Composites: Liquid State Processing *World Academy of Science, Engineering and Technology, World Acad Sci Eng Technol International Journal of Materials and Metallurgical Engineering***15**43-49.

[17] Kawther W A Al-Helal 2013 New Approaches to Casting Hypereutectic Al-Si Alloys to Achieve Simultaneous Refinement of Primary Silicon and Modification of Eutectic Silicon, PhD Thesis, Brunel Centre for Advanced Solidification Technology (BCAST), Brunel University.

[18] Midson S P and Jackson A 2006 A Comparison of Thixocasting and Rheocasting *Proceedings of 67th World foundry congress*vol**1***(June 5 – 7, 2006, Harrogate, UK)* pp 1081-1090

[19] He K, Yu F-X, Zhao D-Z and Zuo L 2012 Characterization of precipitates in a hot-deformed hypereutectic Al–Si alloy*J. Alloys Compd.* **539** 74–81

[20] Annigeri Veeresh Kumar U. K. G. B. 2017 Method of stir casting of Aluminum metal matrix composites: a review *Mater. Today Proc.***4** 1140–1146

[21] Ramesh C S and Prasad T B 2009 Friction and Wear Behavior of Graphite-Carbon Short Fiber Reinforced Al–17%Si Alloy Hybrid composites*J. Tribol.***131** 014501-1-014501-5.

[22] Pal A, Srivastava A, Singh AK, Singh V, Bajpai A, Kumar A and Kandpal B C 2015 A Review on Technological Advancements in Aluminium Based Metal Matrix Composites *Int. j.*

comput. sci. math. sci. **4** 4-9
[23] Alghamdi A S, Ramadan M, Abdel Halim K S and Fathy N 2018Microscopical characterization of cast hypereutectic Al-Si alloys reinforced with grapheme nanosheets *Eng. Technol. Appl. Sci. Res.***8**2514–2519.
[24] Hekmat-Ardakan A, Liu X, Ajersch F and Chen X-G 2010 Wear behaviour of hypereutectic Al–Si–Cu–Mg casting alloys with variable Mg contents , *Wear***269** (2010) 684–692
[25] Campbell J 2015 *Complete casting handbook: metal casting processes, metallurgy, techniques and design* (Oxford: Butterworth-Heinemann Elsevier Ltd) chapter 16 pp 821-882
[26] Huang X, Liu C, Lv X, Liu Gand Li F 2011Aluminum alloy pistons reinforced with SiC fabricated by centrifugal casting*J. Mater. Process. Technol.* **211** 1540–1546
[27] Mallikarjun, Sukruth M, Chethan K S and Dr. Kiran Aithal S 2017 Study of Functionally Graded Al-Si Alloy Fabricated by Centrifuge Casting Technique*International Journal of Advances in Scientific Research and Engineering (IJASRE)* **3** 70-75
[28] Wyatt L M 1994*Mechanical Engineer's Reference Book*12 ed E H Smith(Oxford: Butterworth-Heinemann Elsevier Ltd) chapter 16.5pp 998-1109
[29] Li R, Liu L, Zhang L, Sun J, Shi Y and Yu, B2017 Effect of Squeeze Casting on Microstructure and Mechanical Properties of Hypereutectic Al-xSi Alloys*J. Mater. Sci. Technol.***33**404–410.
[30] Husain N H, Ahmad A H and Rashidi M M 2017 An overview of thixoforming process 4th International Conference on Mechanical Engineering Research vol 257 (1 – 2 August 2017, Kuantan, Malasya) IOP Conf. Ser.: Mater. Sci. Eng. pp 012053
[31] Lasa L and Rodriguez-Ibabe J M 2002 Effect of composition and processing route on the wear behaviour of Al–Si alloys*ScriptaMaterialia***46** 477-481
[32.] Lasa L and Rodriguez-Ibabe J M 2003 Wear behaviour of eutectic and hypereutectic Al–Si–Cu–Mg casting alloys tested against a composite brake pad*Mater. Sci. Eng.:A***363**193-202
[33] Birol Y 2008 Cooling slope casting and thixoforming of hypereutectic A390 alloy*J. Mater. Process. Technol.* **207** 200-203
[34] Wang B, Wang J, Tang L, Qi J and Wang J 2016 Effect of electric pulse modification on microstructure and properties of Ni-rich Al-Si piston alloy*China Foundry***13** 139-142.
[35] Walker J C, Murray J, Narania S and Clare A T 2012 Dry Sliding Friction and Wear Behaviour of an Electron Beam Melted Hypereutectic Al–Si Alloy *Tribol. Lett.* **45**49–58.
[36] Timmermans G and Froyen L1999 Fretting wear behaviour of hypereutectic P/M Al–Si in oil environment*Wear* **230**105-117
[37] Al-Samarai R A, Haftirman Ahmad K R and Al-Douri Y 2012 The influence of roughness on the wear and friction coefficient under dry and lubricated sliding *Int. j. sci. eng. res.***3** 1–6.

Acknowledgments
This paper presents the results obtained during research within the framework of the project TR 35021, supported by the Ministry of Education, Science and Technological Development of the Republic of Serbia.

IOP Conf. Series: Materials Science and Engineering 1271 (2022) 012013 doi:10.1088/1757-899X/1271/1/012013

Topology optimization of a MacPherson control arm

Clio Vossou[1], Andreas Nakos[2], Dimitrios Koulocheris[3]

[1] National Technical University of Athens, School of Mechanical Engineering, Vehicles Laboratory, Greece, Zografou Campus, Iroon Polytexneiou 9, 157 80

[2] National Technical University of Athens, School of Mechanical Engineering, Vehicles Laboratory, Greece, Zografou Campus, Iroon Polytexneiou 9, 157 80

[3] National Technical University of Athens, School of Mechanical Engineering, Vehicles Laboratory, Greece, Zografou Campus, Iroon Polytexneiou 9, 157 80

E-mail: klvossou@mail.ntua.gr

Abstract. One of the most important systems in ground vehicles is the suspension system, which is responsible for the ride comfort, the road holding and handling. A crucial part of a passenger car passive suspension system is the control arm or wishbone arm or A-arm. Its main function is to provide a link between the body of the vehicle and the wheels. In this paper, the studied control arm is a part of a MacPherson strut suspension and its dimensions correspond to a front right suspension of a SUV vehicle. Two geometrical models of the control arm were developed in Solidworks® CAD-CAE Software, one with the measured dimensions and one with the outline dimensions of the control arm. A finite element model of the first one was built, using typical loading conditions, in order to determine the deformation and the stress contours. Then a topology optimization analysis was carried out to the second one in order to obtain a new design for the control arm. The FE results of the existing control arm and the optimized one were compared in terms of weight and structural integrity in different loading conditions leading to conclusions regarding the optimization procedure.

1. Introduction

The suspension system is crucial for the handling and the ride of a vehicle. It contributes to the safe driving of the vehicle and the comfort of the passengers by reducing the vibrations during vehicle's ride. A MacPherson strut suspension is an independent type of suspension which is mainly used in the front axle of four-wheel-drive (FWD) cars and has only one control arm (or wishbone or A-arm) which is located at the lower end of the assembly as presented in Figure 1. The role of a control arm is to provide optimum control in steering and bumps and connect the tire to the chassis [1].

IOP Conf. Series: Materials Science and Engineering 1271 (2022) 012013 doi:10.1088/1757-899X/1271/1/012013

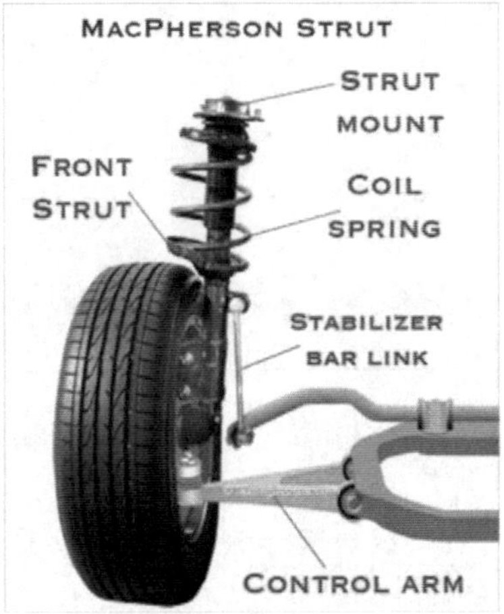

Figure 1. MacPherson suspension [2].

In order to keep up with the rapid progression of the automobile industry design and simulation (CAD - CAE) software is used for the design of new parts or the optimization of existing ones. Using a CAD - CAE software a control arm can be designed and tested in real loading case scenarios in order to evaluate its design without constructing a physical prototype. Furthermore, using such software the engineer has also the possibility to optimize the design using built in topology optimization features. Topology optimization is a methodology that aims to employ the stiffest material distribution in a way that the resulting structure is as light as possible [3]. The most typical topology optimization method embedded in CAD software is the Solid Isotropic Material with Penalization method (SIMP), which predicts an optimal material distribution within a given design space, for given loading and boundary conditions, manufacturing constraints and performance requirements [4].

In this paper the possibilities of the topology optimization analysis embedded in Solidworks ® CAD – CAE software has been explored, using the control arm as a test case. In order define the design space for the control arm and create a baseline FE simulation, a commercial control arm from an SUV vehicle with a MacPherson suspension has been measured and accurately designed (M1). Then, the geometrical model of the outline of the geometrically accurate control arm has been developed (M2). Two topology optimization analyses with different objective functions were performed to M2. In both topology optimization analyses to two typical loading cases and their combination have been used as loading conditions. The outcome models of the topology optimization analysis were, then, redesigned and mechanically analysed using the abovementioned loading conditions and their results were compared to the results of the mechanical analysis of M1 to the same loading conditions.

2. Materials and methods

In Figure 2 the commercial control arm used as a baseline in this paper is presented. The control arm consists of two metal plates welded together and its weight is equal to 3.0 kg.

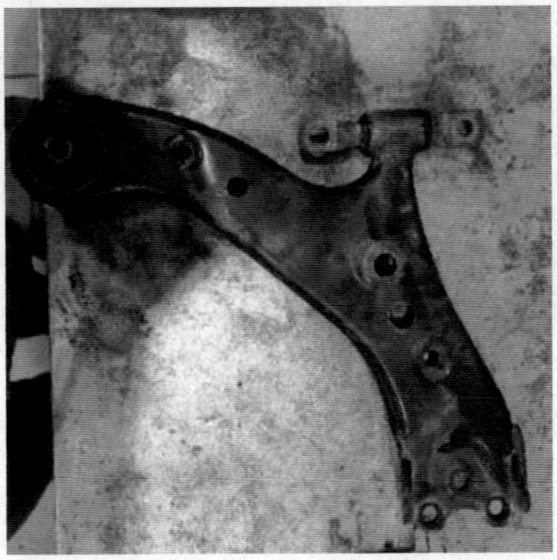

Figure 2. Commercial control arm of the front right suspension of an SUV.

2.1. Geometrical models

In order to build the geometrically accurate model (Figure 3) in Solidworks® software a surface model, consisting of three surface bodies has been created.

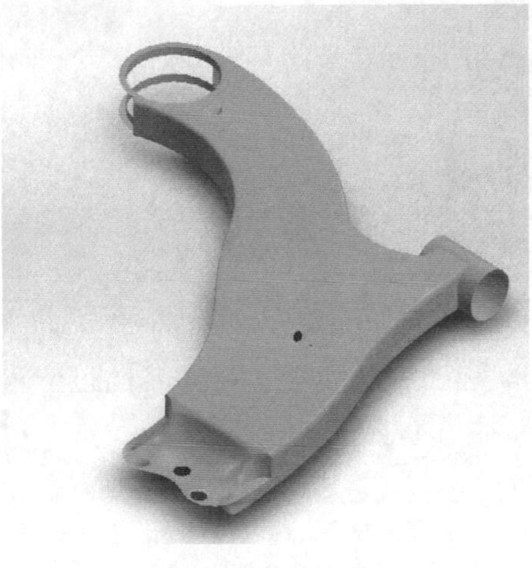

Figure 3. Geometrically accurate 3D CAD model of the control arm (*M*1) in isogeometric view.

Using the outer dimensions of the geometrically accurate model of the control arm (Figure 4) a solid 3D CAD model has been designed (M2).

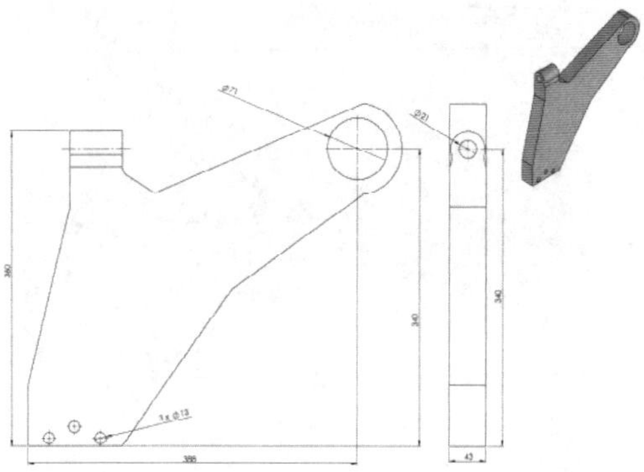

Figure 4. Outline part.

2.2. *Finite elements models*

In order to perform the mechanical analysis on M1 and the topology optimization of M2, two Finite Element (FE) models have been created. In Figure 5 the mesh of both FE models, FE1 and FE2, respectively, is presented.

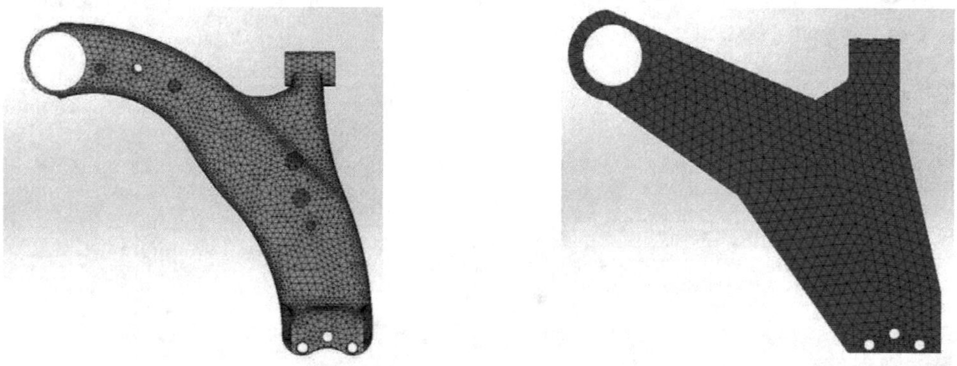

(a) (b)

Figure 5. Mesh of (a) M1 (FE1) and (b) M2 (FE2).

It is important to mention that FE1 consists of shell FE while FE2 consists of solid FE. In more details in FE1 parabolic triangular shell FE are created automatically from the software with the use of the curvature-based mesh controls while for FE2 tetrahedral 3D solid FE are created automatically from the software with the use of the blended curvature-based mesh controls. The size of the FE, in both cases emerged after a convergence study. In Table 1, the characteristics of both FE models are presented.

Table 1. Characteristics of finite elements.

	FE1	FE2
No. of elements	16326	16117
No. of nodes	33371	25441
Max element size [mm]	8.00	13.00
Min element size [mm]	2.61	4.97
Jacobian points	16	16
Max aspect ratio	-	6.08
% of elem. with aspect ratio<3	-	99

For the solution of both FE models the direct sparse solver, which solves the equations using exact numerical techniques, has been used [4], since FE1 has less than 100000 FE.

2.3. Material properties
According to the literature control arms are mostly manufactured with casting iron or casting aluminum or stamped steel [5]. In Table 2 the mechanical properties of the most common materials in the automotive industry are presented, as they are defined in Solidworks software [6], [7].

Table 2. Mechanical properties of materials used in automotive industry.

Material	Modulus of Elasticity [MPa]	Poisson's ratio	Yield Strength [MPa]	Density [kg/m^3]
AISI 1045 Steel	205000	0.29	530	7850
ASTM A36 steel	200000	0.26	250	7850
Aluminum Alloy 7075-T6	72000	0.33	505	2810

However, in this paper the material which is used for both FE1 and FE2 is the Aluminum Alloy 7075-T6, since it has a low density and a high value of yield strength.

2.4. Loading conditions
For both the mechanical analysis of FE1 and the topology optimization of FE2 the same loading conditions were used and they were retrieved from the literature.

Considering the coordinate system presented in Figure 6, on a vehicle the forces generated from the acceleration and braking are developed on x-axis, on y-axis are developed the loads from cornering and on the z-axis are developed the forces generated from the contact of the wheels on the road. For the mechanical analysis of vehicle components such as the control is usual to ignore the forces on z-axis since their value is a lot lower that those generated on x and y axis [8].

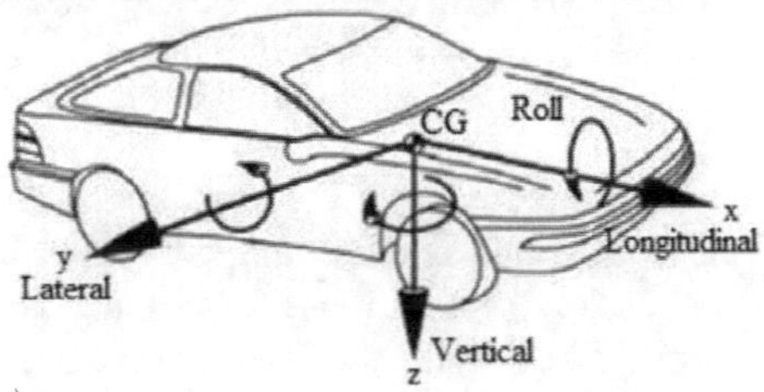

Figure 6. Coordinate system [2].

Having the above in mind, as loading conditions, two typical load case scenarios were assumed and they are presented in Table 3. The first load case (LC1) represents the pot-hole brake case, the second one (LC2) represents typical loading of the control link during the movement of the vehicle.

Table 3. Load case scenarios for static tests.

Load Case No	Fx [N]	Fy [N]
LC1	-24200	8200
LC2	-31460	15730

The abovementioned forces act on the connection of the control arm to the wheel. In Figure 7, with purple color, are depicted the forces acting on the control arm in FE1.

IOP Conf. Series: Materials Science and Engineering 1271 (2022) 012013 doi:10.1088/1757-899X/1271/1/012013

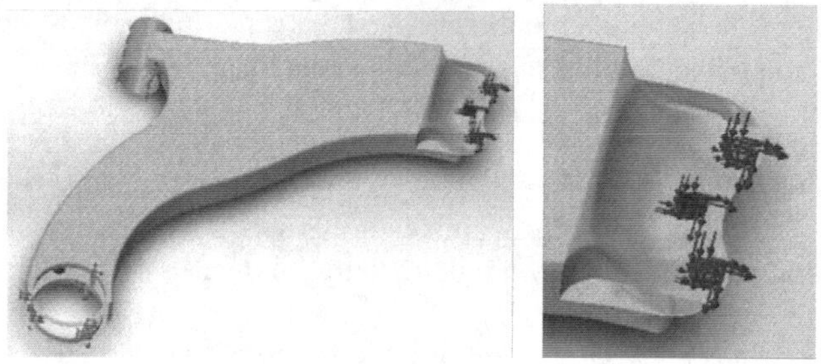

Figure 7. Loading conditions

Concerning the connection of the control arm to the chassis was modeled as a fixed support in the back region and as cylindrical support in the front (Figure 8). In the fixed support all the degrees of freedom of the nodes of the finite elements have been restrained, while in the cylindrical support only the rotation in x axis is allowed.

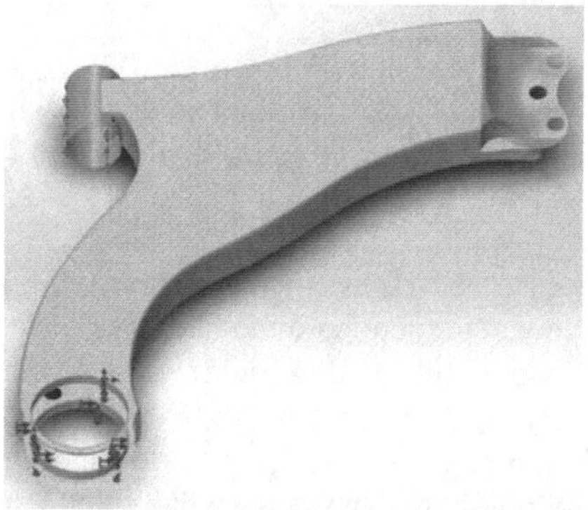

Figure 8. Boundary conditions (orange arrows: fixed support, green arrows: cylindrical support.

2.5. Topology optimization analysis

In order to perform a topology optimization, (a) the goal option and (b) the restrictions have to be set. In this paper two different topology optimization analysis have been performed with different restrictions. In more details as goal option, best stiffness to weight ratio has been used for both analysis forcing the software to create a stiff model with the least possible mass. The topology optimization analyses differ in the restrictions which are dual there is a mass constraint that refers to the maximum reduction of mass compared to the initial model and a stress constraint which refers to the maximum stress value that the optimized model can have with respect to the yield strength of the material. In Table 4 the parameters of both topology optimization analyses are presented.

Table 4. Parameters for the topology optimization (TO) analysis.

	TO1	TO2
Goal option	Best stiffness to weight ratio	Best stiffness to weight ratio
Mass constraint	50%	75%
Stress constraint	40%	40%

For the topology optimization analysis, the direct sparse solver was used when each LC was considered independently and the intel direct sparse solver when the combination of the LC was considered. Also, in the last case the Min Max formulation in order to minimize the maximum compliance and maximize the component's stiffness calculated independently from each LC. The optimized shape of the component has the maximum stiffness for each LC [4]. Moreover, in the solver option of the topology simulations the average stresses at mid-nodes command activated and set to automatically preserve the regions with loads and supports.

3. Results

Firstly, the results of the mechanical analysis of FE1 and then the topology optimization analyses results are presented for both load cases. Finally, the mechanical analysis of the topologically optimized control arm is presented.

3.1. Mechanical analysis of FE1

of the real control arm (FE1) are presented for LC1 and LC2 in terms of equivalent von Mises stress contours (Figure 9). Since FE1 consists of shell elements, the stress contours correspond to the top shell face and represent the total membrane and bending stress on the whole thickness for the control arm [4].

LC1

LC2

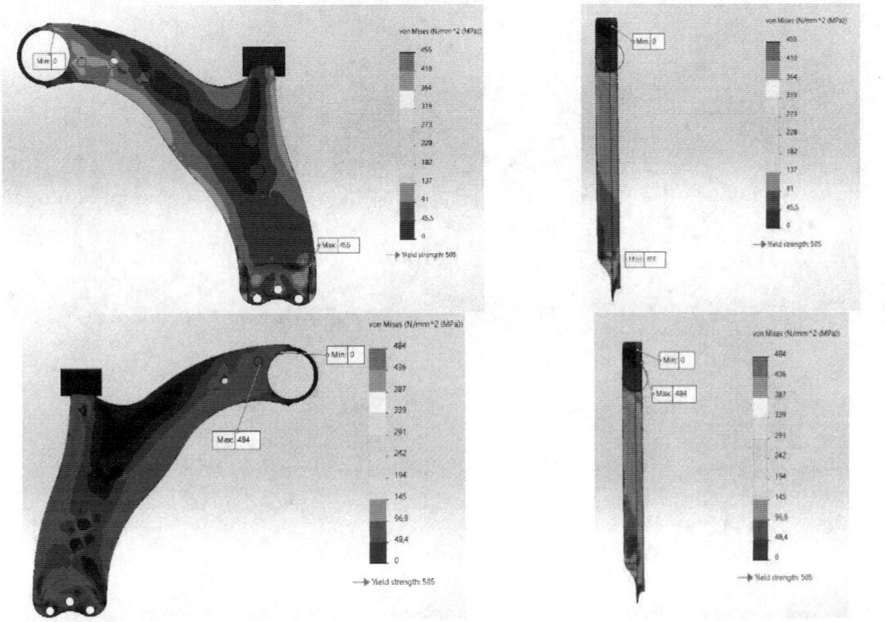

(a) (b)

Figure 9. Stress contour on FE1 for LC1 and LC2 in (a) front and (b) side view.

The control arm is created from two metal plates connected with a MIG weld. This weld has been simulated as a perimetrical bonded contact. This is an area of stress concentration for both load cases. While the contours of the equivalent von Misses stresses are similar in both loading cases, the maximum values of stress differ and are presented in Table 5 that follows along with the maximum displacement values.

Table 5. Stresses and Displacements in Real part.

Load case	Stress [MPa]	Total Displacement [mm]	Displacement on x-axis [mm]	Displacement on y-axis [mm]	Displacement on z-axis [mm]
LC1	455	5.2	3.1	1.3	4.0
LC2	484	5.4	3.9	1.7	3.5

In Table 5 is obvious that LC2 leads to higher value of displacement (total and per axis) as well as for maximum equivalent Von Mises stress.

3.2. Topology optimization of FE2

In Figure 10 the resulting geometry for TO1 and TO2 is presented for the worst-case scenario that is the combination of LC1 and LC2

(a)

(b)

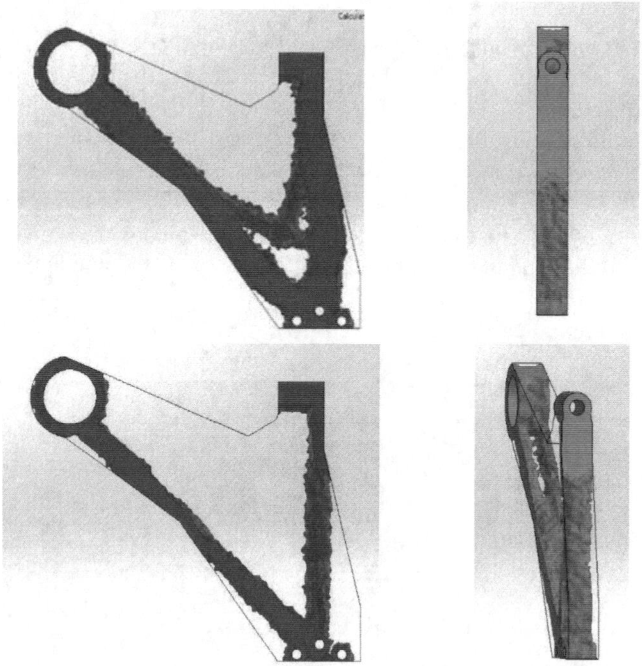

Figure 10. Resulting geometry for (a) TO1 & (b) TO2 in front and side view.

In Figure 11 the convergence graphs for the goal and the mass constraint are presented for both topology optimization analyses.

TO1

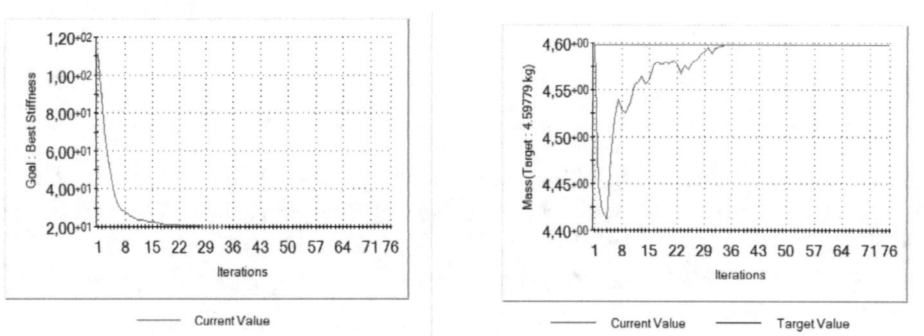

IOP Conf. Series: Materials Science and Engineering 1271 (2022) 012013 doi:10.1088/1757-899X/1271/1/012013

TO2

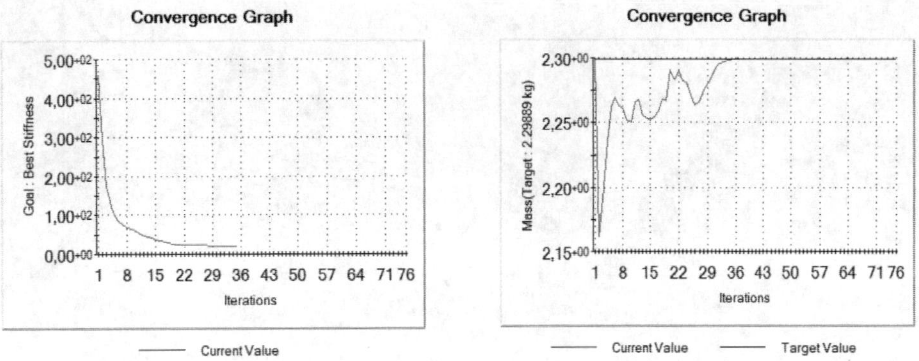

(a) (b)

Figure 11. Best stiffness(a) and Mass reduction (b) convergence graphs for FE3.

The resulting stress contours on FE2 are provided in Figure 12. Reviewing Figure 12 attention should be paid to the fact that the software creates a stress distribution diagram for the optimized model with the loads from the topology optimization simulation. However, the stress values are based on the entire model which, after the optimization it may contain porous elements that are less rigid than fully dense elements. For this reason, these values should be thought of as a rough estimate on how the final model might behave.

TO1

TO2

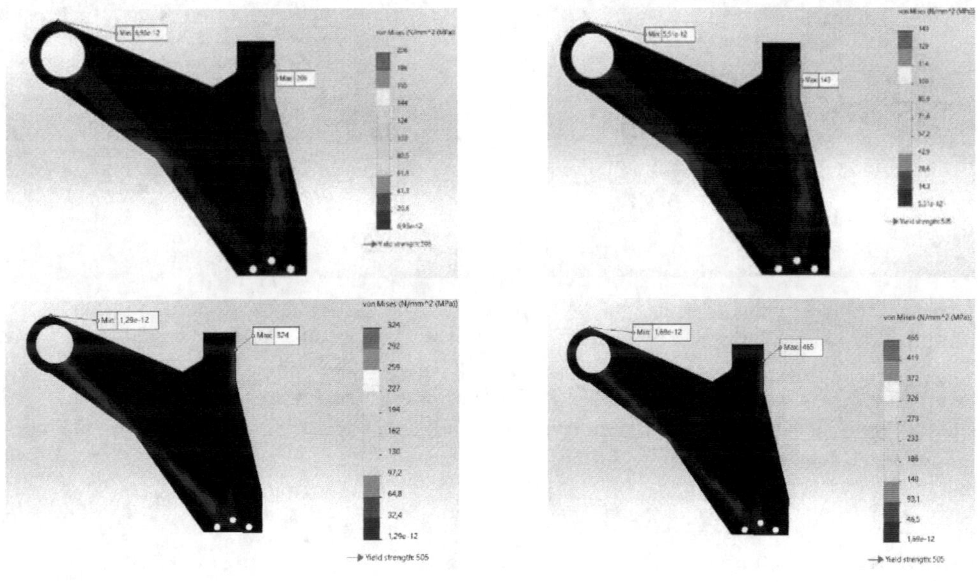

(a) (b)

Figure 12. Stress on FE3 charged with LC1(a) and LC2 (b).

In Figure 13 the optimized geometrical models are presented. It is obvious that the automatically created designs from the software cannot be engineered with conventional manufacturing techniques but in the future, they could be 3D.

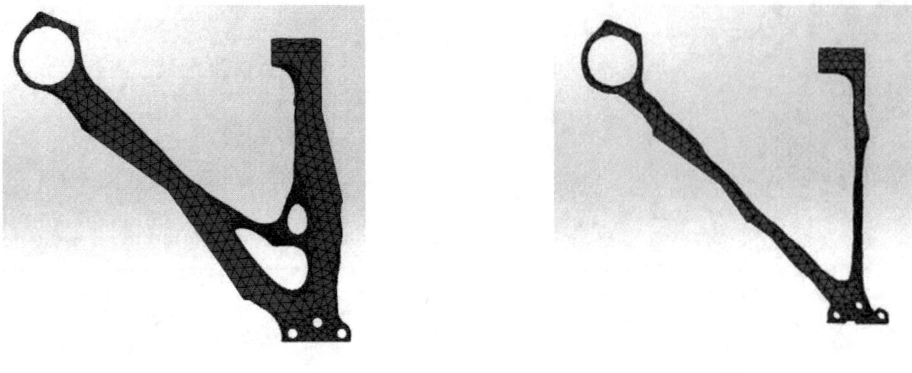

(a) (b)

Figure 13. Optimized geometrical models from (a) TO1 and (b) TO2 produced by the software.

3.3. Mechanical analysis of TO1 and TO2

In order to perform the mechanical analysis of the optimized geometrical models they have been redesigned to consist of typical geometrical features, in order to be manufacturable. In Figure 14 the redesigned geometrical models M3 and M4 are presented.

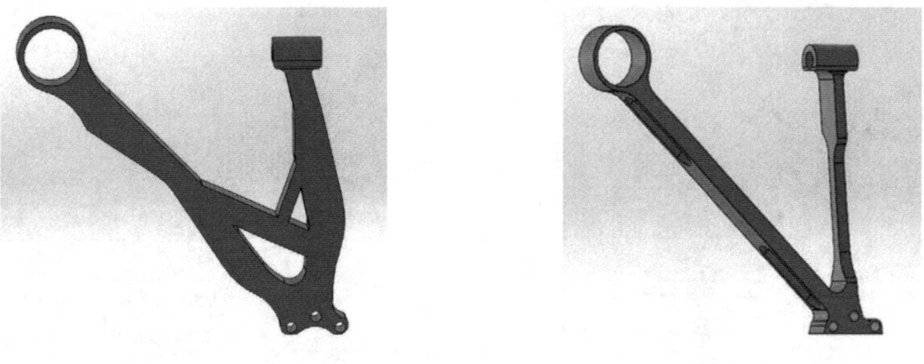

(a) (b)

Figure 14. Redesigned geometrical models from (a) TO1 and (b) TO2 produced by the software.

From M3 & M4 two new FE models (FE3 & FE4, respectively) have been built with the same parameters as FE2. In Figure 15 the equivalent von Misses stress contours for both FE3 and FE4 is presented for the static analysis with LC1.

FE3

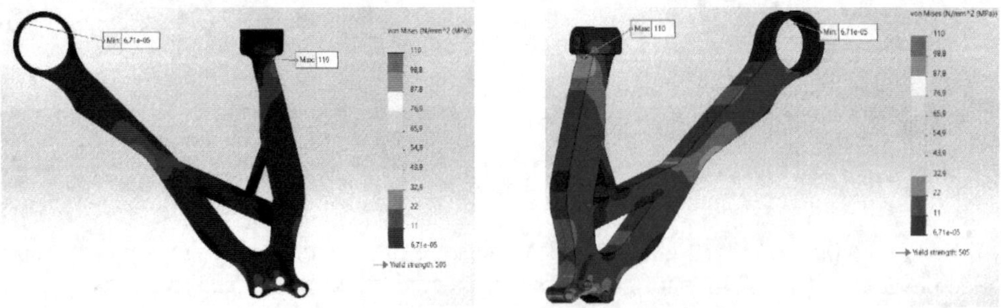

IOP Conf. Series: Materials Science and Engineering 1271 (2022) 012013 doi:10.1088/1757-899X/1271/1/012013

FE4

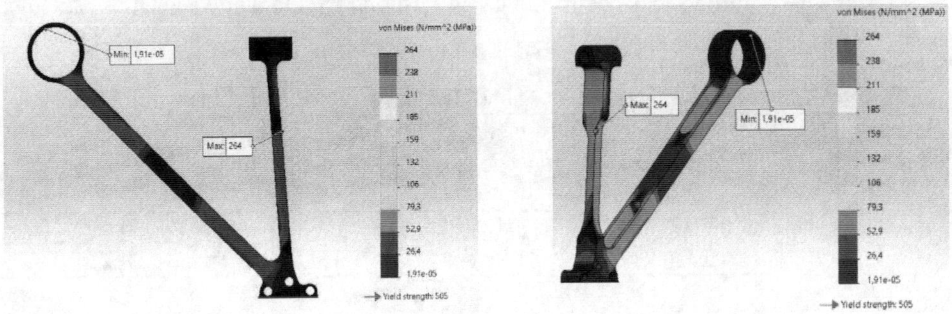

(a) (b)

Figure 15. Stress contour on FE3 & FE4 for LC1 in (a) front and (b) isometric view.

In Table 6 the maximum value of equivalent Von Mises stress and displacement are presented for both FE models.

Table 6. Maximum values of stress and displacements for LC1 in both FE3 & FE4.

Stress [MPa]	Total Displacement [mm]	Displacement on x-axis [mm]	Displacement on y-axis [mm]	Displacement on z-axis [mm]
110	0.34	0.30	0.16	0.01
264	1.57	1.48	0.78	0.16

Finally, in Figure 16 the equivalent von Misses stress contours for both FE3 and FE4 is presented for the static analysis with LC2.

FE3

FE4

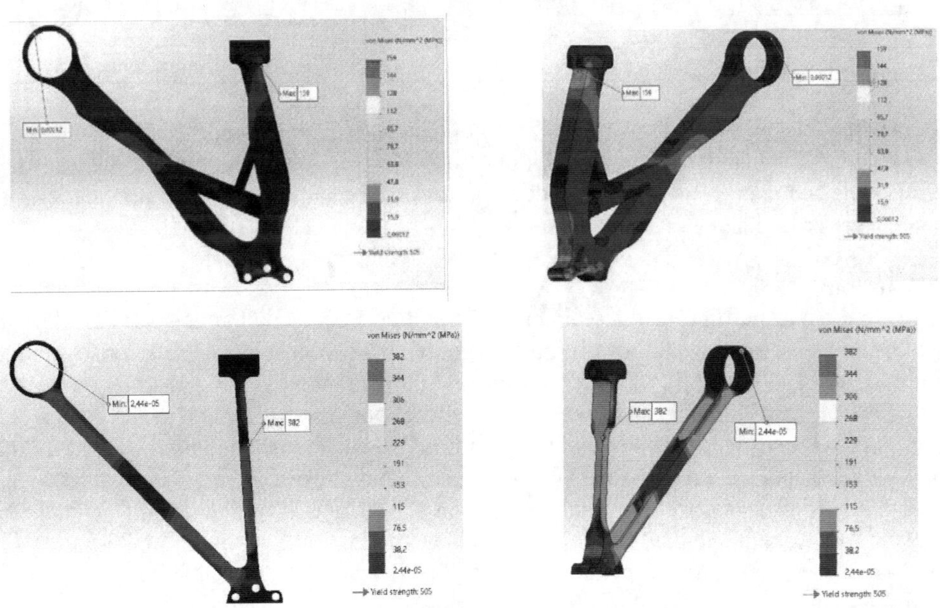

(a) (b)

Figure 16. Stress contour on FE3 & FE4 for LC2 in (a) front and (b) isometric view.

In Table 7 the maximum value of equivalent Von Mises stress and displacement is presented for FE3 and FE 4.

Table 7. Maximum values of stress and displacements for LC2 in both FE3 & FE4.

Stress [MPa]	Total Displacement [mm]	Displacement on x-axis [mm]	Displacement on y-axis [mm]	Displacement on z-axis [mm]
159	0.47	0.42	0.25	0.01
382	2.13	1.99	1.07	0.21

4. Discussion

The real control arm is constructed with the weldment of two metal plates with thickness of 3 mm. FE1 is the geometrically accurate model and it is built as a surface model with shell FE. Both optimized control arms (FE3 and FE4) are considered solid bodies and are built as solid models with solid FE.
In Table 8 and Table 9 the weight of each model, along with the maximum value of Von Mises stress and this of total displacement for each load case are presented, respectively.

Table 8. Results of the mechanical analysis of FE1, FE3 and FE4 in LC1.

Models	FE1	FE3	FE4
Stress [MPa]	445	110	264
Displacement [mm]	5.20	0.34	1.57
Mass [kg]	3.3	4.4	1.5

Table 9. Results of the mechanical analysis of FE1, FE3 and FE4 in LC2.

Models	FE1	FE3	FE4
Stress [MPa]	484	159	382
Displacement [mm]	5.44	0.47	2.13
Mass [kg]	3.3	4.4	1.5

The lighter model is FE4 but it has higher value of Von Mises stress and displacement than FE3 which on the other hand is heavier than both FE4 and FE1. It is important to point out that apart from their mass, also the morphology of the FE models is different requesting for the application of different manufacturing procedures.

5. Conclusions

In the present paper starting from a real control arm of an SUV vehicle one shell FE model (FE1) has been built and two topology optimization procedures have been set up in order to produce lighter structures, leading to two solid FE (FE3 & FE4) in Solidworks ® CAD-CAE software. Comparing the results of the mechanical analysis of these three models one can conclude that the choice of the better control arm for the suspension of a vehicle depends on the requirements the designer has. For example, the FE4 model would fit in an off-road vehicle where higher stresses are developed and its weight is not so important. Another criterion that can be used in this choice is the behavior of the control arm in fatigue and the cycle of life they have. In fact, this would be worth studying in the future. Finally, the cost of manufacturing is a factor of difference for them since FE1 consists of sheets of material while FE3 and FE4 consists of bulk material.

References

[1] Birch W T 1999 *Automotive Suspension and Steering Systems*
[2] *Testing Autos* [Online] Available: www.testingautos.com

[3] Rozvany G I 1997 *Topology Optimization in Structural Mechanics* Springer Science & Business Media
[4] Dassault Systems *Solidworks help* [Online] Available: www.help.solidworks.com
[5] Konieczny L Burdzik R and Lazarz B 2012 *Analysis of properties of automotive vehicle suspension arm depending on different materials used in the MSC.Adams environment*
[6] *Matmatch* [Online] Available: www.matmatch.com/learn/material/7075-t6-aluminium
[7] *REDSTONE Manufacturing* [Online] Available: www.redstonemanufacturing.com/stamped-steel-vs-cast-steel/
[8] Vossou G C Koulocheris V D Provatidis G C and Costopoulos N Th 2014 *Design and Optimization of a Control Link of a Vehicle Suspension System*
[9] Macfarlane A 2016 *Modular Electric Automatic Guided Vehicle Suspension-Drive Unit*

IOP Conf. Series: Materials Science and Engineering 1271 (2022) 012014 doi:10.1088/1757-899X/1271/1/012014

Functionally graded materials in transport vehicles – overview, fabrication, application, modelling

G Tošić[1], G Bogdanović[2], D Čukanović[3], A Radaković[4]

[1] University of Banja Luka, Faculty of Mechanical Engineering, Bulevar vojvode Stepe Stepanovića 71, 78000 Banja Luka
[2] University of Kragujevac, Faculty of Engineering, Sestre Janjić 6, 34000 Kragujevac
[3] University of Priština in Kosovska Mitrovica, Faculty of Technical Sciences, Knjaza Miloša 7, 38220 Kosovska Mitrovica
[4] State University of Novi Pazar, Vuka Karadžića bb, 36300 Novi Pazar

E-mail address: gordana.tosic@mf.unibl.org

Abstract: This paper deals with functionally graded materials (FGM) as modern materials in transport vehicles. Basic methods of FGM production and areas of their application are presented, as well as an example of modelling an FGM plate. Bending of freely supported square FGM plates under the influence of sinusoidal transverse load was analyzed. The plates are made of metal (aluminium) and ceramic (alumina) in which material characteristics change as a function along the thickness of the plate. The static analysis was done by applying finite elements methods in the FEMAP software package for different mesh sizes. The results were compared with those obtained in the MATLAB software package using the classical plate theory, quasi 3D theory as well as high order shear deformation theories. The previously mentioned comparative results were analysed and certain observations and appropriate conclusions were made.

1. Introduction

The importance of materials on the daily life of every single individual is not at all questionable. Going back through the centuries, we can follow the ideas that man received and the ways in which he used materials, from the first primitive forms of objects, all the way to the advanced technologies of today. Materials research is very important for modern society, primarily because the development of industry requires it, but also because of the impact on the living environment with the aspiration to make it better and safer. New materials are used more and more in various disciplines, primarily in medical activities, aviation and automotive industry, construction, etc.

Functionally graded materials (FGM) are modern materials from the family of engineering composites in which there is a continuous or discontinuous change in chemical composition and/or microstructure through a defined geometric distance. They represent a class of attractive materials in which it is possible to create a gradient of properties, such as: modulus of elasticity, density and similarly. Constant progress in processing methods has enabled the concept of graded materials to create new systems and solve various engineering problems. In the family of engineering composites, functionally graded materials are increasingly being used. By combining potentially available materials, the aim is to use their properties in the best possible way.

The idea for FGM actually comes from nature where we can see a gradient structure on a microscopic level in different structures such as bones, teeth, skin tissue or bamboo wood. It is common to combine two isotropic material phases, but they also can contain more. Most often, metal and ceramics are combined, for the reason that ceramics have good heat resistance, while in terms of metals, mechanical strength and toughness are significant [1]. They can be designed in different ways and with different proportions of constituents in the desired direction.

2. Fabrication methods

There are different methods for the production of FGM that allow control of composition and structure. A basic classification can be made based on whether powders or melts techniques are used [2].

As far as powder metallurgy is concerned, layer pressing, powder mixture formation by centrifugal force, sedimentation by gravitational force or centrifugal force are most often applied. This procedure includes the production of powders, forming a mixture of components and sintering. Also, it can be used and some other method of consolidation. This method can be applied to a large number of powders (metal and alloy, ceramics) with a particle size of several nanometers to several hundred micrometers. The main failing of this method is that it is very difficult to get FGM with a continuous change of features, already it mostly gets FGM with a step-type trait gradient [2].

When it comes to melts techniques, the following methods are common: centrifugal casting, centrifugal sediment method and centrifugal powder mixture method. Centrifugal casting is one of the most efficient methods for producing FGM because possibilities of a wide range of composition and microstructure control [3]. In this method centrifugal force is used to form a gradient structure in the molten metal which contains some other solid phase (solid reinforcement particles)[4]. The lack of powder metallurgy in the form of a non-continuous gradient of properties can be overcome by the centrifugal sediment method. The centrifugal method of a mixture of powders is an improved variant of the centrifugal method of solid particles, and can be used to obtain FGM that contain nano particles [2,5].

Vapour deposition (physical and chemical) is a technique used for FGM films (coatings) with a gradient structure at the nano level. Excellent microstructure can be produced using fabrication processes, but only thin surface layer can be obtained [2].

3. Application of FGM

FG materials are increasingly used in demanding conditions of exploitation that require resistance to elevated temperature and heat shocks, oxidation, corrosion, wear and the similarly. Starting from the first applications in spacecraft projects, through the energy and nuclear sectors, and all the way to medical sciences, such composites have an important role. FGM has found its application in many areas as shown in figure 1. For example, in the military industry for the production of parts for armored vehicles (heat resistant door), bulletproof vests, then in the automotive industry for brake systems as well as to exhaust and drive systems, etc.

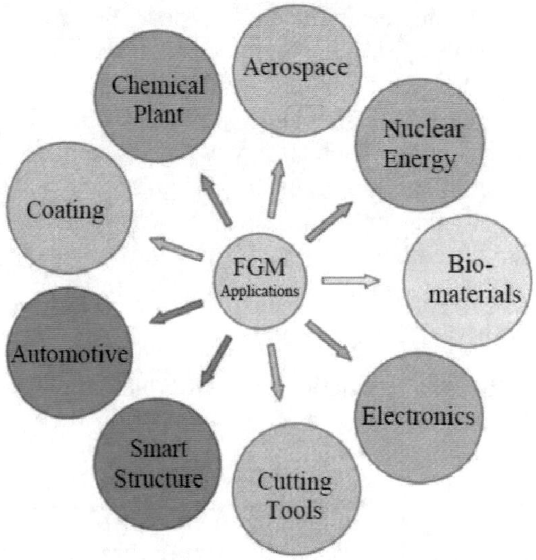

Figure 1. Areas of FGM practical applications.

3.1. *FGM in aerospace industry*

In terms of spacecrafts, there was an effort to develop materials for thermal shields in extreme conditions where the temperature changes even up to several thousand degrees. In that case, thermal stability was achieved through ceramic materials, while mechanical strength was achieved through metal. The first application was carried out at the Japanese space shuttle project in 1983 [6]. Most parts of spacecraft and aircraft now used FGM with gradient properties such as a rocket engine, spacecraft gear structure, heat exchange plates, and some structures, such as reflectors, solar panels, camera bunks, turbine wheels, turbine blade coating, nose covers, as shown in figure 2 [6,7]. FG Al_2O_3/Al alloy with an excellent thermal and corrosion resistance in recent years are used to produce engine parts and rocket nozzles [6].

3.2. *FGM in automotive industry*

Many automotive components, such as diesel engine pistons and cylinder liners, combustion chambers, racing car brakes, driveshafts, and flywheels are demanding in terms of properties such as thermal, pressure, or stress. Therefore, FGM with gradient properties are very suitable for these components, as shown in figure 3. Due to the high cost, the use of FGM is very restrictive in automotive industry and generally only used in for critical parts cars [6, 8].

In [9] FG Al_2O_3/Al_2TiO_5 composites are successfully made using powder metallurgy in order to investigate the influence of graded structure on microstructure behaviour and physical as well as thermal properties for automotive applications. The results showed that the FGM with graded structure improved the mechanical and thermal properties for car brake rot. In another research [10] FG Al–Mg_2Si composites are designed and produced using centrifugal casting in order to study the effect of process parameters on the microstructure and mechanical properties for automotive cylinder block liners. The results indicated an improvement in the properties of FGM composites compared to the conventional composites to serve the required application [6, 10].

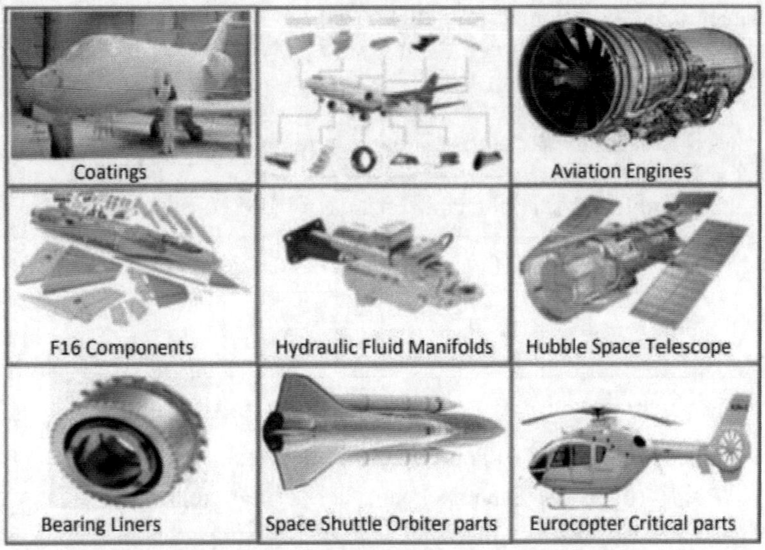

Figure 2. FGM in aerospace industry.

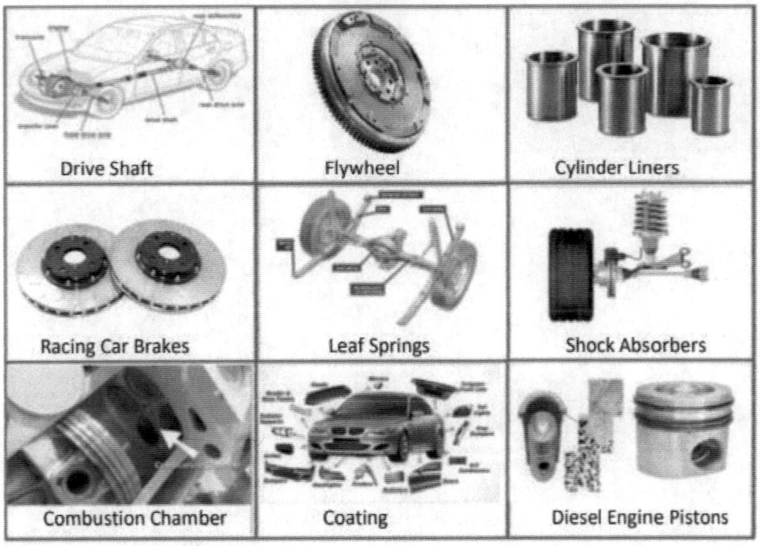

Figure 3. FGM in automotive industry.

4. Finite elements modelling of functionally graded plate

The subject of the finite elements (FE) analysis are freely supported square functionally graded (FG) plates $(a/h=10)$ made of metal and ceramics. For metal, the material characteristics of aluminium (Al) were used, and as for ceramics, it is alumina (Al_2O_3).

Aluminium is a common constituent of FGM because of its toughness, corrosion resistance, low density and non-toxicity. It is characterized by satisfactory mechanical strength, as well as high electrical and thermal conductivity. It is widely distributed in various areas of modern technology.

Alumina is characterized by different properties, from low conductivity to high brittleness. On the other hand, it is extremely resistant to corrosion and high temperature influences, and is therefore often used as a thermal shield.

The table 1 shows the material characteristics of aluminium and alumina used for modelling of FG plate.

Table 1. Material characteristics of FGM constituents.

	Modulus of elasticity E (GPa)	Poisson's coefficient
Aluminium	$E_m = 70$	0.3
Alumina	$E_c = 380$	0.3

The material characteristics change according to the equation (1), where the properties of the metal are $P_b = P_m$ (on lower side of the plate), the properties of the ceramic are $P_t = P_c$ (on the upper side of the plate) and $P_{tb} = P_t - P_b$ (figure 4). The exponent p defines the proportion of constituents, whereby for $p = 0$ the plate is homogeneous, made of ceramics, while for $p = \infty$ a plate would be homogeneous, made of metal. For values in between, the plate has a gradient structure.

$$P(z) = P_b + P_{tb}\left(\frac{1}{2} + \frac{z}{h}\right)^p \quad (1)$$

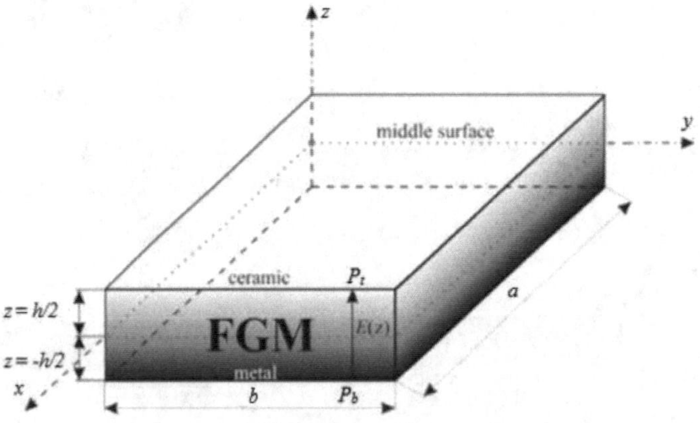

Figure 4. FG plate.

4.1. *Input data for static linear analysis*

Since the subject of the analysis is a square plate $(a/h=10)$, the dimensions of the plate are adopted as input data, so that the defined ratio is maintained $(a=b=10\text{m},\ h=1\text{m})$.

The change in material is expressed by the function (1), where the change in modulus of elasticity as an input parameter was of greatest interest. The following form is obtained for the change of the modulus of elasticity along the thickness of the plate depending on the exponent p:

$$E(z)=E_m+(E_c-E_m)\left(\frac{1}{2}+\frac{z}{h}\right)^p . \tag{2}$$

Due to comparison with other results, the calculation will be done for $p=1,\ 2,\ 4,\ 8,\ 20$.

In the program, the modulus of elasticity will be entered over the layers. For the corresponding layer modulus of elasticity will be calculated according to equation (2) for the level z corresponding to the center of that layer. The analysis will be performed in case of 5, 10 and 20 layers. In the table 2 are shown the modulus of elasticity for the case of dividing the plate along thickness into 5 layers.

Table 2. Modul of elasticity in the case of 5 layers, *E* (GPa).

	$p=1$	$p=2$	$p=4$	$p=8$	$p=20$
5th layer	349.000	321.1	273.391	203.445	107.689
4th layer	287.000	221.9	144.431	87.871	70.247
3rd layer	225.000	147.5	89.375	71.211	70.000
2nd layer	163.000	97.9	72.511	70.020	70.000
1st layer	101.000	73.1	70.031	70.000	70.000

It is interesting that even for $p=20$ the plate is mostly made of metal.

A sinusoidal transverse load acts on the plate:

$$q(x,y)=q_0\sin\left(\frac{\pi x}{a}\right)\sin\left(\frac{\pi y}{b}\right), \tag{3}$$

where q_0 represents the load amplitude.

The contour conditions for a freely supported plate are given as:

$$v=w=\theta_y=0, \text{ on the sides where it is } x=0 \text{ and } x=a, \tag{4}$$

$$u=w=\theta_x=0, \text{ on the sides where it is } y=0 \text{ and } y=b, \tag{5}$$

Eight-node 3D finite elements were used for the modelling of FG plate in case of 5 layers (figure 5) as well as 10 layers (figure 6).

.

IOP Conf. Series: Materials Science and Engineering 1271 (2022) 012014 doi:10.1088/1757-899X/1271/1/012014

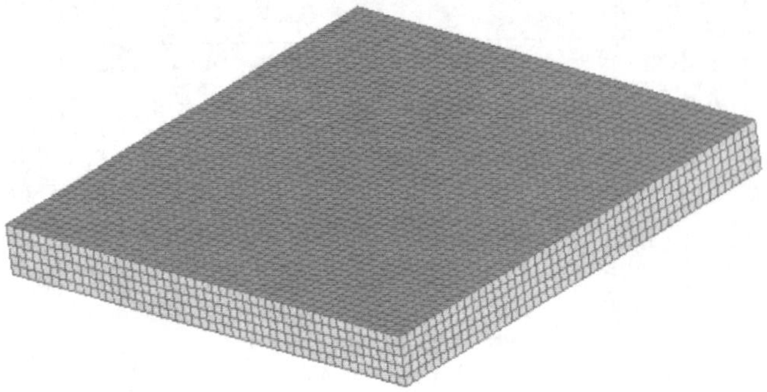

Figure 5. Finite elements model of FG plate in the case of division into 5 layers.

Figure 6. Finite elements model of FG plate in the case of division into 10 layers.

4.2. Results

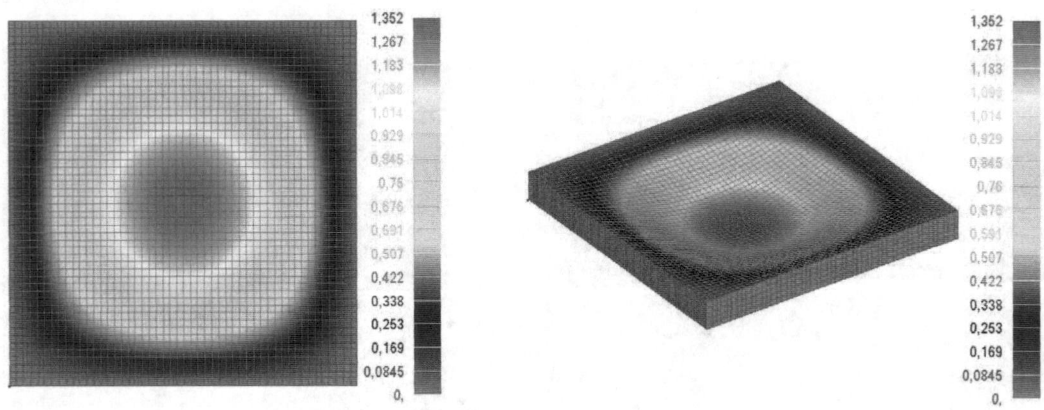

Figure 7. Normalized values of vertical displacement field $\overline{w}$ of the FG plate in case of 5 layers and value of the exponent $p = 20$.

The table 3 shows the normalized displacements values of plate centre $w_s = w\left(\frac{a}{2},\frac{b}{2}\right)$.

Normalization was performed according to:

IOP Conf. Series: Materials Science and Engineering 1271 (2022) 012014 doi:10.1088/1757-899X/1271/1/012014

$$\bar{w} = \frac{10E_c h^3}{q_0 a^4} w\left(\frac{a}{2}, \frac{b}{2}\right). \tag{6}$$

Table 3. Normalized displacement values $\bar{w}$ for different mesh size and values of the exponent p.

p	5 layers, 50×50	10 layers, 100×100	20 layers, 200×200
1	0.5776	0.5852	0.5890
2	0.7638	0.7600	0.7600
4	0.9120	0.8892	0.8854
8	1.0488	0.9956	0.9804
20	1.3528	1.2008	1.1514

The table 4 shows the results where the corresponding values were reached by applying the classical plate theory (CPT), quasi 3D theory, third order shear deformation theory (TSDT), and high order shear deformation theory (HSDT) based on shape functions (SF).

Table 4. Normalized displacement values $\bar{w}$ for different values of the exponent p.

	$p = 1$	$p = 2$	$p = 4$	$p = 8$	$p = 20$
CPT [11]	0.5623	-----	0.8281	-----	-----
Quasi 3D [12]	0.5876	0.7571	0.8823	0.9739	-----
TSDT [13]	0.5890	0.7573	0.8815	0.9747	-----
SF 1 [2]	0.5889	0.7573	0.8814	0.9746	1.1377
SF 2 [2]	0.5889	0.7573	0.8814	0.9746	1.1377
SF 3 [2]	0.5889	0.7573	0.8818	0.9749	1.1374
SF 4 [2]	0.5880	0.7563	0.8815	0.9739	1.1338
SF 5 [2]	0.5889	0.7572	0.8814	0.9745	1.1377
SF 6 [2]	0.5888	0.7568	0.8802	0.9730	1.1375
SF 7 [2]	0.5887	0.7572	0.8820	0.9751	1.1368
SF 8 [2]	0.5887	0.7572	0.8820	0.9751	1.1367
SF 9 [2]	0.5887	0.7572	0.8820	0.9751	1.1367
SF 10 [2]	0.5889	0.7572	0.8814	0.9745	1.1377
SF 11 [2]	0.5887	0.7567	0.8798	0.9727	1.1375
SF 12 [2]	0.5889	0.7573	0.8817	0.9749	1.1375
SF 13 [2]	0.5889	0.7573	0.8815	0.9746	1.1377

5. Conclusions

The tables 5-8 show the relative errors obtained by comparing FE results to results achieved using CPT, quasi 3D, TSDT as well as HSDT based on SF 13.

As was to be expected, results based on finer mesh of finite elements are closer to the values obtained by applying shear deformation theories in the mentioned references. A significant deviation is obtained in relation to CPT [2]. It is a consequence of the fact that the mentioned theory gives worse results in the analysis of thick and moderately-thick plates.

Table 5. Relative errors [%], comparison FE results with CPT results [11].

	$p = 1$	$p = 2$	$p = 4$	$p = 8$	$p = 20$
5 layers, 50×50	2.72	----	10.13	----	----
10 layers, 100×100	4.07	----	7.38	----	----
20 layers, 200×200	4.74	----	6.92	----	----

Table 6. Relative errors [%], comparison FE results with quasi 3D results [12].

	$p = 1$	$p = 2$	$p = 4$	$p = 8$	$p = 20$
5 layers, 50×50	1.70	0.88	3.37	7.69	----
10 layers, 100×100	0.41	0.38	0.78	2.23	----
20 layers, 200×200	0.24	0.38	0.35	0.67	----

Table 7. Relative errors [%], comparison FE results with TSDT [13].

	$p = 1$	$p = 2$	$p = 4$	$p = 8$	$p = 20$
5 layers, 50×50	1.93	0.86	3.46	7.60	----
10 layers, 100×100	0.64	0.36	0.87	2.14	----
20 layers, 200×200	0	0.36	0.44	0.58	----

Table 8. Relative errors [%], comparison FE results with HSDT results based on SF 13 [2].

	$p = 1$	$p = 2$	$p = 4$	$p = 8$	$p = 20$
5 layers, 50×50	1.92	0.86	3.46	7.61	18.91
10 layers, 100×100	0.63	0.36	0.87	2.15	5.55
20 layers, 200×200	0.02	0.36	0.44	0.59	1.20

When increasing the values of the exponent p from 2 to 20, it is evident that the error also increased, regardless of the mesh density. When comparing the obtained values with e.g. SF 13 [2], it can be observed that the largest relative error is obtained in the case of larger volume fraction of metal (aluminium) in FG plate. The best match of the results was achieved for the linear change of the modulus of elasticity in the case that the plate is divided into 20 layers.

References

[1] Bharti I, Gupta N and Gupta K M 2013 Novel Applications of Functionally Graded Nano, Optoelectronic and Thermoelectric Materials, *International Journal of Materials, Mechanics and Manufacturing* **1(3)** pp 221-224

[2] Čukanović D 2017 Statička i dinamička analiza ploča od funkcionalno gradijentno raspoređenih materijala, Doctoral Dissertation, University of Kragujevac, Serbia

[3] EL-Wazery M S and EL-Desouky A R 2015 *A review on Functionally Graded Ceramic-Metal Materials*, Journal of Materials and Environmental Science **6(5)** pp 1369-1376

[4] Saiyathibrahim A, Nazirudeen M S S and Dhanapal P 2015 Processing Techniques of Functionally Graded Materials – A Review, *International Conference on Systems, Science, Control, Communication Engineering and Technology* **1** pp 98-105

[5] Watanabe Y and Sato H, 2011 *Review Fabrication of Functionally Graded Materials under a Centrifugal Force*, Nanocomposites with Unique Properties and Applications in Medicine and Industry pp 133-150

[6] Saleh B, Jiang J, Fathi R, Al-hababi Tareq, Xu Q, Wang L, Song D and Ma A 2020 30 Years of functionally graded materials: An overview of manufacturing methods, Applications and Future Challenges, Composites part B **201** pp 1-46

[7] Udupa G, Shrikantha S R and Gangadharan K V 2014 *Functionally graded composite materials: An overview*, Procedia Materials Science **5** pp 1291-1299

[8] Udupa G, Shrikantha Rao S and Gangadharan K V 2012 Future applications of carbon nanotube reinforced functionally graded composite materials, *IEEE-International Conference on Adances in Engineering, Science and Management* pp 399-404

[9] Kimberly R and Oo Z, Sujan D 2015 *Microstructure analysis, physical and thermal properties of Al_2O_3 - Al_2TiO_5 functionally graded ceramics for the application of car brake rot*, Pertanika Journal of Science and Technology **23(1)** pp 153-161

[10] Ram S C, Chattopadhyay K and Chakrabarty I 2017 *High temperature tensile properties of centrifugally cast in-situ Al-Mg2Si functionally graded composites for automotive cylinder block liners*, Journal of Alloys and Compounds **724** pp 84-97

[11] Carrera E, Brischetto S and Robaldo A 2008 *Variable kinematic model for the analysis of functionally graded material plates*, AIAA Journal **46(1)** pp 194-203

[12] Wu C P, Chiu K H and Wang Y M 2011 RMVT-based meshless collocation and element-free Galerkin methods for the quasi-3D analysis of multilayered composite and FGM plates, Composite Structures **93(2)** pp 923-943

[13] Wu C P and Li H Y 2010 An RMVT-based third-order shear deformation theory of multilayered functionally graded material plates, Composite Structures **92(10)** pp 2591-2605

ANALYSIS OF THE IMPACT OF THE MACHINE ON THE FORGING PROCESS OF A CONNECTING ROD USING NUMERICAL SIMULATION

Marko Delić[1], Vesna Mandić[1], Filip Pavlović[2]

[1]Faculty of Engineering University of Kragujevac
[2]IMW institute

Corresponding author's e-mail: marko.delic@kg.ac.rs

Abstract. The application of numerical simulations of production technologies, using the finite element method or the finite volume method, is a widely used approach in advanced engineering design and research. Unlike industrial experiments, this approach allows the analysis of the influence of different process parameters on the output performance of a workpiece. The paper analyzes the influence of the choice of the machine on the forging process of the connecting rod, which is made, among other things, for the needs of the auto industry. For this purpose, numerical experiments were carried out, using the finite volume method and the Simufact.forming software package, with variations of three types of forging machines: drop forging hammer, crank press and screw press. Finally, the results of the numerical experiments and conducted analyses are presented, with concluding considerations and recommendations.

Keywords: Numerical simulation, Finite volume method, Forging

1. Introduction

Hot forging technology has an important role in modern industry. Forgings made by hot forging have good mechanical properties and are therefore widely used in the automotive and military industry.In hot forging processes, machines are used that differ according to their construction and mode of action. Hot forging machines are divided into several groups: forging hammers, crank, screw and hydraulic presses. Within this paper, the impact of the forging machine type on the hot forging process will be determined using numerical simulations. The impact of the forging hammer, crank press and screw press will be discussed.

Each of the listed types of forging machines has its own characteristics.The action of the forging hammer is impact, the velocity of the forging mallet can be up to 10m/s, and the filling of the forging engraving is better in the vertical direction due to inertial forces. According to their properties, screw presses are between forging hammers and crank presses and are used only when one forging operation is required for forging a part - final forging. Crank presses do not have an impact effect, so complex engravings are more difficult to fill than usinga forging hammer, because the material flow is more intense in the zone of the parting plane and more difficult in the vertical direction.

The hot forging process is the subject of numerous researches.The analysis of the hot forging process of aluminum 5083, previously deformed by the ECAE (Equal Channel Angular Extrusion)

process, using the finite element method (FEM) is presented in [1]. In [2], two ways of determining the forming load were compared, analytically and using FEM. The process of hot forging of the connecting rod on the crank press was analyzed. The impact of the workpiece dimensions on the forming load and the effective stress in the process of hot forging of the pin on the screw press (1500kN) was discussed in the paper [3]. The aim of the research, which was carried out experimentally and numerically, is to reduce the forming load and the effective stress, and it was concluded that by reducing the volume of the workpiece, i.e. determining the optimal billet volume can reduce the value of the forming load. A detailed review of the materials processed by hot forging as well as the materials used to make dies for hot forging is given in [4]. Engineers have the goal to reduce material consumption by optimally determining the workpiece dimensions and thus preventing excessive filling of the flash [5], and FEM analysis plays a significant role in this process. The finite element method is used when defining the forging technology, primarily the workpiece dimensions and the choice of the forging machine [6]. For the successful carrying outof the forging process, the reliability of the press is very important, and the FEM analysis of the loaded componentsof the press and the heat impact can be important [7, 8]. The calculation of the workpiece dimensions and other production parameters can be automated using special software [9].

2. Description of the research methodology

The aim of the research presented in this paper is to determine the impact of machine selection on the process of hot forging of connecting rod, that is an important component for parts in the auto industry.In order to verify the obtained results, an analysis of the industrial process was carried out, so that CAD models of the connecting rod and corresponding engravings of the forging dies for two operations were obtained from the partner company Zastava Kovacnica.

In order to analyze the impact of the choice of forging machine on the forging process and the performance of connecting rod forging, three numerical experiments were performed on different machines (forging hammer, crank press, screw press). Numerical simulations of all analyzed forging processes were realized using Simufact.forming software and finite volume method (FVM) [10]. For all numerical simulations, the same parameters of the process were followed (filling of the engraving, material flow into the flash, homogeneity of the temperature field which affects the microstructure of the forging, required forming load, i.e. forging energy, as well as strainand stress fields in the forging).

The process of hot forging of the connecting rod consists of two operations, preliminary and final forging, and the design of the technology is based on the calculation of reduced forging for the preliminary forging operation, in order to perform an optimal redistribution of the mass of the workpiece and for the correct filling of the final engraving. The billet dimensions for forging the connecting rod is ϕ65x111mm, made of 42CrMo4 material, which is heated to a temperature of 1225°C. The dies were heated to 150°C, and the ambient temperature for process simulations was set to 50°C.

The design of the reduced forging is based on the shape and dimensions of the final forging. A preform (reduced) forging is a shaped perform of regular circular cross-section that is the same length as thefinal forging, where the cross-sectional area is equal to the sum of the cross-sectional area of the forging and the flash at a certain cross section.CAD models of the connecting rod forging and the calculated and modeled perform forging (shape after preforging) are shown in figure 1. The virtual perform forging for FVM analysis is discretized with 4mm facets.

Forging die engravings for both forging operations are shown in figure 2, as CAD models of hammer forging dies. The same engravings were used for the other two numerical experiments on other forging machines, in order to compare and analyze the results. The die material is H13, and rigid dies with heat conduction were used in the simulations.

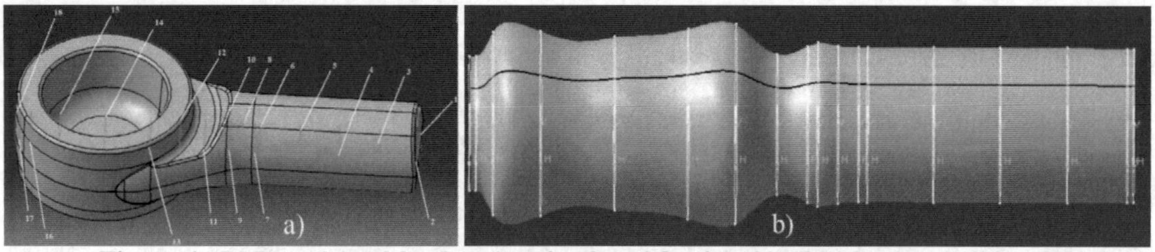

Figure 1. CAD models of forging a) and reduced forging, b) with characteristic sections

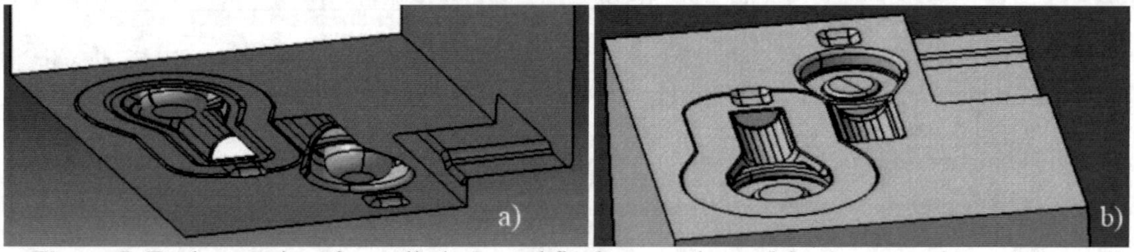

Figure 2. Tool engravings for preliminary and final connecting rod forging operations, a) upper die b) lower die

The properties of the considered presses (forging hammer, crank and screw press) are given in table 1. According to the data listed intable 1, in the Simufact.forming software i.e. in its press definition module "virtual presses" with given properties are defined.

Table 1. Press properties

Drop forging hammer	Crank press	Screw press
Max. impact energy: 9500 kGm Weight of falling part:3570kg Number of blows: 3	Crank radius: 300 mm Rod length: 1200 mm Revolution: 30 Rpm	Gross energy: 40000 [J] Maximum ram speed: 340 mm/s Efficiency during stroke [0-1]: 0,9

3. Research results and comparative analysis

The results of numerical simulations of the connecting rod forging process in the final forging operation were considered, namely the fields of effective plastic strain, effective stress and temperature in the forging, then forming load diagrams, and the material flow into the flash and filling of the engravings via the *Die contact option* (two-color display) were monitored.

The results of the numerical experiment when a forging hammer was used are shown in figure 3. The effective plastic strain is the highest in hammer forging process and in that case the strainfield is the least homogeneous compared to other used forging presses. The maximum value of the effective plastic strain is 2.93, but the area belonging to higher values is located on the flash of the forging, which is subsequently removed, while the maximum values on the forging itself have smaller values and amount to 2.2. The maximum value of effective stress is 315MPa. The highest values of the effective stress are inside the forging and such values are expected because the stress in the material depends on the strain rate because the viscoplastic behavior of the material is present. The temperature field is divided into several areas. The highest temperature is 1178°C and belongs to the area located on the surfaces of the forging that were least in contact with the die (top of the forging and engraving). One of the most important things in the analysis of the forging process is to check the fulfillment of the die engraving, which is checked in numerical simulations with the *die contact option*.Figure 3.d shows that the engraving is completely filled, that confirms the success of the process.The protrusion of the material into the flash is uniform in the dividing plane. The maximum value of the forming load is large and amounts to 32000kN. The value of the formingload increases sharply after reaching the forming stroke of 46 mm in the third blow of the hammer and corresponds to the moment when the material starts to protrude into the flash and the die closes.

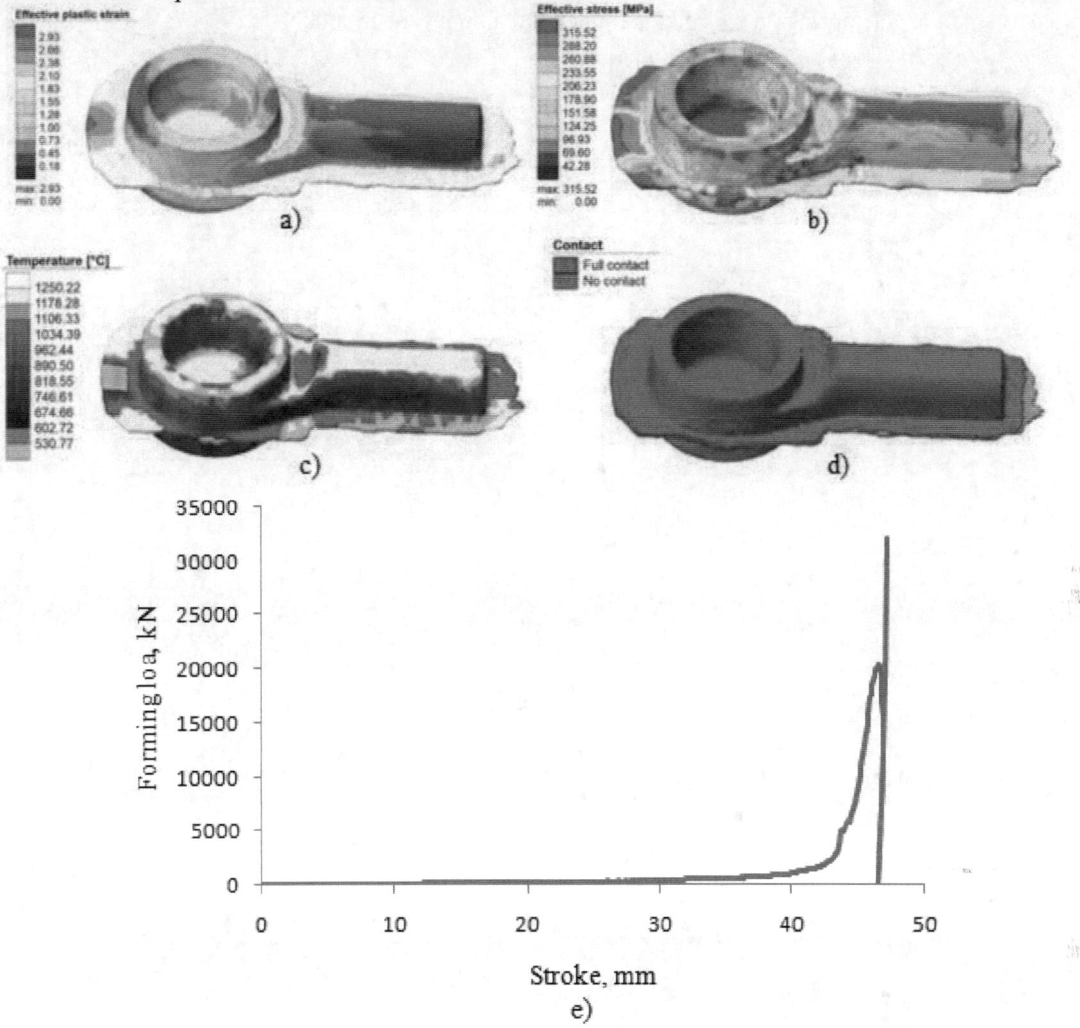

Figure. 3. Forging ondrop forging hammer a) effective plastic strain b) effective stress c) temperature d) die contact e) forming load diagram

The results of numerical simulation of the forging on crank press are shown in figure 4. The maximum value of the effective plastic strain is 3.88, which is a significantly higher value than in the case of forging ona forging hammer, and that in the zone of greater protrusion of the material into the flash in the part of the parting plane of the forging. The maximum value of the effective stress is 125MPa, significantly lower than for hammer forging, which is expected because the formingvelocityon a crank press is significantly lower than onforging hammer. The temperatureof the forging is lower than when forgingon a hammer, due to longer forging time and cooling of the forging in longer contact with the tool. The impact of the forming velocity on the temperature distribution in the forging is evident, so higher temperatures are achieved in hammer forging, because the process takes less time and the impact loads are the highest, so additional heat is generated due to forming. On the crank press, forming velocityis significantly lower, so the material in contact with the tool and the environment cools down during the longer duration of the forging process. Even in the case of forging on a crank press, the engraving of the die is adequately filled, but the shape of the flash is different than in forging on a hammer, namely in the cylindrical part of the connecting rod handle, which could have a lack of material for filling the connecting rod hub. The maximum forming load is more than twice as small as for forging on a forging hammer and amounts to 15000kN.

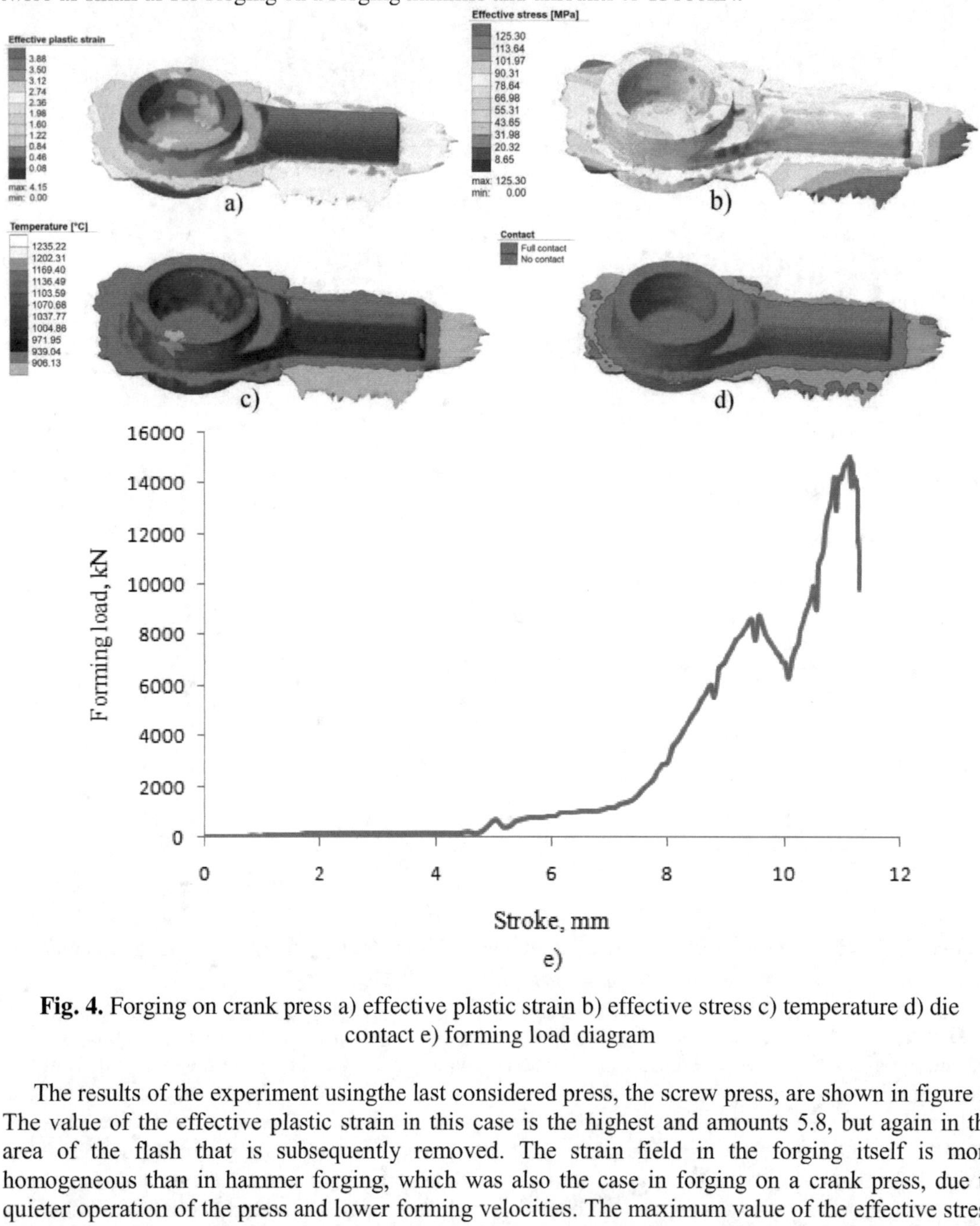

Fig. 4. Forging on crank press a) effective plastic strain b) effective stress c) temperature d) die contact e) forming load diagram

The results of the experiment usingthe last considered press, the screw press, are shown in figure 5. The value of the effective plastic strain in this case is the highest and amounts 5.8, but again in the area of the flash that is subsequently removed. The strain field in the forging itself is more homogeneous than in hammer forging, which was also the case in forging on a crank press, due to quieter operation of the press and lower forming velocities. The maximum value of the effective stress is 257 MPa, which is higher than for forging on a crank press and lower when using forging hammer. In the case of forging on a screw press, the temperature distribution is similar to that of a forging hammer.As in the previous two cases, the filling of the engraving is good, so the conclusion is that this forging can also be forged on a screw press.The extrusion of excess material into the flash is more even, as with a forging hammer, unlike forging on a crank press. The highest value of forming load was recorded in forging on ascrew press and amounts 42000kN at the moment of closing the tool. Otherwise, the courseof the forming load diagram is different on all considered forging machines, which is expected and confirmed.

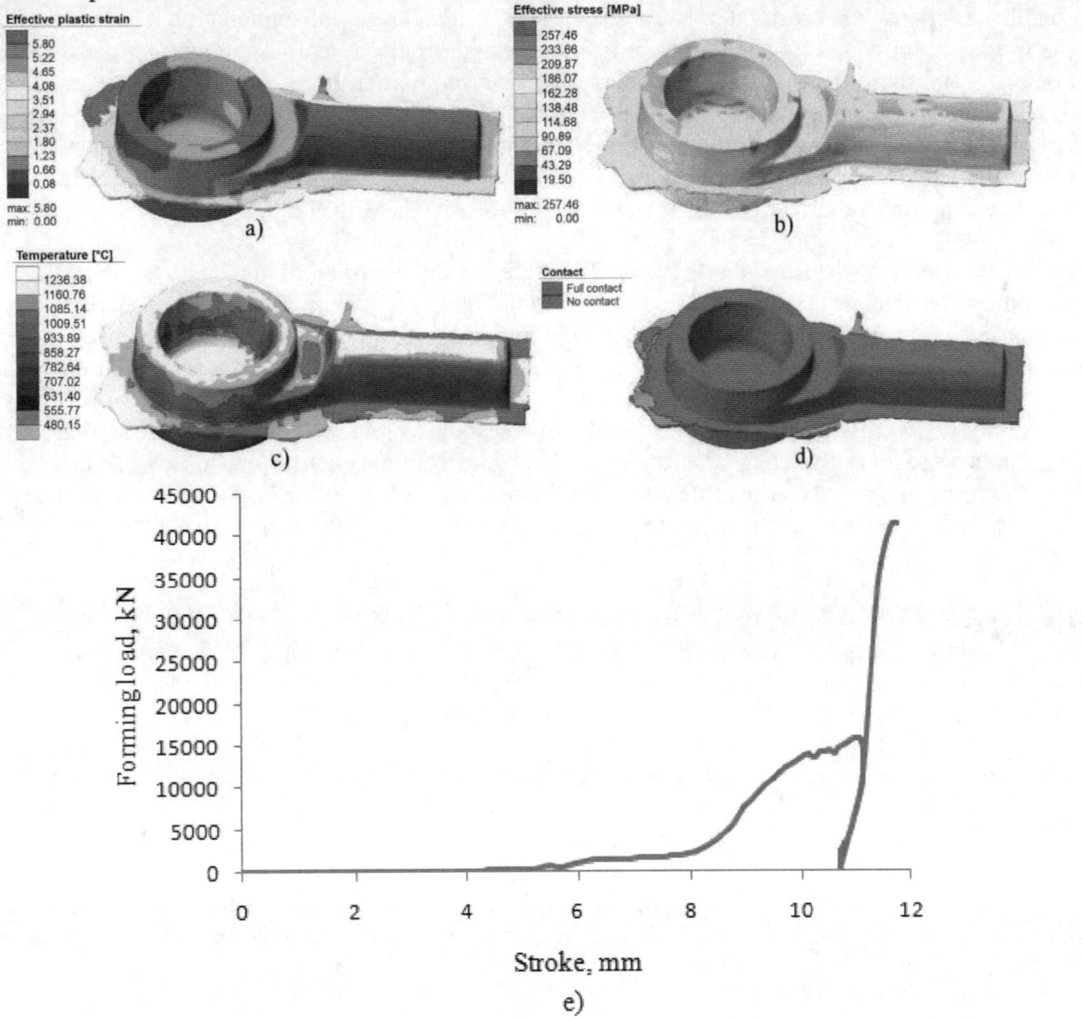

Figure. 5. Forging on screw press a) effective plastic strain b) effective stress c) temperature d) die contact e) forming load diagram

4. Conclusion

The impact of the choice of the forging machineon the forging properties and on the forging process itself can be analyzed using numerical methods and process simulations, as shown in this paper.Due to the exploitation propertiesof the forging, its microstructure, it is important that the strain field in the forging be as homogeneous as possible, and that the amounts of maximum strainsbe within acceptable limits for a certain material.

The results of numerical simulations showed that the highest effective stress is in the inner part of the forging, and it has significantly higher values in hammer forging than in crank and screw press forging. Such values of effective stress in the forging are expected, because the forming velocityisthe highest onthe hammer, then on the screw press, and the least onthe crank press. The stress in the forging material depends on the forming velocity, i.e. the achieved strain rate, because the viscoplastic behaviour of the material is presentat elevated forging temperatures. By comparative analysis of the field of effective plastic strain, maximum values were observed in the flash zone, which is not crucial from the aspect of forging quality, considering that it is removed after forging. However, for the quality of the forging, it is important to have as homogeneous a strain field as possible, which gives smaller variations in microstricture and resistance characteristics during exploitation. Highest strainsin the forging occurred during hammer forging, due to high forming velocities and impact loads. A more homogeneous strain field was obtained during forging on a crank press, and the most unfavourable distribution was obtained during forging on a hammer, due to impact loads and a high forming velocity. On the other hand, filling of the die and uniform protrusion of the material into the flash is evident in hammer forging, while in forging on the crank press, the protrusion into the flash is more prominent in the cylindrical part of the connecting rod shank, with the possibility of not filling the hub.

The temperature distributions in the forging are in similar proportions for these three forging machines, the highest temperatures are during forging on the hammer, and the lowest during forging on the crank press, again due to the different forming velocities of these forging machines, the duration of the forging process, and especially the time of contact with the die and environment and cooling of the workpiece.

Numerical process simulations provide the opportunity to review and analyze all aspects of the machine's impact on the complete forging process and forging quality, outside of the production environment, and to make correct decisions on the process parameters or on the selection of the most favorable machine based on the results.

References

[1] Luria R, Luisa C, Salcedoa D, Leóna J, Fuertesa J and Puertasa I 2013 *The Manufacturing Engineering Society International Conference, MESIC 2013*, FEM analysis of the isothermal forging of a connecting rod from material previously deformed by ECAE 63 540-546

[2] Zahálka M and Staněk J 2014*25th DAAAM International Symposium on Intelligent Manufacturing and Automation, DAAAM 2014,*Options of Advanced Simulations of Crank Presses Loading 891-898

[3] Vishal P and George P 2021*Materials Today: Proceedings*, Effect of preform design on forging load and effective stress duringclosed die hot forging process of pin 44 106-112

[4] Madhankumar S, Narayanan H, Harini V, Gokulraj K, Selvakumar S, Dharshini R, Dharshini K and Harikrishnan T 2021 *Materials Today: Proceedings*, Study and selection of hot forging die materials and hardness 45 6563–6566

[5] Vishal P and George P2021 *Materials Today: Proceedings*, Preform optimization for the anchor shackle during closed die forging process on one ton hammer 47 3256–3262

[6] Jayanthi A, Anilkumar M and Ravisankar B 2022 *Materials Today: Proceedings* Selection of forging process for compressor disc for aero engine using finite element analysis 60 2111–2116

[7] Zdenek C and Cechura M 2015 *ProcediaEngineering*Monitoring Extremely Stressed Points on Stands of Forging Presses 100 841-846

[8] Zdenek C 2014 *ProcediaEngineering* Effect of Heat Load on a Forging Press 69 897 – 901

[9] Bogoyavlensky A and Bogoyavlensky M 2019 *Materials Today: Proceedings*Calculation technological parameters of forging on hammers and presses

[10] Mandic V 2020 *Lecture Notes in Mechanical Engineering* Model-Based Manufacturing System Supported by Virtual Technologies in an Industry 4.0 Context, Springer, APM2020,ISBN 978-3-030-46211-6.

FE mesh density influence on blast loading analysis

Miloš Pešić[1*], Nikola Jović[2], Vladimir Milovanović[2], Mladen Pantić[3] and Miroslav Živković[2]

[1]University of Kragujevac, Institute for Information Technologies – National Institute of the Republic of Serbia, Jovana Cvijica bb, Kragujevac, Serbia

[2]University of Kragujevac, Faculty of Engineering, Sestre Janjić 6, 34000 Kragujevac, Serbia

[3]Faculty of Maritime Academic Studies, Bulevar vojvode Putnika 7, 11000 Belgrade, Serbia

*Corresponding author: Miloš Pešić, email: milospesic@uni.kg.ac.rs

Abstract. To examine the influence of mesh density on the blast loads of anti-landmines (ALMs) and improvised explosive devices (IEDs) under an armored vehicle, models with different FE mesh densities were numerically analysed. The multi-material arbitrary Lagrangian-Eulerian (MM-ALE) algorithm was used to simulate under armored vehicle explosions. Explicit dynamic analysis was performed in LS-DYNA. Blast loading simulation was performed using the ConWep (conventional weapon) loading model, which is implemented in LS-DYNA. The material characteristics of high-strength steel were used to model the protective plates on the armored vehicle. Protective plates were modeled in a V-shape and as a deformable solid with the Johnson-Cook plasticity model. In this paper, the FE model with different mesh densities was numerically tested to analyse the influence of mesh density on the obtained results. For the purposes of the modeling, different types of finite elements were used for the FE models. The floor of the vehicle was modeled with 3D hexahedral eight-noded finite elements, and the protective plates were modeled with four-noded plate finite elements. The maximal value of plastic strain and the maximal value of the displacement of the central node on the protective plate was chosen as the parameters to compare the obtained results.

1. Introduction

The bottom of the vehicle has become crucial in preventing explosion damage as a result of the threat posed by ALMs and IEDs on modern battlefields. The bottom of the vehicle has a key role in resisting the effect of the blast and thus protecting the troops aboard the armored vehicle. Within 4 ms after the explosion, the structure (armored vehicle) is exposed to large elastoplastic deformation as a result of landmine detonation. In paper [1] Kartikeya et al. conducted a mesh convergence study to identify the optimal element size. Maximal vertical displacement of the central node and values of plastic strain at the end of the numerical simulation were monitored in order to compare numerical and experimental results. Makwana et al. [2] proved that increased stand-off, decreased internal angles of V-plates, or providing sharp V-angles can all be used to reduce blast momentum. Counter momentum measures can also be used to reduce overall momentum. Trajkovski et al. [3] investigated the influence of blast loading on light armored vehicles with flat and V-hull protective plates. Their results proved that the V-hull protective plates offer greater protection for armored vehicles and troops aboard. For this research, they combined Smooth Particle Hydrodynamics – Finite Element (SPH-FE) methods. In paper [4] blast loading simulation on V-shaped protective plates was conducted using commercial software ABAQUS Explicit for the numerical simulation. The parameters chosen to monitor obtained results were plastic strain and generated impulse due to the blast loading. Ivanco et al. [5] investigated how blast waves propagation influence on kinetic energy acting on armored vehicle chassis and the displacement generated by the energy of blast loading. In this paper, FE models with different mesh densities were investigated in order to obtain convergence. Based on V-shaped hull protection research was completed. To determine the convergence, the protective effect of the FE models with different mesh density under the influence of 6-kg trinitrotoluene (TNT) blast were examined. The mass of TNT in this research is in accordance with NATO AEP-55 STANAG 4569 standard – level 2 [6]. Slavik [7] gave specifics on how ConWep loading model was implemented, show how it was used in the simulation environment, and compared obtained results with those of previous techniques for modeling air blast loads on a structure.

2. Problem description and FE model

The armored vehicle is equipped with a V-shaped hull to adequately protect its troops from the anti-landmine and improvised explosive device blasts. The hull is made of a metal plate that is installed beneath the floor of the armored vehicle and bent into a V shape. Four edges of the protective plate are merged to the body of the armored vehicle, thus simulating welded joints. The ability of this plate to withstand or divert the blast wave away from the armored vehicle determines how blast resistant the armored vehicle is. STRENX700 (S690QL) steel is used as the material for the protective plate.

In order to determine FE mesh density influence and to achieve convergence the floor and V-shaped hull of the armored vehicle is modeled with five different mesh densities. The number of 3D hexahedral eight-noded finite elements and four-noded plate elements which are used to model all five FE models and the number of the nodes are shown in table 1.

Table 1. Number of finite elements and nodes

Finite element size	Number of 3D finite elements	Number of plate elements	Number of nodes
FE model 1 10mm x 10mm	60000	79800	199411
FE model 2 20mm x 20mm	15000	17100	47867
FE model 3 30mm x 30mm	6800	7600	21715
FE model 4 40mm x 40mm	3800	4636	12543
FE model 5 50mm x 50mm	2400	2760	7869

Eight-noded 3D hexahedral finite elements are used to model the supporting structure of the vehicle's floor, while four-noded plate finite elements are used to model the protective plate. The thickness of the protective plate is 10 mm for all FE models presented in table 1. Using the CAD model for creating the FE model in the FEMAP v2021.2 software [8] the input file for the LS-DYNA software [9] was exported. All additional FE model preparations and explicit dynamic analysis setups were completed using the LS-DYNA software. A hemispherical incident wave was applied to all protective plates. In every FE model, there was a 0.5 m gap between the TNT and the protective plate. The software FEMAP v2021.2 was used for pre-processing and post-processing. The V-shaped geometry for ALMs and IEDs protection of armored vehicles and troops are shown in figure 1. Suhaimi et al. [10] conducted research on how blast loading affects troops in armored vehicles using a dummy in experimental research.

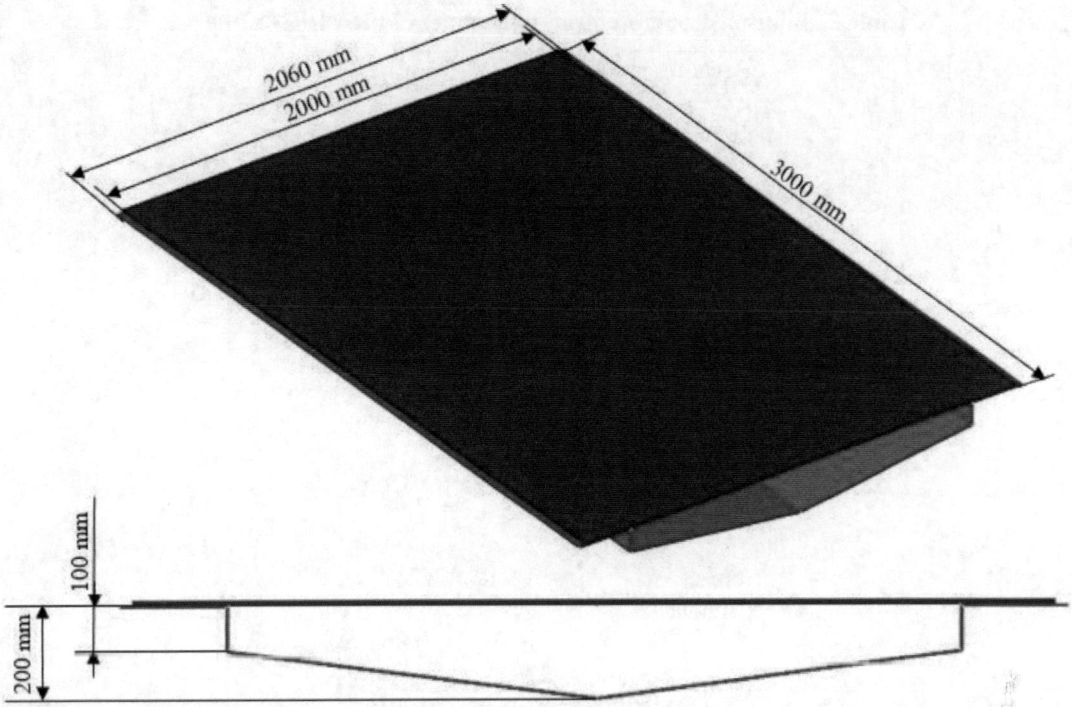

Figure 1. The floor and V-shaped hull protection of the armored vehicle

2.1 *Material model*

STRENX 700 belongs to the class of high-strength steel, and it can be used for military and civil applications where high protection is needed. For the purpose of numerical presentation of STRENX700, the Johnson-Cook strength model was used:

$$\bar{\sigma} = \left(A + B\left(\bar{\varepsilon}_{pl}\right)^n\right)\left(1 + C\ln\dot{\bar{\varepsilon}}_{pl}^*\right)\left(1 - T^{*m}\right) \tag{1}$$

where A, B, C, n, and m are material constants, $\bar{\varepsilon}_{pl}$ is the equivalent plastic strain, $\dot{\bar{\varepsilon}}_{pl}^* = \frac{\dot{\bar{\varepsilon}}_p}{\dot{\varepsilon}_0}$ dimensionless strain rate, where $\dot{\bar{\varepsilon}}_p$ is the equivalent plastic strain rate and $\dot{\varepsilon}_0$ is the reference strain rate. Dimensionless temperature is given by equation:

$$T^* = \frac{T - T_r}{T_m - T_r} \tag{2}$$

where T is the current temperature, T_r is the reference temperature and T_m is the melting temperature of the material.

The Johnson-Cook damage model is given by equation:

$$\dot{\bar{\varepsilon}}_p^f = \left(D_1 + D_2\exp\left(D_3\sigma^*\right)\right)\left(1 + D_4\ln\dot{\bar{\varepsilon}}_{pl}^*\right)\left(1 + D_5T^*\right) \tag{3}$$

where $D_1 - D_5$ are material damage parameters and $\sigma^* = \frac{\sigma_H}{\bar{\sigma}}$ is a stress triaxiality ratio and σ_H is hydrostatic pressure.

STRENX700 is used for this study due to the availability of relevant data. Johnson-Cook material parameters for STRENX700 are given in table 2.

Table 2. Johnson-Cook material parameters for STRENX700 [11]

Parameters		Values
Density	ρ [t/mm^3]	7.85E-9
Young's Modulus	E [MPa]	228368.9
Poisson's ratio	ν	0.33
Yield stress	A [MPa]	767.38
Proportionality coefficient	B [MPa]	445.13
Reinforcement exponent	n	0.5075
Strain rate Impact parameter	C	0.0265
Temperature impact parameter	m	1.354
Damage parameters	D_1	-0.066
	D_2	3.028
	D_3	-1.408
	D_4	-0.00851
	D_5	0.633

2.2 *TNT equivalent method*

According to the shortened distance between the chosen object and the explosion site, the TNT equivalent method – which is frequently used in explosion resistance modeling – estimates the pressure caused by the explosion. The US Navy and US Air Force invented this technique, which is still employed by armies all over the world. Based on a review of numerous military experiments, this method uses correlations between explosives and damage. After calculating the combustion heat of the escaping gas, the combustion energy of the explosion can be set to the same level as the equivalent of the TNT charge, and then the distance of the overpressure is estimated. Because of its simplicity and low computational time, the TNT equivalent method is widely used.

The equivalent weight of TNT for a flammable gas can be calculated using the following equation:

$$m_{TNT} = \frac{\eta m \Delta H_c}{E_{TNT}} \tag{4}$$

where m_{TNT} represents equivalent of TNT, η is the empirical explosion efficiency, m is the mass of hydrocarbon, ΔH_c is the energy of explosion of flammable gas and E_{TNT} is the energy of explosion of TNT.

3. Results and discussion

All FE models were examined in accordance with the tasks established in Chapter 2 in order to monitor vertical displacements of the central node on the protective plate in dependence of time. Central node is located on the longitudinal plane of symmetry in the direction perpendicular to the protective plate. Vertical displacement of the central node on the protective plate in dependence of time for all five FE models are shown in figure 2. In figure 3 magnified results for vertical displacement of the central node on the protective plate in dependence of time for all five FE models are shown.

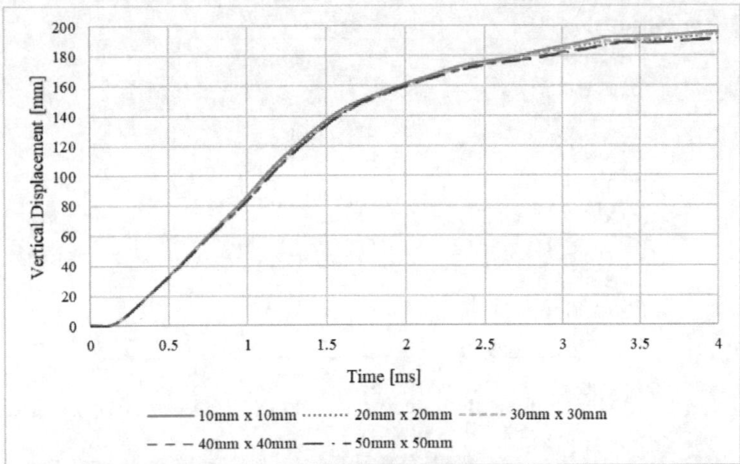

Figure 2. Vertical displacement of the central node on the protective plate in dependence of time

As can be seen from figure 2, with the increase in the size of the finite elements, the vertical displacement of the central node on the protective plate decreases. Also, it can be seen that the difference in the value of the vertical displacement of the central node on the protective plate between FE model 4 and FE model 5 is very small (figure 3, table 3), which means that the result converges.

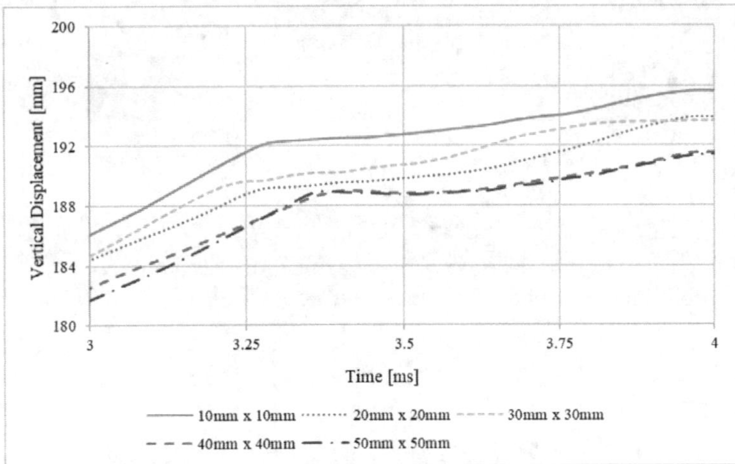

Figure 3. Vertical displacement of the central node on the protective plate in dependence of time – magnified

Given that convergence has occurred, it can be concluded that a further change in the size of the finite elements does not affect the result, that is, the value of the vertical displacement of the central node on the protective plate remains the same. In table 3 obtained results for vertical displacement of the central node on the protective plate for all five FE models are shown.

Table 3. Values of vertical displacement of the central node on the protective plate for all five FE models

Finite element size [mm]	Vertical displacement [mm]
FE model 1 10 x 10	195.6473
FE model 2 20 x 20	193.892
FE model 3 30 x 30	193.6688
FE model 4 40 x 40	191.5166
FE model 5 50 x 50	191.4053

An example of vertical displacement of the central node on the protective plate for FE model 3 is shown in figure 4.

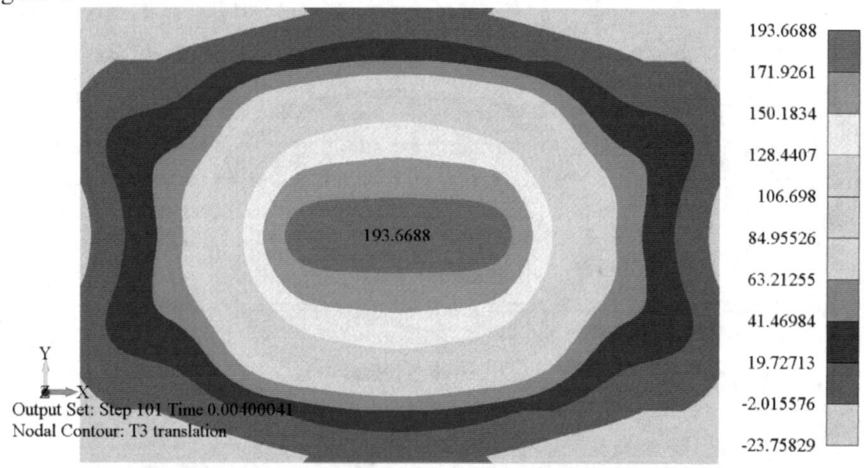

Figure 4. Vertical displacement distribution

Plastic strain distribution of FE model 3 is shown on figure 5.

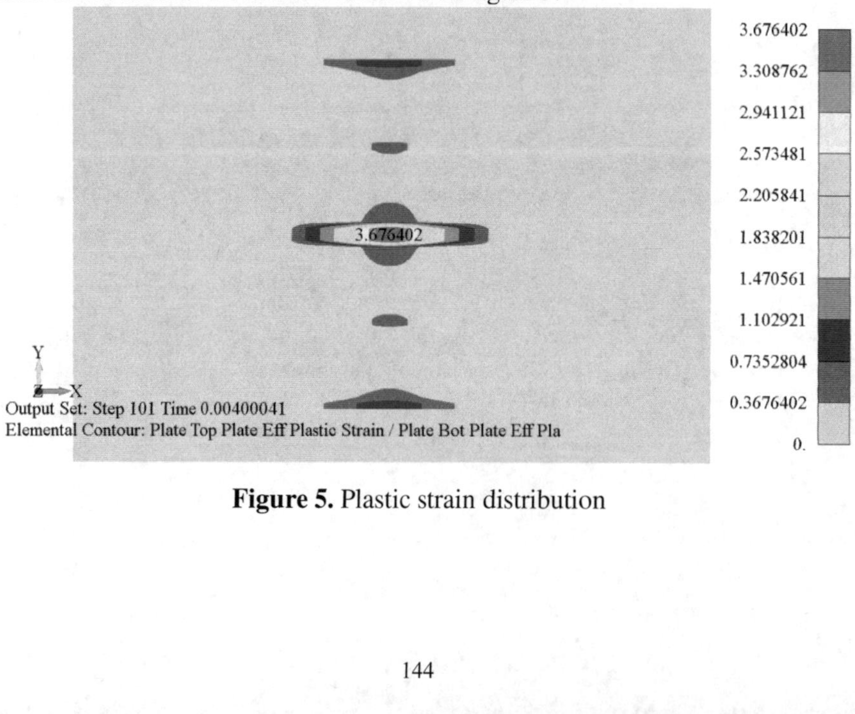

Figure 5. Plastic strain distribution

4. Conclusion

From specialized mine-resistant vehicles to main battle tanks, the threat posed by ALMs and IEDs is a significant consideration in the design of armored vehicles. In this paper, floor of the armored vehicle with a V-shaped hull was numerically tested. The floor and protective plates of the armored vehicle are modeled with five different mesh densities. The objective was to achieve convergence. For that purpose, maximal vertical displacements of the central node on the protective plate, which is located on the longitudinal plane of symmetry in the direction perpendicular to the protective plate were monitored.

With the increase in the size of the finite elements, the difference in the analysis values for vertical displacements and plastic deformation due to the impact of the blast wave as a result of the anti-landmine explosion, decreases. Five different FE models were tested, i.e. five FE models with different sizes of finite elements. The differences in the values of vertical displacements in the FE models 4 and 5 are minimal, so it can be confirmed that convergence has occurred. Considering the accuracy of the results in FE models 4 and 5, there is no need to further increase the mesh density. In this way, engineering and computational time is saved.

Further research will be focused on the protection of troops, armored vehicles and military equipment from the effects of ALMs and IEDs by using different geometries of protective plates, changing the angles of V-shaped hulls, as well as using different types of high-strength steels, sandwich constructions and composite materials.

Acknowledgments

This research is partly supported by the Ministry of Education and Science, Republic of Serbia, Grant 451-03-68/2022-14/ 200378, and Grant TR32036.

References

[1] Kartikeya *et al.*, 2019. Finite Element Simulation of Armor Steel used for Blast Protection. *Procedia Structural Integrity*, vol. 14, pp. 514–520.
[2] Makwana DR et al., 2019. Numerical Analysis and Effects on Rigidity of Combat Vehicle Structure Due to Blast Load. Procedia Structural Integrity, vol. 14, pp. 44–52.
[3] Trajkovski J et al., 2018. Blast response of Light Armoured Vehicles (LAVs) with flat and V-hull floor. Thin-Walled Structures, vol. 131, pp. 238–244.
[4] Markose A and Rao CL, 2017. Mechanical response of V shaped plates under blast loading. Thin-Walled Structures, vol. 115, pp. 12–20.
[5] Ivančo M et al., 2019. Simulation of detonation and blast waves propagation. Transportation Research Procedia, vol. 40, pp. 1356–1363.
[6] NATO, 2014. AEP-55 Vol2 Procedures for Evaluating the Protection Level of Armoured Vehicles – Mine Threat, NATO Standardization Agency.
[7] Slavik TP 2010. A coupling of empirical explosive blast loads to ALE air domains in LS-DYNA®. IOP Conf. Ser.: Mater. Sci. Eng., vol. 10, pp. 012146.
[8] Femap, 2021. Finite Element Modeling and PostProcessing Application FEMAP v2021.2, Siemens.
[9] LS-DYNA, 2014. LS-PrePost-4.5-A New Post Processor for Use LSDYNA, California: Livermore.
[10] Suhaimi K et al., 2016. Simulation on the Shock Response of Vehicle Occupant Subjected to Underbelly Blast Loading. Procedia Computer Science, vol. 80, pp. 1223–1231.
[11] Dišić A, 2018. Development of methodology and devices for dynamic testing of materials and welded joints with application in numerical calculations of structures at high deformation rates – PhD Thesis, Faculty of Engineering University of Kragujevac, Serbia.

HOW THE CORE TECHNOLOGIES OF INDUSTRY 4.0 ARE CHANGING THE AUTOMOTIVE INDUSTRY IN THE WORLD, WITH A FOCUS ON CHINA

Isak Karabegović[1], Ermin Husak[2], Edina Karabegović[2], Mehmed Mahmić[2]

[1]Academy of Sciences and Arts of Bosnia and Herzegovina, Bistrik 7.,Sarajevo, Bosnia and Herzegovina
[2] Univesity of Bihać, Irfana Ljubijankića bb, Bihać, Bosnia and Herzegovina

E-mail: isak1910@hotmail.com

Abstract.: Industry 4.0 has a significant impact on the automation of production processes, by causing numerous changes in three segments: companies, technology and workers. Developed countries worldwide have their own strategies on Industry 4.0, which offer guidelines on its implementation in production processes, with the aim of their complete flexible automation. The core technologies on which Industry 4.0 rests have led to a complete transformation in production processes, especially in the automotive industry. The basic technology of Industry 4.0 is robot technology, i.e., the implementation of industrial and service robots in production processes. The paper provides an analysis of the implementation of industrial robots and service robots in the automation of production processes in the automotive industry with a focus on China. The analysis of the automation of production processes of the automobile industry in China was carried out for two reasons. The first is that China has a growing middle-class population, so demographic trends are encouraging the growing demand for certain products, such as cars. Another reason is that in China (as in Japan, Russia and Western Europe) the average age of factory workers is increasing (the labor force is older), thus the performance of certain tasks becomes more difficult so greater efficiency is not achieved. The paper analyzes vehicle production in China, as well as the readiness of production processes in the automotive industry for the implementation of Industry 4.0.
Keywords: Industry 4.0, automation, robot, vehicle, production process

1. Introduction

In the past ten years, new technologies have been rapidly changing various branches of industry, including the automotive industry. It is well-known that the automotive industry was the first to implement the latest technologies due to global competition on the market. The automation system in the automotive industry is changing the way vehicles are produced. By introducing automation into production processes in the automotive industry, tedious and time-consuming processes are performed in less time with increased quality. This is achieved by the introduction of industrial robots into production processes, which improve the very aspects of the production process, such as design technology, aerodynamic efficiency, fuel economy, determination of chassis strength, etc. [1,2,3,]. It is known that the term Industry 4.0 is already widely used, and it is related to various production concepts. From a technological perspective, Industry 4.0 should be understood as the increasing digitization and automation of production processes, that is, the production environment and the design of the digital value chain from the product to the customer. Some authors call Industry 4.0 a new disruptive technology. Industry 4.0 is inevitable because the world today is in the dynamic era, and it changes the very aspect of production processes that are currently implemented in most companies [5-9]. The concept of Industry 4.0 is defined by many technologies, some of which are: robotics and automation, Internet of Things (IoT), Big Data, Cloud Computing, 3D printing, smart sensors, Radio Frequency Identification (RFID), virtual and augmented reality (AR), Artificial Intelligence (AI), advanced security systems, Cyber-Physical Systems (CPS), etc. The implementation of the above technologies allows car manufacturers to simulate dangerous scenarios in order to test and develop new security measures. IoT technology enables access to information about safety operations and driving time, among others. Automotive telematics insurance and surveillance vehicles are deployed to accurately assess risk by observing mileage, driver behavior and vehicle type. The inclusion of IoT in the automotive industry makes it easier to prevent accidents on the highway. Tracking devices and software are new technologies that will allow automotive companies to make the most of their available assets, while its optimization reduces costs, increases profits, and facilitates the growth of the company and the automotive industry as a whole [2,12-14].

2. Industry 4.0 transforms production processes in the automotive industry to its limits

Competition on the global market and the rapid development of demand forces the companies of the automotive industry to implement modern technologies, i.e., Industry 4.0, which leads to progressive automation and information exchange in the production and technological sectors. Industry 4.0 in the automotive industry extends modern digital technologies to a comprehensive scope, which includes Cyber-Physical Systems (CPS), Internet of Things (IoT) and Cloud Computing. Automotive industry companies perceive the implementation of Industry 4.0 as a big challenge they need to overcome because these are disruptive changes, given the constantly changing conditions for adapting products and services. Value chains need to be continuously analyzed and improved using modern technologies [15-18]. The concepts of Industry 4.0 are continuously researched with the aim of ensuring smooth production in the automotive industry. With the implementation of Industry 4.0 in companies, more transitive business models are expected: at all levels of the company, production process, products, and supply chain, success in measuring data in relation to clients, as shown in Table 1. As depicted in Table 1, a current way of management and production is changed in a way that we acquire: companies become agile and automated, decentralization of the company to a lower level, personalization and adaptability, the company becomes dynamic and predictive, high return on invested capital, connection with clients is high and direct, etc. We can state that Industry 4.0 represents a change in the paradigm of the previous production, which was rigid and its very efficiency focused on manual production, to a dynamic production that is automated, agile and adaptable for any product. In other words, we can say that there will be a transformation that changes focus from mass production to mass customization, enabled by flexible production and shorter product delivery times.

The implementation of Industry 4.0 supports the transition of companies, which become companies with high-tech equipment that are able to produce more different products at competitive prices. The basic technologies of Industry 4.0 presented in the make production processes in the company more flexible because there is the possibility of working remotely. The very indicator of success is, for example, achieving low production costs, increasing profitability by facilitating greater adaptation, reducing labor costs and reduction of complexity costs, as well as reduction of invested capital thus enabling greater flexibility and use of resources [19-24].

***Table 1** Transitions expected with the implementation of Industry 4.0 [4]*

	Production before Industry 4.0	Production with the implementation of Industry 4.0
SCALE OF FACTORIES	Large companies in centralized locations	Small companies in centralized locations
PROCESS	Rigid and manual	Agile and automated
PRODUCT	Standardized	Customized and personalized
SUPPLY CHAIN	Stock bazed planning	Dynamic and predictive
SUCCESS METRIC	Low cost, high efficiency	High return an capital employed
CLIENT RELATIONSHIP	Low and indirect	High and direct

The key prerequisites for the implementation of Industry 4.0 are:

- Networked production systems and vertical integration with the aim of ensuring an intelligent and integrated production environment, all thanks to the integration of information and communication technology and its consistency within the company, from the development through the production process and logistics to sales;
- Horizontal integration by ensuring the networking of information technologies, processes and data flows among companies, suppliers, business partners in such a way as to promote intensive cooperation based on partnership within the value chain;
- Within the process of the complete value chain, continuous and digital integration of engineering is necessary, thus enabling a high degree of design of end products that are oriented towards the customer, whose goal is to enable cost savings in the company [10].

The implementation of Industry 4.0 in the automotive industry is reflected in the implementation of the following elements [11]:

- Collaborative robots and machines, which are based on machine-machine (M2M) and human-machine (H2M) communication, which releases workers from risky jobs;
- Augmented reality, i.e., using virtual simulations to design workplaces, production and adjustment processes, as well as repair machines;
- Greater flexibility in all segments, which gives a faster production response to global market fluctuations; digital production facilitates the production of complicated products;
- 360-degree networking;
- Virtual assembly allows employees to master and evaluate how best to perform a task using the control module;
- Increased efficiency, rational use of resources and energy consumption;
- Increased speed, flexible production processes, simplified adaptation of existing production processes and installation of new equipment enable easier and more efficient production processes;
- The goal is a smart factory where, by incorporating the real world into the digital world, it is possible to create a so-called digital twin that enables a real-time display of the process itself, the system and the entire production.

The very decision to implement Industry 4.0 in the company lies in its ability to act as a change in the economic game, which would open countless opportunities for companies to remodel or create completely new offers and business models. Industry 4.0 enables mass customization that offers greater flexibility for companies to meet the diverse demands of their customers, who increasingly prefer more flexible and customizable results compared to standard product offerings. Various industrial and research reports indicate a huge potential in the transformation of industrial production.

3. The trend of implementation of industrial robots in vehicle production in the world with a reference on the representation of robots in the automotive industry

The technologies of Industry 4.0 are rapidly changing all sectors of industrial branches worldwide, with main focus on the automotive industry. The automotive industry is the first in the implementation of most modern technologies compared to other industries, the reason being the competition on the global market. Robotic technology is one of the technologies that bring changes in the production processes of the automotive industry in the world. In order to explain this, an analysis of the implementation of industrial robots in the world was conducted. The statistical data on the number of industrial robots were obtained from the International Federation of Robotics (IFR), the UN Economic Commission for Europe (UNECE) and the Organization for Economic Cooperation and Development (OECD) [25-33]. The data on the number of manufactured vehicles in the world were obtained from the Verband Deutscher Verkehrsunternehmen (VDV) [21,35,36], which are graphically shown in the following figures. Figure 1 shows the trend of the implementation of industrial robots in the world on an annual basis in the period 2010-2020 and the predictions of application until 2023.

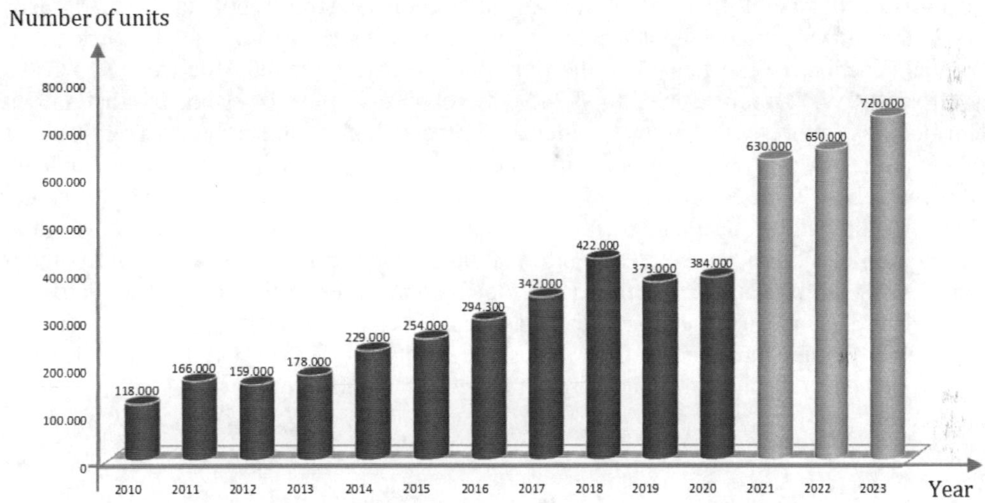

Figure 1 *The trend of the implementation of industrial robots in the world on an annual basis in the period 2010-2020 and predictions of application until 2023 [25-34]*

Based on Figure 1, we come to the conclusion that since 2010, the trend of the implementation of industrial robots in the world had a growing character until 2018. The growth trend increased from 118.000 industrial robot units used in 2010 to 422.000 robot units used in 2020. In other words, in just eight years, the implementation has increased about 3.5 times. In 2019 and 2020, the implementation was reduced, so that in 2020 it amounted to around 384.000 robot units. The reason that the trend of implementing robots has decreased is that the world has been affected by the pandemic of the COVID-19 virus. After the end of the pandemic, there was an increase in the trend of implementation of industrial robots in the world, and the predictions are that in 2023 about 720.000 robot units will be implemented. The trend of the overall implementation of industrial robots in the world is shown in Figure 2.

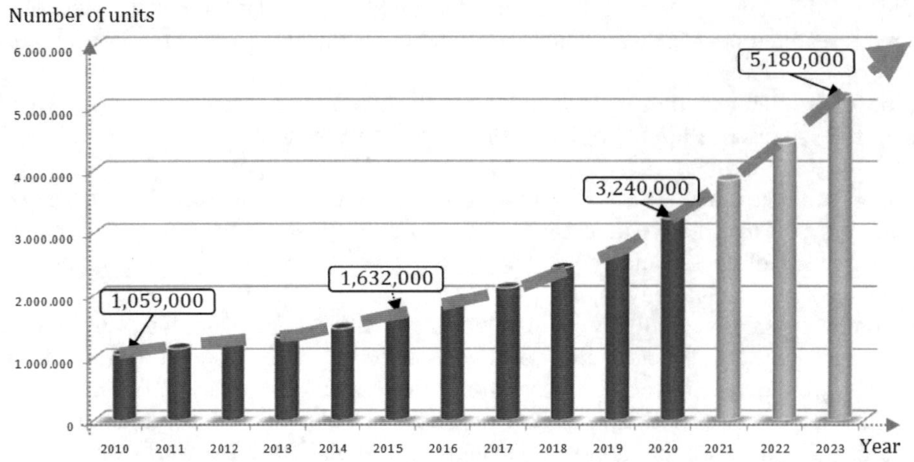

Figure 2 *Total implementation of industrial robots in the world in the period 2010-2020 and predictions of application until 2023 [25-34]*

The analysis of the overall trend of the implementation of industrial robots in the world depicted in Figure 2 has shown that the implementation has a growing character, and is carried out by an exponential function. In just ten years, the total implementation of industrial robots increased by 3 times (from 1.059.000 robot units to 3.240.000 robot units). It is estimated that the trend of implementation of robots will continue to increase in the coming years, and it is expected that in 2023 it will amount to about 5.180.000 robot units. The reason for the growing trend in the implementation of industrial robots is the implementation of Industry 4.0 in companies that have to maintain their positions on the global market. It is interesting to note which industries have the largest implementation of industrial robots. This information is given in Figure 3, which shows the trend of the implementation of industrial robots in the top five industries in the world: automotive industry, electrical/electronic industry, metal industry, plastic and chemical industry and food industry. The first place in terms of the implementation of industrial robots in the world is held by the automotive industry, followed by the electrical/electronic industry and the metal industry.

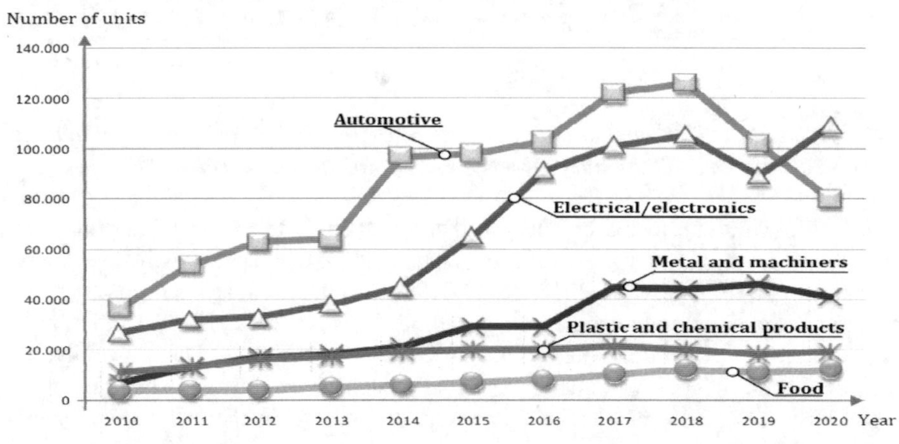

Figure 3 *The trend of implementation of industrial robots in five top industrial branches in the world in the period 2010-2020*

The plastic and chemical industry ranks fourth in the world in terms of the use of industrial robots, while the food industry occupies the fifth place. Considering that the automotive industry is in first place, an analysis of the trend of the implementation of robots in the world on an annual basis was conducted and shown in Figure 4.

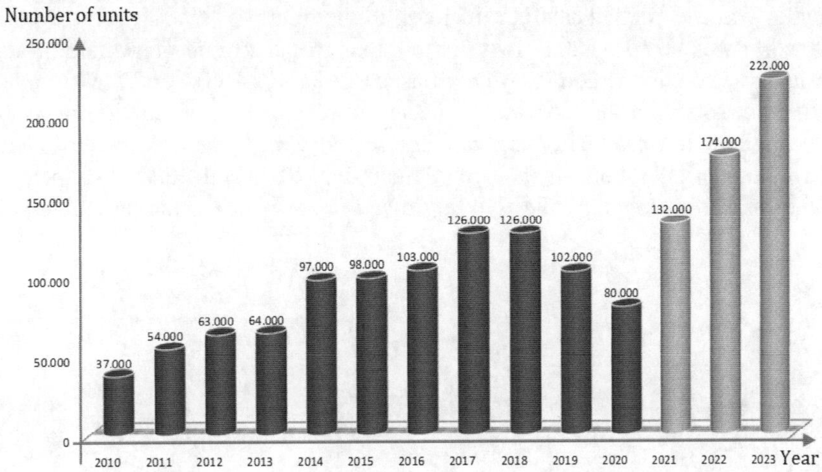

Figure 4 *An annual implementation of industrial robots in the automotive industry in the world in the period 2010-2020 and predictions of application until 2023 [25-34]*

The analysis of the trend of the implementation of industrial robots in the automotive industry in the last ten years has provided us with the conclusion that it had a growing trend until 2018. In the period from 2010 to 2013 the implementation increased from 32.000 robot units to 64.000 robot units, which is almost double increase. In 2014, the implementation increased by 51% compared to 2013. The trend of implementation has increased until 2018, when around 126.000 robots were implemented in the automotive industry. Due to the pandemic of the COVID 19 virus in the world, the trend of the use of industrial robots in the automotive industry has declined, so that in 2020 it amounted to about 80.000 robot units. At the end of the COVID 19 virus pandemic, the growing trend of the implementation of robots in the automotive industry is predicted, and the implementation of around 222.000 robot units is predicted in 2023.

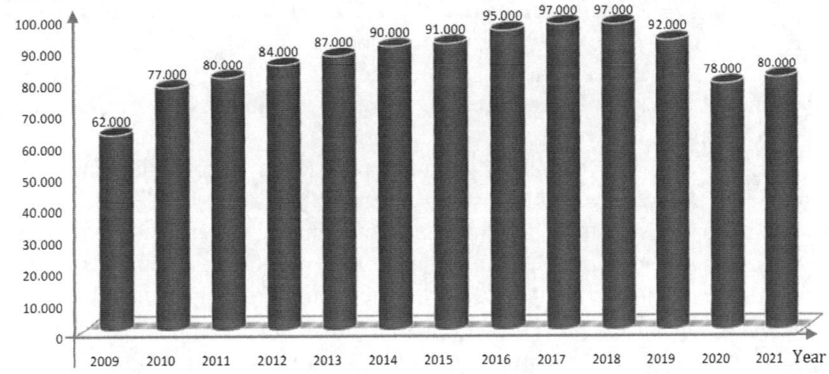

Figure 5 *The trend of vehicle production in the world in the period 2009-2021 [35-36]*

As a result of the implementation of industrial robots in the automotive industry, the trend of vehicle production in the world in the period 2009-2021 was created and shown in Figure 5. We come to the conclusion that the trend of vehicle production has continuously grown since 2009, increasing from 62 million vehicles to 97 million vehicles produced in 2018. Due to the pandemic of the COVID 19 virus, the production of vehicles in the world had declined in 2019 and 2020, when the lowest production of 78 million vehicles was recorded. The slight increase in the trend of vehicle production of 80 million vehicles in the world was noted in 2021, with predictions that the trend of vehicle production in the world will continue to grow in the coming years. Based on Figures 1, 3, 4 and 5, we can conclude that the implementation of robots in the world is connected to the work in production processes, so when there was a decrease in the production of vehicles in the world, there was also adecrease in the implementation of industrial robots in that year, including 2019 and 2020. The percentage of the implementation of industrial robots in the five top industrial branches in the last three years is shown in Figure 6.

.

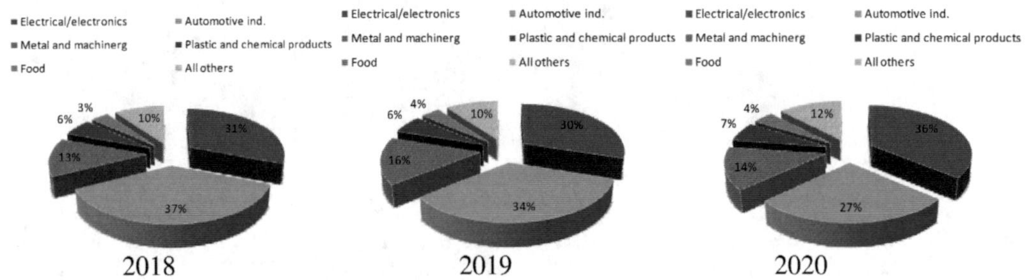

Figure 6 *Percentage of implementation of industrial robots in five top industrial branches in the world in the period 2018-2020 [25-27]*

The analysis of the diagrams shown in Figure 6 has provided the conclusion that the highest percentage of industrial robot implementation is recorded in two industries: the automotive industry and the electrical/electronic industry, in which over 60 % of the total amount of robots used that year is implemented. The first place is held by the automotive industry with 37 % of industrial robots implemented in 2018 and 34 % of industrial robots implemented in 2019. In 2020, due to the pandemic of the COVID 19 virus, the electrical/electronics industry has taken the primacy with 36 %. The reason is that many vehicle manufacturing companies reduced production, and therefore the implementation of industrial robots decreased. At the same time, there was an increase in the production of devices in the electrical/electronic industry that can reduce and mitigate the effects of the pandemic, which was reflected in the increase in the implementation of industrial robots. In third place is the metal industry, where the percentage of robots implemented in the last three years has ranged from 13 % to 16 %. We must note that part of the percentage of robot implementation from this industry can be joined to the automotive industry, as many components for vehicles are produced for the automotive industry, such as gearboxes, engines and etc.

In fourth place in terms of the robot implementation is the plastic and chemical industry, where in the last three years the percentage ratio has been in the range of 6-7 %. The fifth place in the implementation of industrial robots is held by the food industry, where 3-4 % of robots have been implemented in the last three years. We hope that this trend of implementation in these industrial branches will continue in the coming years. In the automotive industry, industrial robots are implemented in many operating procedures, such as: welding, painting, control, assembly, etc. However, the development and implementation of collaborative robots leads to expansion of the sites of robot implementation, which perform various tasks and automate those tasks that could not be automated so far. We must note that the largest number of industrial robots in the automotive industry is implemented in the processes of welding vehicle bodies and the production of vehicle parts. For this reason, an analysis of the application of industrial robots in welding processes in the world was conducted and shown in Figure 7.

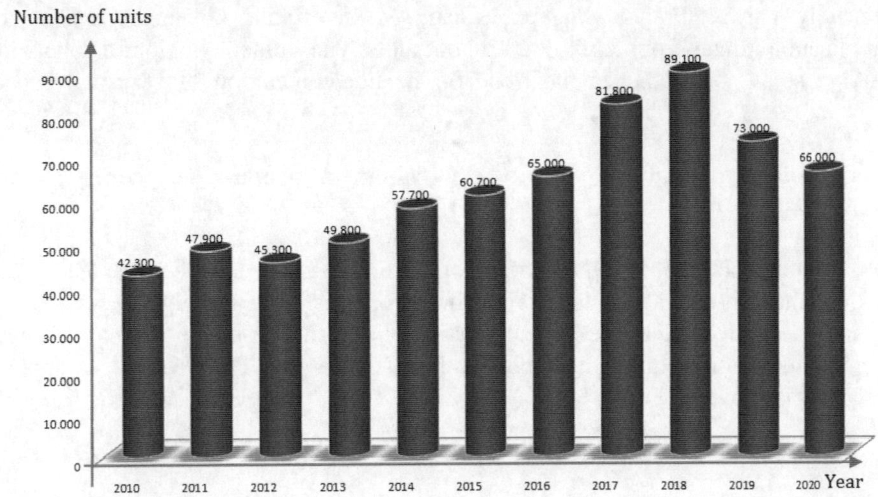

Figure 7 *Annual implementation of industrial robots in welding processes in the world in the period 2010-2020 [25-34]*

The analysis of Figure 7 has shown that the trend of implementation of industrial robots in the welding process had a growing character from 2010 to 2018, with the increase of 47.300 robot units in 2010 to 89.100 robot units in 2018. In 2019 and 2020, the trend of implementation has decreased due to the pandemic of the COVID virus, so that in 2020, the implementation amounted to about 66.000 robot units. The decline in the implementation of robots has also been recorded in the automotive industry and vehicle production worldwide, including the welding process. The trend of increased implementation of industrial robots is expected in all segments in the coming years, including the welding process.It is interesting to note which countries were the top countries in the implementation of industrial robots in 2020, as shown in Figure 8.

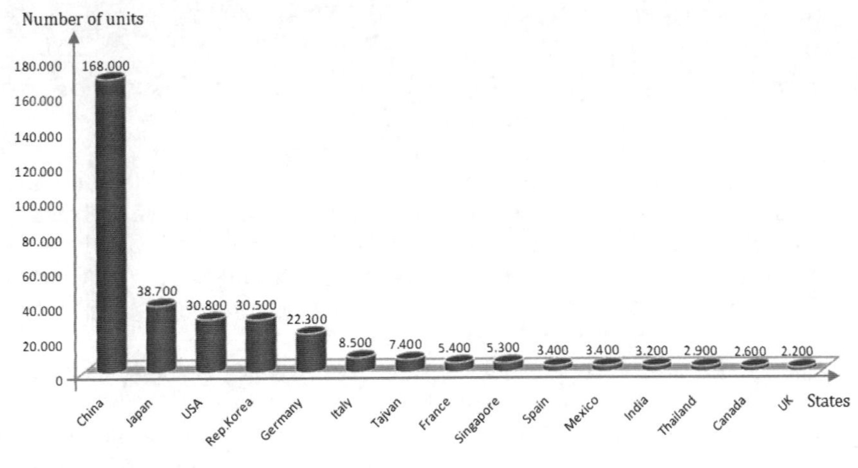

Figure 8 *Implementation of industrial robots in top fifteen countries of the world in 2020 [25]*

Figure 8 shows the implementation of industrial robots in top fifteen countries in the world in 2020. As can be seen, China stands out in first place with the implementation of around 168.000 robot units.

IOP Conf. Series: Materials Science and Engineering 1271 (2022) 012017 doi:10.1088/1757-899X/1271/1/012017

As of 2014, China has been the first in the world to implement industrial robots, which can be seen from [25-31]. Following China, Japan is in second place with 38.700 robot units, followed by the USA, the Republic of Korea, and Germany, which implemented 22.300 robot units. The next top ten countries are: Italy, Taiwan, France, Singapore, Spain, Mexico, India, Thailand, Canada and the UK, where the application ranges from 8.500-2.200 robot units. The implementation of robots in China is most interesting, so we will analyze the trend of the implementation of industrial robots in that country.

4. The trend of implementation of industrial robots in production processes and vehicle production in China

As could be concluded and confirmed by the reports [25-34], China has been the leading country in the world to implement industrial robots in production processes in the last eight years.
In order to get an accurate state of the implementation of industrial robots in China, it is necessary to analyze the implementation of industrial robots in the last ten years. The trend of the implementation of industrial robots is shown in Figure 9.

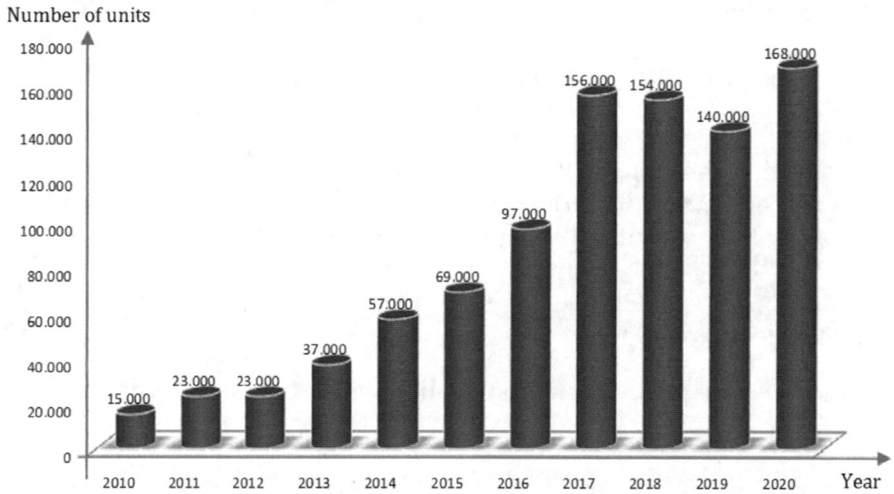

Figure 9 *The trend of the implementation of industrial robots in the production processes of China in the period 2010-2020 [25-34]*

The analysis of the data shown in Figure 9 has shown that from 2010 to 2013, the trend of robot implementation was increasing linearly. As of 2014, there has been a sudden increase in the use of robots in China, and with about 57.000 units it has become the leading country in the world in the use of robots, and has remained until today. In 2020, 168.000 robot units were implemented in China. There are several reasons why China is the leader in the implementation of industrial robots in the world:

- China has adopted a national strategy (in May 2015) named "Made in China 2025'', which implements specific and long-term general goals until 2035 and 2049. The goal of the national strategy is the transformation of China into the world's leading manufacturing power.
- China is the first in the world in terms of vehicle production, and industrial robots are mostly implemented in the automotive industry.
- China, like other developed countries, is in the stage of implementation of Industry 4.0, where robot technology is the basic technology, and the automation of production processes cannot be achieved without the use of industrial robots.

- The price of the working hour in China is continuously increasing. The companies are automating production processes where they implement industrial robots, as shown in Figure 10.

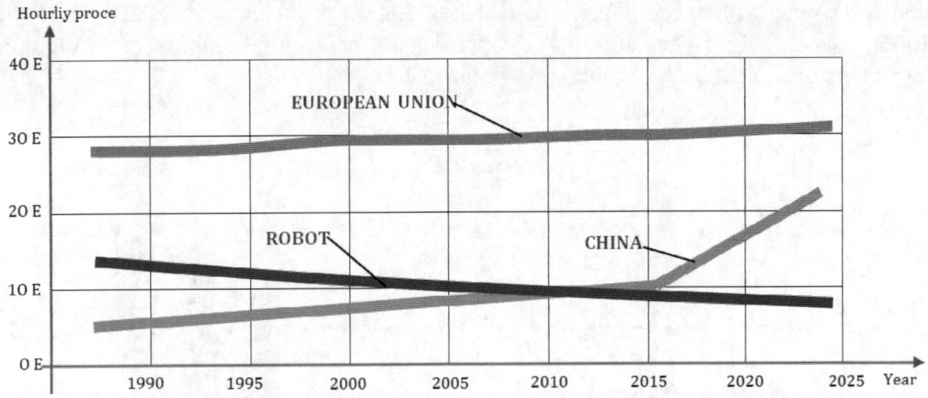

***Figure 10** Cost of robot working hour in the European Union and*

Based on Figure 10, we are able to compare the price of a working hour of a robot in China and the European Union. The price of a robot's working hour is continuously decreasing, while the trend of the labor price in the European Union is slightly increasing every year. Meanwhile,the price of labor in China has been skyrocketing since 2015, which justifies the introduction of industrial robots into the production processes in China. China is the first in the world in terms of vehicle production, as shown in Figure 11.

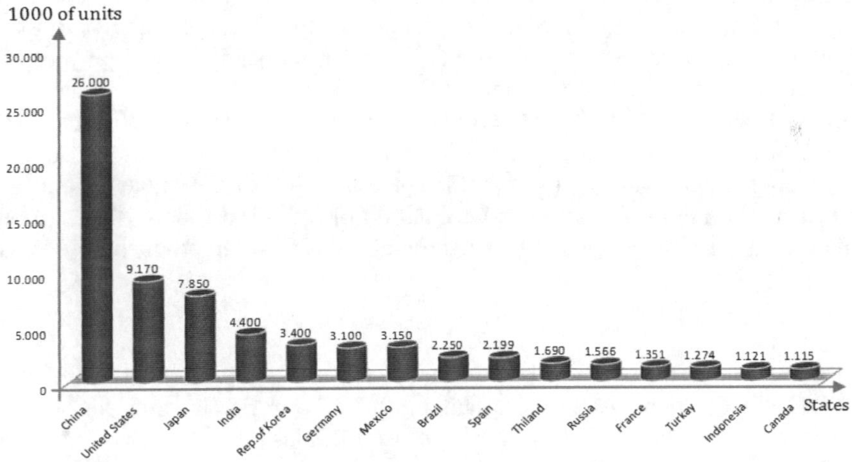

***Figure 11** The trend of vehicle production in fifteen top countries in the world in 2021 [35,36]*

As of 2009, China has been the first in the world in terms of vehicle production (in 2009 it produced 10.3 million vehicles). Every year it is the top country in the world and is far ahead of other countries in terms of vehicle production. This is depicted in Figure 11, which shows that in 2021 China produced about 26 million vehicles, which is about 16.8 million more than the USA, which produces about 9.17 million vehicles. This trend of vehicle production justifies the trend of increased implementation of industrial robots in China. The number of installed industrial robots can be a misleading measure in the analysis, because countries have different economic sizes and it does not reflect the size of a country's economy. In order to obtain the real state of affairs, the robot density is introduced, which represents the number of operating industrial robots in relation to the number of employees. The robot density coefficient is the number of multipurpose industrial robots per 10.000 employees. The robot density coefficient is a measure of the penetration of industrial robots in the industry. There are two countries in the world that stand out: the Republic of Korea with robot density of 932, and Singapore with robot density of 921 per 10.000 robot units in 2020, while the world average robot density is 126. The ratio of the robot density per 10.000 employees in the industry in Chinais increasing on annual basis, as shown in Figure 12.

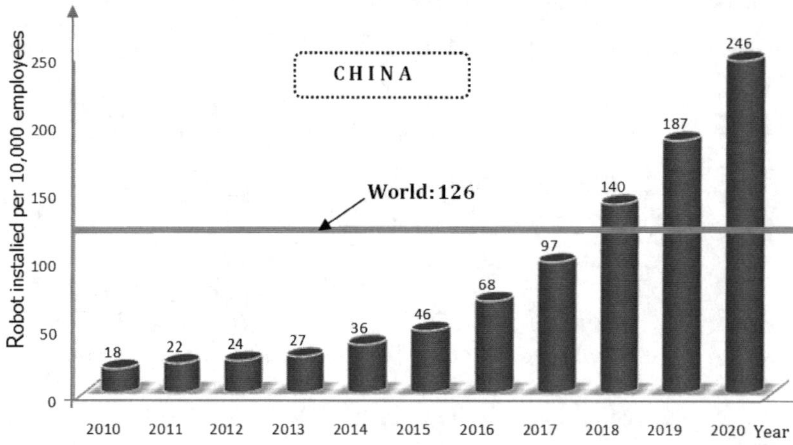

Figure 12 *The trend of the robot density per 10.000 employees in production processes in China in the period 2010-2020*

Up until 2017, China's robot density per 10.000 employees was below the world average, so in 2010, the robot density in China was 18, while the world average was 50 robot units. With the continuous implementation of industrial robots, the ratio of the robot density per 10.000 employees in China was continuously increasing on annual basis, so that in 2017 it exceeded the world average, which was 85, and China's robot density was 97. In 2020 the robot density in China was far above the world average, as shown in Figure 12. The robot density coefficient per 10.000 employees in China follows an exponential function, which confirms the fact that industrial robots are implemented in production processes in China. The reason for such implementation of industrial robots is the national strategy ''Made in China 2025'', as well as the implementation of Industry 4.0 in production processes.

5. Conclusion

The automotive industry was the first to implement the latest technologies due to global competition on the market. The automation system in the automotive industry is changing the way vehicles are produced, which would not be possible without the use of industrial robots. The implementation of Industry 4.0 must enable modern production systems in the automotive industry to become flexible/agile, integrated, reactive, and profitable at the same time in order to empower industrial companies to remain globally competitive. From a large number of fundamental technologies (over 40 according to some authors) of Industry 4.0, we will single out the basic ones: robotics, Internet of Things (IoT), Industrial Internet (IIoT), data storage in the cloud, automation, intelligent sensors, 3D printing, radio frequency identification (RFID), cognitive computing, artificial intelligence, etc. The basic technology of Industry 4.0 is robotic technology, i.e., the implementation of industrial and service robots in production processes.

The trend of the implementation of industrial robots in the world has been increasing in the last ten years, as shown in Figures 1 and 2, and will continue to increase in the coming years due to the implementation of Industry 4.0. Two leading industries in the application of industrial robots in the world are the automotive industry and the electrical/electronic industry, which in 2020 used 50 % of industrial robots. The first place in terms of application is held by China with about 168.400 robot units. In the recent years, China has always been the first in the world to implement robots, and the reasons are:

- The cost of labor in China has been increasing in the recent years, and many companies are opting for automation;
- The Chinese government is implementing a strategy called ''Made in China 2025'' which aims to make China the most technologically developed country in the world by 2025;
- China holds the first place in the world in terms of car production, where most industrial robots are used.

The analysis of the robot density per 10.000 employees in production processes shows that the world average in 2020 was 126 robots. The greatest progress was recorded in China, which in 2010 had robot density of only 18 robots per 10.000 employed workers in industry, which was below world average, as shown in Figure 12. In ten years, in 2020, China recorded an increase of robot densityof 246 robots per 10.000 employees, which was above the world average. In the past four years, there has been an increasing trend of the application of II generation industrial robots – collaborative robots, with the assumption that the trend will grow in the future due to the implementation of Industry 4.0. Robotic technology is the basic technology of Industry 4.0,which is indicated by the research that shows that the automotive industry production processes will become ''smart'' production processes in the years to come.

References

[1] Schwab, K.: The Fourth Industrial Revolution, World Economic Forum, Geneva, Switzerland, (2016)

[2] Karabegović, I.,Kovačević, A., Banjanović-Mehmedović,L.,Dašić,P.:Integrating Industry 4.0 in Busniess and Manufacturing, IGI Global,Hershey, PA, USA, (2020) https://www.igi-global.com/book/handbook-research-integrating-industry-business/237834

[3] Karabegović Isak, Karabegović Edina, Mahmić Mehmed, Husak Ermin, The Implementation of Industry 4.0 in the Function of Application of Industrial and Service Robots in Production Processes, International Scientific Conference, Application of Industriy 4.0 on Opportunity for a New Step Forward in All Industrial Branches, 14 April, 2022, Special Editons, Volume CCII, Department of Technical Sciences, Volume 20, ANU B&H, pp.103-117; DOI:10.5644/PI2022.202.24;
https://www.anubih.ba/images/news/PDF/20220526_Aplikacija_industrije_4.pdf

[4] Wilfried Albur, Arvind CJ, Rishi Bigghe: Skill Development for Industry 4.0, Ronald Berger, 2016, Munch, Germany, pp.7-12.; https://www.globalskillsummit.com/Whitepaper-Summary.pdf

[5] Karabegović Isak, Karabegović Edina, Mahmić Mehmed, Husak Ermin, The Implementation of Industry 4.0 by Using Industrial and Service Robots in the Production Processes**,** Chapter 1. Handbook of Research on Integrating Industry 4.0 in Business and Manufacturing, IGI Global, USA, pp.1-30, (2020), DOI: 10.4018/978-1-7998-2725-2.ch001

[6] Chryssolouris, G., Mavrikios, D., Papakostas, N., Mourtzis, D., Michalos, G., Georgoulias, K.: Digital manufacturing: History, perspectives, and outlook**,** Journal of Engineering Manufacture, Proceedings of the Institution of Mechanical Engineers, Part B, pp.451–462(2009) http://dx.doi.org/10.1243/09544054JEM1241

[7] Karabegović Isak and Karabegović E**.,** The Role of Collaborative Service Robots in the Implementation of Industry 4.0,*International Journal of Robotics and Automation Technology*,Vol.6.,pp. 40-46.; (2019) https://www.zealpress.com/;

[8] Kagermann, H., Wahlster, W. and Helbig, J. (2013) Securing the Future of German Manufacturing Industry: Recommendations for Implementing the Strategic Initiative INDUSTRIE 4.0. Final Report of the INDUSTRIE 4.0 Working Group, Vol. 4.

[9] Lasi, H., Fettke, P., Kemper, H-G.(2014) Industry 4.0, Follow Journal, Business & Information Systems Engineering 6(4) pp.249-242; DOI 10.1007/s12599-014-0334-4 (2014)

[10] Oesterreich, T.D. and Teuteberg, F. (2016) Understanding the Implications of Digitisation and Automation in the Context of Industry 4.0: A Triangulation Approach and Elements of a Research Agenda for the Construction Industry. Computers in Industry, 83, 121-139. https://doi.org/10.1016/j.compind.2016.09.006

[11] Juraj SINAY, Zuzana Kotianová, Automotive Industry in the Context of Industry 4.0 Strategy Transactions of the VSB - Technical university of Ostrava, Safety Engineering Series, Vol. XIII, No. 2, 2018, pp. 61 – 65.; DOI 10.2478/tvsbses-2018-0014

[12] Wang, K.: Intelligent Predictive Maintenance (IPdM) system—Industry 4.0 scenario. WIT Transactions on Engineering Sciences, Vol. 113, pp.259–268 (2016) www.witpress.com,doi:10.2495/IWAMA150301

[13] Muller, J.M., Buliga, O., Voigt, K.I.: Fortune favors the prepared: How SMEs approach business model innovations in Industry 4.0. Technological Forecasting and Social Change, Vol.132,Iss.C, pp.2–7 (2018)
DOI: 10.1016/j.techfore.2017.12.019

[14] Isak Karabegović The Role of Industrial and Service Robots in Fourth Industrial Revolution with Focus on China ,Journal of Engineering and Architecture ,Vol. 5, No. 2, pp. 110-117 , (2017) DOI: 10.15640/jea.v5n2a9 URL: https://doi.org/10.15640/jea.v5n2a9

[15] Edina Karabegovic, Isak Karabegovic, Edit Hadzalic (2012),Industrial Robot Application Trend in World's Metal Industry *Inzinerine Ekonomika-Engineering Economics,* , 23(4),pp.368-378, http://dx.doi.org/10.5755/j01.ee.23.4.2567

[16] Hermann, M.; Pentek, T.; Otto, B. Design principles for industrie 4.0 scenarios. In Proceedings of the IEEE 2016, 49th Hawaii International Conference on System Sciences (HICSS), Koloa, HI, USA, pp. 3928–3937 (2016), http://dx.doi.org/10.1109/HICSS.2016.488

[17] Thoben, K.D.; Wiesner, S.; Wuest, T. Industrie 4.0 and smart manufacturing-a review of research issues and application examples. International Journal Automation Technology , Vol.11, pp.4–16 (2017)
https://doi.org/10.20965/ijat.2017.p0004

[18] Karabegović,I.: The Role of Industrial and Service Robots in the Fourt Industrial Revolution, ACTE Technica Corviniensis-Bulletin of Engineering,University Politehnica Timisoara,Tome XI, Fascicule 2. April 2018. Hunedoara,Romania, pp.11-16 (2018) http://acta.fih.upt.ro/pdf/2018-2/ACTA-2018-2-01.pdf

[19] Karabegović. I., Karabegović, E., Mahmić, M., Husak, E.: The application of service robots for logistics in manufacturing processes, Advances in Production Engineering & Management, Vol. 10. No. 4. 2015,Maribor, Slovenia, EU, pp. 185-194 (2015)https://www.apem-journal.org/Archives/2015/APEM10-4_185-194.pdf

[20] Karabegović, I., Karabegović , E., Mahmić, M., Husak, E.: Implementation of Industry 4.0 and Industrial Robots in Production Processes In: Isak Karabegović (eds) New Technologies,Development and Application II 2019. Lecture Notes in Networks and Systems. Springer Nature Switzerland AG 76, pp.96-102 (2020)

[21] Karabegović, I., Husak, E.: Industry 4.0 based on Industrial and Service Robots with Application in China, Journal Mobility and Vehicle,44(4), pp.59-71 (2018)

[22] Karabegović, I.: The Role of Industrial and Service Robots in Fourth Industrial Revolution with Focus on China,Journal of Engineering and Architecture, 5(2), pp.110-117 (2017)

[23] Ostrgaard, E.: Collaborative Robot Technology and Applications, International Collaborative Robots, Workshop, Columbia, (2015)

[24] Karabegović I.,Husak E.,Predrag D.:The Role of Service Robots in Industry 4.0 - Smart Automation of Transport,International Scientific Journal Industry 4.0, Vo.4. Iss.6. pp.290.292 (2019)https://stumejournals.com/journals/i4/2019/6/290
[25] World Robotics 2020,Industrial Robots, Service Robots, The International Federation of Robotics, Statistical Department, Frankfurt am Main, Germany ,(2020) https://ifr.org/
[26] World Robotics 2019,Industrial Robots, Service Robots, The International Federation of Robotics, Statistical Department, Frankfurt am Main, Germany ,(2019) https://ifr.org/
[27] World Robotics 2018,Industrial Robots, Service Robots, The International Federation of Robotics, Statistical Department, Frankfurt am Main, Germany ,(2018) https://ifr.org/
[28] World Robotics 2017,Industrial Robots, Service Robots, The International Federation of Robotics, Statistical Department, Frankfurt am Main, Germany ,(2017) https://ifr.org/
[29] World Robotics 2016,Industrial Robots, Service Robots, The International Federation of Robotics, Statistical Department, Frankfurt am Main, Germany ,(2016) https://ifr.org/
[30] World Robotics 2015,Industrial Robots, Service Robots, The International Federation of Robotics, Statistical Department, Frankfurt am Main, Germany ,(2015) https://ifr.org/
[31] World Robotics 2014,Industrial Robots, Service Robots, The International Federation of Robotics, Statistical Department, Frankfurt am Main, Germany ,(2014) https://ifr.org/
[32] World Robotics 2012,Industrial Robots, Service Robots, The International Federation of Robotics, Statistical Department, Frankfurt am Main, Germany ,(2012) https://ifr.org/
[33] World Robotics 2011,Industrial Robots, Service Robots, The International Federation of Robotics, Statistical Department, Frankfurt am Main, Germany ,(2011) https://ifr.org/
[34] World Robotics 2010,Industrial Robots, Service Robots, The International Federation of Robotics, Statistical Department, Frankfurt am Main, Germany ,(2010) https://ifr.org/
[35] Verband Deutscher Verkehrsunternehmen VDA: „AutoJahresbericht 2021“,Koln, Deutschland,2021
[36] Verband Deutscher Verkehrsunternehmen VDV:"Jahresbericht 2020/2021, Koln,Deutschland,2021.

Measurement noise level of E-bus HIGER KLQ6125GEV3 on the polygon

S Misanovic[1], D Taranovic[2], M Maljkovic[3] and B Milicic[4]

[1] PhD, Project manager, City Public Transport Company "Belgrade", Kneginje Ljubice 29, 11000 Belgrade, Serbia
[2] PhD, Associate prof. Faculty of Engineering University of Kragujevac, Serbia
[3] MSc, Assistant, Faculty of Mechanical Engineering University of Belgrade, Serbia
[4] MSc, Assistant, Faculty of Mechanical Engineering University of Belgrade, Serbia

corresponding author[1], e.mail: slobodan.misanovic@gsp.co.rs

Abstract. Buses for public city transport in cities represent a significant source of noise, especially in the central city zones, on the traffic corridors with the highest hourly frequencies of public transport vehicles. The introduction of electric buses is one of the possible ways to reduce noise. The paper will present the results of the experimental measurement of noise levels of e-bus Higer KLQ6125GEV3 on the polygon, using the Lutron SL-4001 measuring instrument, following UN ECE 51.02 regulations and a comparison with the results of noise levels measurement with buses on diesel and CNG driveline.

Keywords: Noise level, E-bus, Environment

1. Introducing

Conventional public transport buses are a significant source of noise, especially in central urban areas. The term noise means an unwanted sound, ie a sound phenomenon that interferes with human activity. Noise in vehicles occurs as a result of work: drive unit, transmission, exhaust system, tires, brakes, aerodynamics, auxiliary systems on the vehicle. Noise emissions from buses can be viewed in terms of environmental impact (external noise) or driver impact and passenger comfort (internal noise) [1].

Noise affects a person through the senses of hearing because it transmits all stimuli and transforms them into auditory feelings to which a person subjectively reacts. From the total frequency domain, a healthy person hears sounds in the frequency range of 20-20000 Hz. The level of air pressure at which a person begins to hear is called the "threshold of audibility" (0 dB), while the upper limit is defined as the "threshold of pain" (120 dB). The harmful effect of noise on humans is manifested on the psychological and physiological level, and its impact depends on several factors, the most important of which are: the level (intensity) of noise, frequency spectrum, and exposure time [2].

The noise level of motor vehicles used in road transport is defined by the rules of permissible noise level, UN ECE Regulation 51.02 [3] or Directive 2007/34/EC [4]. For buses of standard length or articulated, which are mostly used in public city transport, whose power of the propulsion system exceeds 150 kW, the maximum value of the level of emitted external noise is 80 dB. Reducing the level of noise emitted by public transport vehicles is one of the goals of improving environmental performance and environmental impact. The introduction of electric buses is one of the possible ways to reduce noise.

2. Literature Rewiew

A study of the impact of internal noise on the driver of different types of diesel buses (standard, articulated, mini) was conducted at the public transport company in Curtiba (Brazil). The average measured value of the internal noise level for a standard bus was 80.2 dB. Significant variation in the measured level was observed depending on the age of the vehicle. The range of noise levels was between 68.5 dB and 90.6 dB [5]. A similar study was conducted in India in the city of Burdwan (West Bengal) where the impact of internal noise levels on city bus drivers was investigated. Since the buses were extremely old, most of them were between 10÷25. year, the level of the internal noise in the driver's space zone of 88.6÷102.4 dB was measured. As many as 65% of the surveyed drivers had been diagnosed with some the health problems caused by noise [6].

The study of the internal noise level in the bus with the hybrid-series configuration Volvo 7900 and the tram Konstal 105N HF was done at an inter-station distance between the stations Zawodzie Laczna and Zawodzie Petla, in Katowice (Poland). The range of emitted internal noise in the hybrid bus was between 54÷78 dB, while in the tram the range was in the range of 75÷85 dB [7]. Measurement of the level of external noise generated by city buses of standard length when the vehicle is in motion and when at rest, was done in the study Urban buses - Alternative powertrains for Europe. The values of the emitted external noise in purely electric buses for the case when the vehicle is "standing" are 63 dB. Compared to a diesel-powered bus, where the noise level reaches 80 dB, it is clear that electric vehicles have a 17 dB lower noise level. If the level of emitted noise is measured for a moving vehicle, electric buses reach the level of 69 dB, which is 8 dB less than diesel buses [8].

Measurement of external noise levels at the test site for diesel and CNG buses was reported in the paper by Milojevic et.al [9]. Measurement of external noise levels for a standard city bus with hybrid and fully electric drive was performed by Volvo bus manufacturer. For the characteristic speeds of 30 $km{\cdot}h^{-1}$ and 40 $km{\cdot}h^{-1}$, which are representative of buses in public transport, the level of emitted external noise in the hybrid bus was observed at 69 dB, 72 dB, while the level of 64 dB and 66 dB was measured for fully electric buses [10]. Measurement of external noise levels at the test site for three standard city buses with different propulsion systems (diesel- Mercedes Citaro O 530; Hybrid-electric drive Volvo 7900HA; Fuel cell - Mercedes Citaro) was reported in the work Laib et.al [11]. In the standby mode, the level of external noise was measured at 65 dB, 61 dB, 57 dB. For a speed of 20 $km{\cdot}h^{-1}$, values of 76 dB, 69 dB, 62 dB were measured. For a speed of 30 $km{\cdot}h^{-1}$, the measured noise values are 76 dB, 71 dB, 66 dB, while for a speed of 40 $km{\cdot}h^{-1}$, the measured noise level is 78 dB, 75 dB, 74 dB. At a speed of 50 $km{\cdot}h^{-1}$, the level of measured external noise for all three tested buses is 78 dB. Polygon research where four buses were tested, two with diesel drive of different ages Marcopolo Mercedes Benz Euro 5 model 1721 LE from 2020, Marcopolo Mercedes Benz, model OH 1318/51 from 1999 and two fully electric buses BYD K9W from 2020. and ANKAI model G9 10.5 from 2020, was performed in Montevideo (Uruguay) through the MOVES project [12]. The external noise level for the speed of 20 $km{\cdot}h^{-1}$ is respectable: 69.9 dB, 77.4 dB, 56.1 dB, 56.3 dB. At a speed of 30 $km{\cdot}h^{-1}$, noise was measured; 71.4 dB, 76.3 dB, 60.6 dB, 62.1 dB, while at a speed of 40 $km{\cdot}h^{-1}$, noise values are: 70.5 dB, 75.4 dB, 64.6 dB, 66.7 dB.

3. Methodology for measuring external noise level

Measurement of emission noise emitted by E-bus Higer KLQ6125GEV3 performed in accordance with UN ECE 51P Regulations, at the prepared polygon within the tram depot "Sava" JKP GSP "Belgrade", Figure 1, using a digital measuring device Lutron (Lutron SL-4001 Digital sound level meter) Figure 2.

Figure 1 Polygon for measuring external noise levels

The operating characteristics of the Lutron SL-4001 measuring device are shown in Table 1 [13].

Table 1. Characteristics of the measuring device Lutron SL-4001

Description	Characteristics
Display	18 mm, LCD
Noise measurement range	30-80 dB, 50-100 dB, 80-130 dB
Resolution	0.1 dB
Frequency	31.5 Hz – 8000 Hz
Microphone	Electronic-capacitor
Response time	"fast" 200 ms, "slow" 500 ms
Temperature range	0 °C - 50 °C
Power supply	Battery, 9 V DC
Dimensions	205 mm x 85 mm x 35 mm
Mass	250 g

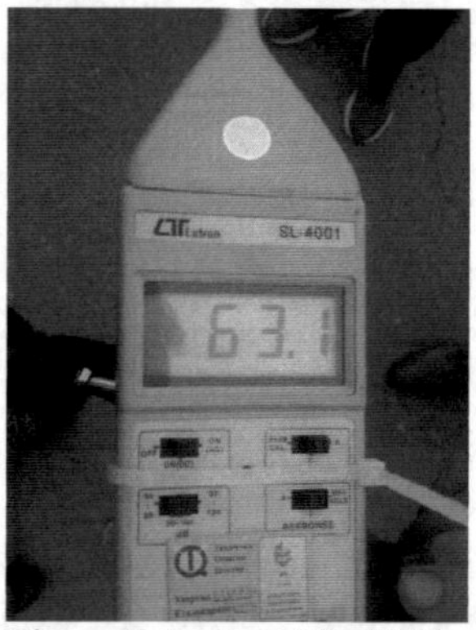

Figure 2 Lutron SL-4001

The measuring device Lutron SL-4001 measures the emitted noise level in decibels (dB) which gives the relative sound pressure ratios as presented in equation [14].

$$L(\text{dB})=20log\frac{p_z}{p_o} \tag{1.1}$$

L (dB) – measured noise level, dB,
p_z – measured sound pressure, N·m^{-2},
p_o – zero sound pressure level, $p_o = 2\cdot10^{-5}$ Nm^{-2}, which corresponds to the threshold ear audibility at an altitude of 0 m, at 1000 Hz.

The Lutron SL-4001 measuring device is placed at a distance of 7.5 m from the reference line K-K " which represents the direction of the E-bus, as shown in Figure 3. The device is placed on a tripod at a height of 1.2 m above the ground and looks in the horizontal direction towards the reference line of motion (Figure 4).

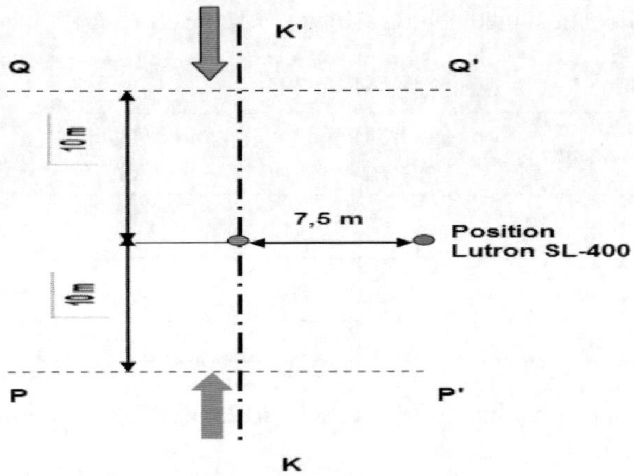

Figure 3 Sketch of a polygon for measuring noise levels

Figure 4 Lutron SL-400, prepared for measurement

Two lines P-P and Q-Q 'at a distance of 10 m in front of and behind the position of the measuring device are marked on the polygon. The e-bus accelerates to the desired speed to the P-P line, after which it maintains a constant speed until the rear of the bus passes the Q-Q line. Measurement of the emitted noise level is performed in the opposite direction K-K '(blue arrow) in an identical way in order to obtain the values of emitted noise for the left and right sides of the E-bus.

4. Results of noise level measurements and discussion

The measured value of the emitted noise when the vehicle is stationary is 63.1 dB. The measuring device was placed at a distance of 0.5 m from the middle of the vehicle, observed from the left side of the vehicle. The values of emitted noise from the moving E-bus at speeds of 30 km·h^{-1} and 40 km·h^{-1} were measured. The values of the measured noise levels that had the greatest mutual homogeneity are shown in Table 2 [15].

Table 2 The noise emission level of the E-bus Higer KLQ6125GEV3

Side of the E-bus	**Measurements**	**Speed, 30 km·h^{-1} dB**	**Speed, 40 km·h^{-1} dB**
left	1.	67,4	70,4
	2.	67,5	70,5
	3.	67,4	70,3
right	1.	67,2	69,8
	2.	67,1	69,8
	3.	67,2	69,6
Average value of measurements , dB		**67,3**	**70,0**

The table concludes that the level of emitted noise of the E-bus in the direction K'-K (left side of the E-bus) has a slightly higher value. In the E-bus, electric motors (two electric motors) a summation transmission, Cardan shaft, and the main gearbox with a differential are centrally located in the rear part of the wheelbase before the drive axle. Noise from these components is emitted evenly on both sides. The position of the auxiliary electric motors for the compressor and hydraulic drive is on the left side of the vehicle in boxes located in front of the driveshaft, so their work affected the noise level in the K-K direction (left side of the E-bus) to be slightly higher. The average noise level emitted at a speed of 40 km·h^{-1} is higher by 2.7 dB compared to a speed of 30 km·h^{-1}. To compare the results of the measured noise emission of electric buses Higer KLQ6125GEV3 with buses powered by diesel fuel, the results of noise emission measurements measured on a bus powered by diesel fuel MAZ-203 (EURO 5) will be used according to the identical methodology shown in Table 3 [9].

Table 3 The noise emission level of the diesel bus MAZ 203

Side of the E-bus	**Measurements**	**Speed, 30 km·h^{-1} dB**	**Speed, 40 km·h^{-1} dB**
left	1.	76,6	76,6
	2.	76,3	77,0
right	1.	74,5	73,5
	2.	75,9	74,4
Average value of measurements, dB		**75,6**	**75,4**

Comparison of the results of measured noise emission of Higer KLQ6125GEV3 electric buses with MAZ-203 CNG (EURO 6) bus is shown in Table 4 [9].

Table 4 The noise emission level of the CNG bus MAZ 203 CNG

Side of the E-bus	**Measurements**	**Speed, 30 km·h^{-1} dB**	**Speed, 40 km·h^{-1} dB**
left	1.	75,8	71,0
	2.	76,3	72,6
right	1.	69,4	70,7
	2.	71,2	71,8
Average value of measurements, dB		**73,1**	**71,5**

5. Conclusion

The emission of noise from public transport vehicles is a significant disrupter of the environment in cities, which is given increasing attention and importance. The introduction of an increasing number of fully electric buses in the public passenger transport system can contribute to the reduction of emitted noise, especially on city roads where public transport vehicles pass. In the research shown in the paper, the measured noise levels of the E-bus Higer KLQ6125GEV3 at speeds of 30 km·h^{-1} and 40 km·h^{-1} have significantly lower values than the noise emission value of the MAZ-203 diesel bus. At a speed of 30 km·h^{-1} the level of measurement noise in electric buses is lower by 10.9% and at a speed of 40 km·h^{-1} by 7.1% compared to a bus with diesel drive, which is proven that electric buses have a lower noise level compared to diesel buses. When comparing the measured noise emission level of the E-bus Higer KLQ6125GEV3 and the MAZ-203 CNG bus at speeds of 30 km·h^{-1} and 40 km·h^{-1}, the E-bus has a lower noise emission level. At a speed of 30 km·h^{-1}, the measured noise level of electric buses is lower by 8.6% and at a speed of 40 km·h^{-1} by 2.1% compared to a bus with CNG drive, which has been proven that electric buses have a lower level of emitted noise compared to CNG buses.

References

[1] Ćućuz N., *Automobilski motori*, Saobraćajni fakultet Beograd,1985
[2] Milosevic S., *Čovek i saobraćajna buka*, Naučna knjiga, Beograd, 1987
[3] *** UN ECE Regulation 51.02
[4] ***Directive 2007/34/EC, EU of 14 June 2007
[5] Portela B., Zannin P., *Analysis of factors that influence moise levels inside urban buses*, Journal of Scientific&Industrial Research, Vol.69, September 2010, pp.684-687
[6] Mondal N., Dey M., Datta J., *Vulnerability of bus and truck drivers affected from vehicle engine noise*, International Journal of Sustainable Bulit Environment, 2014, Vol.3, pp.199-206, http://dx.doi.org/10.1016/j.ijsbe.2014.10.001

[7] Michata A., Haniszewski T., *Traffic Noise Experienced on Buses,Trams and Car in the Urban Agglomeration of the City of Katowice*, Scientific Journal of Silesian University of Technology, Series Transport, 2018,98, 101-109, ISSN:0209-3324, DOI:https://doi.org/10.208/sjsutst.2018.98.10

[8] European Fuel Cells and Hydrogen Joint Undertaking: *Urban buses-Alternative powertrains for Europe, A fact based analysis of the role od diesel,hybrid,hydrogen, fuel cel, trolly and electric powertrains*, november 2012.

[9] Milojevic S., Gročić S., Dragojlovic D., *CNG propulsion system for reducing noise of existing city buses*, JAES,14 (2016)3, 392, pp 377-382, doi:10.5937/jaes14-10991

[10] Turcsany J., *Electric bus and Noise*, Volvo buses, 2016.

[11] Laib F., Braun A., Rid W., *Modelling noise reductions using electric buses in urban trafic. A case study from Stuttgart,* Germany, Transportation Research Procedia 37 (2019) 377-384.

[12] Proyecto MOVES, *Estudio comparativo de nivel de ruido generado por el transporte publico convencional y electrico*, PNUD-SDP URU/17/G32-950, Enero 2021.

[13] ***,http://www.meterindia.com/Catalog/Sound%20Level%20Meter/SL-4001.PDF

[14] Demić M., Diligenski Đ.: *Teorijske osnove projektovanja autobusa*, Mašinski fakultet Kragujevac, 2003.

[15] Mišanović S., *Energetske i ekološke performanse autobusa na električni pogon u sistemu transporta putnika, doktorska diseratcija*, Fakultet inženjerskih nauka Univerziteta u Kragujevcu, 2021.

IOP Conf. Series: Materials Science and Engineering 1271 (2022) 012019 doi:10.1088/1757-899X/1271/1/012019

The ergonomic aspects of a passenger car climate control panel

D Ružić

University of Novi Sad, Faculty of Technical Sciences, Department of Mechanization and Design Engineering, Dositej Obradović square 6, 21000 Novi Sad, Serbia, ruzic@uns.ac.rs

Abstract. In order to make a vehicle cabin environment thermally comfortable, substantial amount of the energy could be needed. This problem is especially notable in electric (EV) and hybrid vehicles (HV) because of the influence on the driving range. Therefore the EV and HV must have an efficient heating, ventilation and air-conditioning system (HVAC). Correct setting and use of the system is also one of preconditions for rational use of the energy and to obtain the efficient energy consumption. Despite of the automation of the system, a driver must be able to set the HVAC while driving. This is secondary task and must not interfere with primary task, the vehicle control. Therefore, design of the climate control panel (CCP) is important in terms of driving safety and energy consumption. An ergonomic analysis of a sample of typical compact/middle class family cars CCPs is given and discussed. The results indicate that the approaches in the CCP design are very different and there is an absence of standardization in terms of general control layout and functions. A user-centred design of the CCP would result in reduced number of the controls and fewer steps needed to set the system.

1. Introduction

To achieve and maintain comfortable thermal environment inside the passenger compartment, the motor vehicles must have efficient heating, ventilation and air-conditioning system (HVAC). This problem is especially notable in electric (EV) and hybrid vehicles (HV) because of the influence on the driving range. Therefore the EV and HV must have properly controlled efficient HVAC. Ambient conditions as well as individual thermal preferences could be very different, making necessary wide range of possible HVAC settings in terms of thermal conditions control inside the vehicle cabin. A driver must be able to control the HVAC system while driving, depending on the current thermal conditions, sensation and preference. The HVAC system adjustment is done via the climate control panel (CCP) as a part of driver-vehicle-interface (DVI) [1], [2], [3]. The CCP configuration is characterized by the control type, layout and location. The HVAC control panel could be very complex, in close relation with the complexity of the vehicle thermal management system. In passenger cars, the HVAC control panel is usually located on lower part of the central console (center stack) as a non-primary function interface. The parameters that can be adjusted are air flow temperature, fan speed, air flow direction, ventilation intake air mode (outside or recirculated), switching the air-conditioning on or off, windshield demisting, rear window heating as well as other options, depending on the system configuration.

It is often necessary to adjust the HVAC system while the vehicle is in motion. The CCP should be designed to be easy for use in such a way that frequency of driver's glances, the duration as well as hand movements do not impair driving safety [4], [5], [6]. A change of the HVAC settings while driving sometimes can increase a driver's workload. In order to reduce the workload, the driver is obliged to be familiar with all relevant controls, signals and information on instrument panel prior to driving the car. A research carried out by Lee and Ji [7] indicates on increased complexity of the control interfaces in modern cars. The results are based on user experiment conducted on 10 vehicles and 91 participants. For example, a "MODE" controller used to change air direction and a double wheel button used to control air volume and temperature were confusing to some participants. A study performed on Volvo XC90 simulator involving two different types of multifunctional interfaces (Rydström et al. [8]) showed that the optimal interface is dependent on the task being performed. Innovative DVI's for HVAC control were proposed in the works of Ackerman et al. [9] and Dvorak et al. [10]. A thermal sensation preference in a passenger compartment is commonly expressed by the temperature value, although there are several heat transfer modes that influence a person's current thermal sensation. Those proposed principles of HVAC setting are based on the subjective thermal comfort of the passengers instead on the air temperature or the fan speed. Schneegass et al. [11] demonstrated models and tools for rapid prototyping and the evaluation of user interfaces in car, implementing an adapted Keystroke-Level Model (KLM). The model is based on empirically collected data for typical operations in the car and it could be used in the early stage of the control panel design process. A software Distract-R presented in [5] allows designers to effectively prototype new basic in-vehicle interfaces and evaluate them using simulation and visualization. A model of driver performance is integrated with models of behaviour on the prototyped interfaces to generate predictions of distraction. In the study presented in [12], 18 different CCPs of a sample of 10 passenger cars were investigated and evaluated in terms of the type and layout of the controls. The evaluation method was based on the Keystroke-Level Model (KLM). The results showed that the time required for a HVAC setup could vary greatly depending on the CCP design. The designs of the CCPs are very different and there is an absence of standardization in terms of general control layout and functions even in the vehicles of the same brand.

In this paper, the CCPs of several typical compact/middle class family cars are analysed and discussed in terms of the ergonomic characteristics: findability, reachability, identification, interpretability, operability and understandability. Although the touch screens are widely used as the DVI, the physical (mechanical) controls are still present in new cars. There is a evident tendency to replace secondary controls and displays with the touch screens, regarding the design and content of the touch screen, but a vehicle is not (and must not be) a smart phone on four wheels. Moreover, the touch screens could be considered as less suitable for complex tasks like HVAC adjustment while driving than use of physical controls [12], [13]. From that reason, the focus of this study is on the CCP realized mostly as a cluster of individual physical controls. The results of the analysis and recommendations from the available literature are compiled here in form of general recommendation for user-centred design of the CCP by considering the ergonomically suitable solutions.

2. The automotive climate system control

Controls of the modern vehicle are designed according to the safety and ergonomics requirements and recommendations. An automotive HVAC system design and a way of control of its operation influence complexity, design and layout of climate control panel. There are three general types of HVAC control, depending on complexity of the system and its management:

- Manually controlled HVAC: a driver adjusts air flow temperature, air flow rate, mode of air flow and air flow direction from individual vents.
- Automatic climate control (ACC): a driver presets desired thermal conditions by setting of preferred temperature and mode of operation. Air flow rate and air distribution mode are controlled automatically. The manual override is possible. The air flow direction from individual vents must be controlled manually.
- Multizone ACC: thermal conditions can be set independently (within the certain limits) for individual seats and/or seat rows and automatically controlled. The manual control and synchronized mode are possible.

The manual climate controls are intended to control individual components of the HVAC system: turning compressor on or off, adjust fan speed, set the air distribution and so on. The more complex is the system, the more complex would be the controls. Automatic HVAC control operates according to programmed algorithm for typical scenarios and depending on parameters from the sensors in the system and in the cabin. Although automated, user still needs to take back control occasionally, making the DVI important even when a high level of automation is applied.

The design principles of driver-vehicle interface (DVI) are based on the following fundamental assumptions [6], [13], [14], [15]:

- a driver prefer to operate the control quickly with minimal mental and physical effort,
- a man or a woman usually prefer not to use the system/controls they do not understand.

A passenger car HVAC system and its control have following general characteristics [12]:

- The controls are labelled by standardised pictograms and/or acronyms (standard ISO 2575: Road vehicles - Symbols for controls, indicators and tell-tales [16]).
- The CCP is complex, consisting of setting labels and/or displays and therefore controls can not be used blindly (without looking at it). The activities must be visually guided.
- The layout of the controls is not standardized or unified. Even the panels of the same brand can have very different layouts.
- The CCP's control shapes are mostly symmetrically arranged due to aesthetic reasons.
- There is no direct visible or sensible feedback to the state change, except of the fan speed. The HVAC system response is relatively slow.
- Correct combination of controls is sometimes necessary (e.g. temperature and mode of air flow).
- A driver does not have to react quickly, except in the case of sudden windshield or rear window fogging or to activate air recirculation to prevent contaminant ingress.
- An airflow direction is not controlled via CCP but by tilting and swivelling the air vents' grilles by hand. Opening and closing of the vents is done by the thumbwheel.
- There are no hazardous or damage potentials from control error or wrong use neither for occupants nor vehicle. The only adverse effect could be windows fogging under some circumstances, or thermal discomfort.
- Since the HVAC control is not considered as a control essential to the safe operation of vehicles, the CCP location does not have to meet the requirements regarding the hand reach and visibility given in the standard ISO 4040: Road vehicles - Location of hand controls, indicators and tell-tales in motor vehicles [17].

3. The CCP design analysis

Since the HVAC control is used while driving, in general its ergonomic suitability can be evaluated in terms of following [14], [18]:

- Interference with the driving task (driver distraction): adverse influence of the DVI on the driver's ability to deal with the vehicle and the environment.
- Efficiency: driver's workload (mental and physical capacity) expended to achieve intended objectives related to desired thermal state.
- Controllability: mode and degree to which drivers can influence the HVAC function.
- Ease of use while learning about the system: acquiring knowledge and developing skills contribute to the reducing of distraction while using DVI.

From the ergonomic and safety point of view, controls should have following general characteristics in order to enable proper and easy setting of the system [13], [14], [15], [19]: findability (visibility, grouping), reachability, identification (logical grouping, labelling, graphics, tactility, discrimination), interpretability, operability and understandability. However, according to the studies [7] and [12] those principles are not fully applied to the automotive CCP as a non-primary DVI. In this section, the CCPs of typical compact/middle class family are analysed and discussed in terms of the ergonomic characteristics using the heuristic approach.

The passenger cars whose CCPs are used in the study are Fiat Tipo, MY 2015 (manual and ACC), Ford C-Max HEV MY 2013 (ACC), Mercedes Benz A-class, MY 2020 (2-zone and 3-zone ACC), Opel Astra MY 2016, (manual and ACC), Renault Captur, MY 2020 (manual and ACC), Volkswagen Golf, MY 2016 (manual and ACC), Toyota Corolla hybrid, HEV, MY 2020 (ACC), BMW i3, BEV, MY 2018 (manual and ACC), Kia E-Niro, BEV, MY 2020 (ACC) and Nissan Leaf, BEV, MY 2020 (ACC).

3.1. Visibility and grouping

A typical location of the CCP on the lower part of the central console (center stack) is outside of the driver's centre field of view. Although the CCP is visible from the driver's seat, the controls must be often used with looking at it. In addition, on CCPs with digital control, an infinitely rotary knob or button is used to control temperature or fan speed. In those cases, the temperature or fan set value must be read on display which could be positioned away from driver's direct field of view. The better option is a design where an each change of settings is shown in the main info-display located closer to the driver's view line [12]. Figure 1.

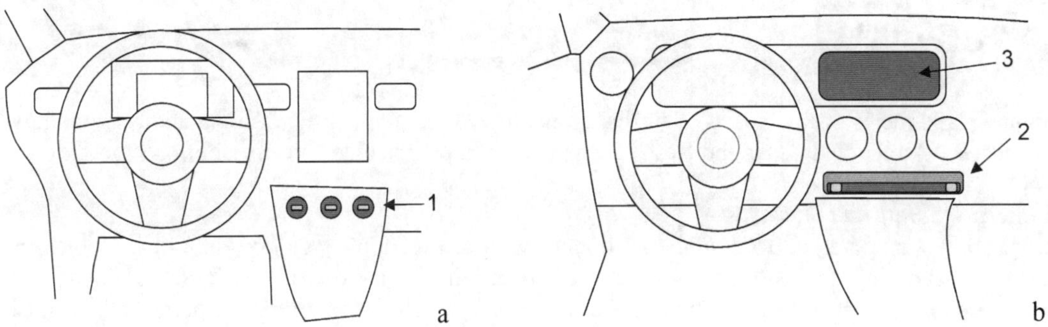

Figure 1. A CCP position in the car cockpit. a: Low positioned CCP (1) with integrated controls and displays (Renault Capture MY 2020). b: The CCP (2) with settings shown in the main info-display (3) (MB A-class, MY 2020)

3.2. Identification

Use of symbols instead of text for describing control's function is more preferable way [19]. However, increasing of a symbol complexity within the available control space may present a problem in the control identification, without previous familiarization [7]. Beside of the labelling according to the standard [16], a shape coding is also used, although a "typical" shape for some control actually does not exist. Type of a control used for setting of a parameter and a function activation that are usually encountered are [12]:

- temperature settings: rotary knob, +/- push button, rocker, +/- icon on a touch screen, slider on a touch screen,
- setting of fan speed: rotary knob, +/- push button, rocker, +/- icon on a touch screen, slider on a touch screen,
- setting of air vents: rotary knob, push button, icon on a touch screen,
- other function activation (AC, automatic control, air recirculation etc.) push button, icon on a touch screen.

The better results for the continuous adjustment (temperature, fan speed) are supposed to be achieved with rotary knobs, compared to the push buttons [11]. However, there are CCPs that have only push buttons (Figure 1, b).

The ergonomic recommendation is that controls that are to be used quickly should be placed closer to the driver and to be distinctive in shape or appearance. Those are controls for activation of demisting or air recirculation. On the most of the CPPs this is not a case and they are blended with the rest of controls, probably due to the aesthetic reasons [12]. Figure 2.

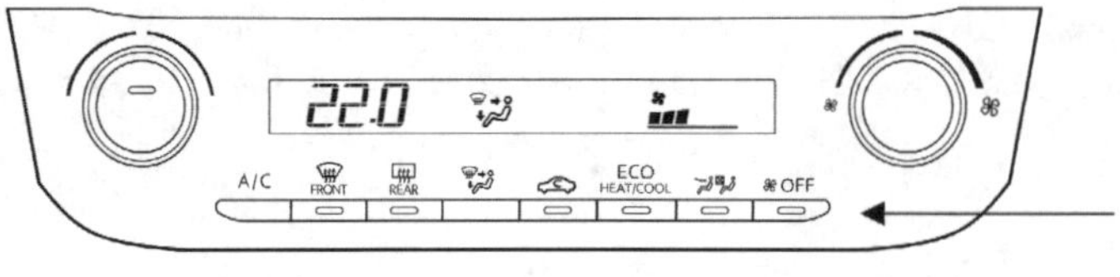

Figure 2. An example of non-distinctive controls for functions of different importance, arrow (Toyota Corolla HEV, MY 2020)

Some text labels or pictograms could look similar, demanding longer glances and drawing more attention to decide which to activate. This problem is more evident when the controls are cluttered and/or placed next to each other. Figure 3.

Figure 3. Examples of three pairs of similar standardized pictograms: windshield and rear window heater (a), recirculated air and fresh air (b), selection of air vents / air distribution mode (c)

3.3. Interpretability and operability

Interpretability is addressed to a single command whose direction of motion should be logical and suggestive according to its function and the operation. Actuation of the climate controls for temperature and fan speed is according to the direction of motion stereotypes: clockwise/up/right for "increase" (the standard ISO 12214: Road vehicles - Direction-of-motion stereotypes for automotive hand controls [20]). However, in order to set the air flow distribution mode using either rotary or push control, a driver must look on the control and/or on the visual indication, because of the size and similarity of the pictograms. This is especially evident if single push button is used for sequential mode selection (Figure 2). There are the cases of rotary buttons where lower vents are activated in upper position of the button. Figure 4.

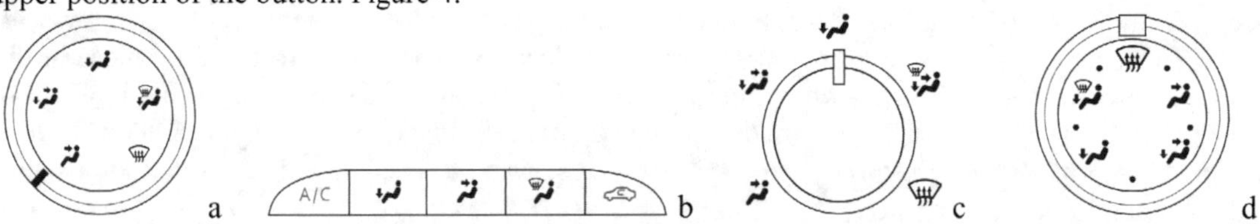

Figure 4. Examples of non suggestive design of air distribution mode control (Renault Captur 2020 (a), Ford C-Max HEV 2013 (b), Fiat Tipo 2020 (c), VW Golf 2016 (d))

Important issue regarding the operability is a climate control for the rear seat row. Typical solution in small-, compact- and middle-class cars is that driver has full control over the climate in the whole passenger compartment. Therefore, the climate setting relies on the feedback from the passengers from the rear seat row as well. The HVAC system has one evaporator and one blower while the air distribution is a way to achieve desired thermal conditions on different points in the cabin space. Four zones HVAC is a feature of high-class vehicles, which in turns have more complex HVAC system (several heat exchangers and evaporators, more than one blower).

3.4. Understandability

The adjustment time will differ depending on that is the driver familiar with the vehicle controls or not [14], [15], [21]. Previous knowledge about the basic controls is expected (temperature, fan speed), usually originated from the use of home air-conditioning devices. Without the automation, a driver must be able to adjust car's HVAC using at least three or four controls (temperature, fan speed, air distribution mode, and air direction from the individual vents) and to change settings while driving. Sometimes he or she must operate additional controls such are temperatures in passenger zones, electric heater(s), seat ventilation etc.

Although the use of standardized labels, pictograms and shapes facilitate the initial understandability of individual functions, the proper setting and the use of a HVAC system is described and explained in the car's owner handbook. The designation of the controls is not suggestive in a way of a proper setting according the current conditions and desired thermal sensation. The user could not be familiar with suitable settings of all parameters which should result in thermal comfort inside the cabin. For example, typical misunderstanding of the HVAC setting is that when the AC is turned on, an air recirculation must be activated all the time. Furthermore, the CCP does not suggest how to properly adjust an air flow. An influence of the air flow over the human body surface, depending on air velocity, turbulence intensity and target surface is complex [2]:

- it changes heat transfer between the air and the body - can result in either comfort or discomfort, depending on the other microclimatic parameters,
- it provides fresh air in face zone,
- it can cause an eye discomfort,
- it can be felt as an air movement.

Other example is a manual CCP that does not have single control for windshield demisting/defrosting. This mode of operation is activated by combination of at least three actions: maximum temperature, maximum fan speed and air flow directed to the windshield. In the case of complex environmental conditions, a driver have to shift from driving task to non-driving task of HVAC setting (driving in busy traffic in night conditions with sudden start of heavy rain combined with glass misting etc.). The "What-to-do?" situation happens if there is no single separated command on the CCP, distracting the driver from driving task in combination with impaired vision in addition.

The often encountered problem in the climate settings (applies on the automatic systems too) is an overshooting [9]: system response in temperature change is slow, compared to the air velocity which reacts almost immediately after the fan speed change. The consequence is a large difference in the temperature settings and the temperature preference (i.e. the preferred heat transfer between the body and the cabin interior). As soon as the system generates too low or too high temperature there will be need for a corrective action. However, a large difference will not increase the speed of change, which is limited by the cooling or the heating power.

4. General CCP design recommendations

The CCP should be designed in such a way that it is to be used by end user (common driver), not by HVAC engineer. The past and current CCPs offer a control of every part of the HVAC system, making the CCP very complex: for example, Toyota Corolla HEV 2020 has 15 controls; VW Golf 2016 has 17 controls. Of course, automatic mode makes unnecessary use of the most of the controls, except of the temperature control.

As is mentioned before, the CCP of today cars are within the driver's visual field, reachable and labelled by standardized pictograms. However, although the HVAC system is configured on very similar way, it is evident that there is a great variety in control layout and type of controls, in contrast to the primary vehicle controls.

In order to facilitate learning of use, improve interpretability, understandability, reduce driver's distraction and improve efficiency by mitigating the wrong settings, the CCP should have following ergonomic features:

- Standardized layout of the controls for temperature, fan speed and air distribution mode as the most used controls,
- Spatial and shape/tactile control distinction.
- Rotary knob for thermal sensation preference / temperature.
- Rotary knob for fan speed.
- Rotary knob combined with push button for related and/or consecutive operation: for example, push button for automatic mode activation on the rotary knob for temperature adjustment.
- Air direction selection by logically positioned push buttons: upper for windshield and windows vents, centre for central vents, lower for foot well vents, with all combination possible.
- Controls for so-called fast actions like windows demisting and air recirculation should be visually and tactile distinctive and easy to find blindly.
- Illuminated indication on activated control and the set values change shown on the main display in order to mitigate delay in the system feedback and the thermal sensation change.
- Automatic "Driver only" mode activated when only the driver is in the cabin. In this mode, the conditioned air mostly blows in the driver's seat. This would lead to less energy consumption, which is especially important for BEV and HEV.
- Passenger seats equipped with essential HVAC controls (temperature, fan speed), to reduce the driver's workload and distraction.

5. Conclusions

This paper presents a study of ergonomic characteristics of different climate control panels in several middle-sized passenger cars. The CCP's visibility, grouping, identification, interpretability, operability and understandability are described using the heuristic approach. As a main ergonomic downsides of the analyzed CCPs are lack of standardized layout, too many cluttered controls and that the controls demand visual guidance for the operation, to mention a few. Recommendations for ergonomically improved and more user-centred CCP features are withdrawn from the presented analysis as well as from general ergonomics benchmarking. Some of the recommendations would require changes in the HVAC system or cabin interior design. The proposed CCP solutions must be tested using the subjects in a driving simulator to analyse time required to adjust the system, to collect subjective responses and to determine the influence on the driver's performance and workload compared to the conventional CCP.

6. References

[1] Daly S 2006 *Automotive air-conditioning and climate control systems* (Oxford: Butterworth-Heinemann)

[2] Ružić D and Časnji F 2012 Thermal interaction between a human body and a vehicle cabin *Heat Transfer Phenomena and Applications* ed Kazi S (Rijeka: InTech) chapter 11 pp 295-318

[3] Wellings T 2012 Designing the Interface *Design for Transport: A User-Centred Approach to Vehicle Design and Travel* ed Tovey M (Aldershot: Gower) chapter 11 pp 317-334

[4] Broström R, Ljung M, Wahlberg L and Källgren L 2013 What drives off-road glance durations during multitasking – capacity, practice or strategy? *Proc. of the Third International Conference on Driver Distraction and Inattention* 04-06.09.2013 Göteborg, paper 08-P

[5] Salvucci D, Zuber M, Beregovaia E and Markley D 2005 Distract-R: Rapid prototyping and evaluation of in-vehicle interfaces, *Proc. of the 2005 Conference on Human Factors in Computing Systems - CHI 2005* 02-07.04.2005 Portland

[6] Kern D and Schmidt A 2009 Design space for driver-based automotive user interfaces *Proc. of 1st International Conference on Automotive User Interfaces and Interactive Vehicular Applications - AutomotiveUI 2009* 21-22.09.2009 Essen

[7] Lee S C and Ji Y G 2018 Complexity of in-vehicle controllers and their effect on task performance, *Int. J. of Human–Computer Interaction* **35** pp 65-74
[8] Rydström A, Broström R and Bengtsson P 2012 A comparison of two contemporary types of in-car multifunctional interfaces, *Appl. Ergonomics* **43** pp 507-514
[9] Ackermann J, Junior C, Wysocki T and Ritter J 2017 Minimale Klimabedienung für maximalen Komfort, *ATZ* **5** pp 44-48
[10] Dvorak D, Krüger V and Wang J 2022 Innovative HMI and Control Concept for Efficient Thermal Management of Electric Vehicles *IEEE Transactions on Intelligent Transportation Systems*, doi: 10.1109/TITS.2022.3177761
[11] Schneegaß S, Pfleging B, Kern D and Schmidt A 2011 Support for modeling interaction with automotive user interfaces *Proc. of the Automotive UI'11* 29.11-02.12.2011 Salzburg
[12] Ružić D 2022 An evaluation of HVAC control panel in passenger cars *Tehnika* **72** pp 203-211
[13] Bhise V D 2012 *Ergonomics in the Automotive Design Process* (Boca Raton: Taylor & Francis Group)
[14] Campbell J L, Brown J L, Graving J S, Richard C M, Lichty M G, Sanquist T, Bacon L P, Woods R, Li H, Williams D N and Morgan J F 2016 *Human factors design guidance for driver-vehicle interfaces* Report DOT HS 812 360 *National Highway Traffic Safety Administration* Washington
[15] Wickens C and Hollands J 2000 *Engineering Psychology and Human Performance* (New Jersey: Prentice Hall)
[16] *ISO 2575:2010 Road vehicles - Symbols for controls, indicators and tell-tales* (Geneva: International Organization for Standardization)
[17] *ISO 4040:2009 Road vehicles - Location of hand controls, indicators and tell-tales in motor vehicles* (Geneva: International Organization for Standardization)
[18] *ISO 17287:2003 Road vehicles - Ergonomic aspects of transport information and control systems - Procedure for assessing suitability for use while driving* (Geneva: International Organization for Standardization)
[19] Sanders M and McCormick E 1993 *Human Factors in Engineering and Design* (New York McGraw-Hill)
[20] *ISO 12214:2018 Road vehicles - Direction-of-motion stereotypes for automotive hand controls* (Geneva: International Organization for Standardization)
[21] Wickens C, Gordon S and Liu Y 1997 *An Introduction to Human Factors Engineering* (New York: Addison Welsley Longman Inc)

Acknowledgments
This paper has been supported by the Ministry of Education, Science and Technological Development through project No. 451-03-68/2022-14/200156 “Innovative scientific and artistic research from the Faculty of Technical Sciences activity domain”.

IOP Conf. Series: Materials Science and Engineering 1271 (2022) 012020 doi:10.1088/1757-899X/1271/1/012020

Investigation of the permitted level of damage of vehicle safety glass

S Petković[1], V Golubović-Bugarski[1], G Globočki-Lakić[1], Ž Đurić[1]

[1]University of Banja Luka, Faculty of Mechanical Engineering, Stepe Stepanovica 71, 78000 Banja Luka, Bosnia and Herzegovina

E-mail: snezana.petkovic@mf.unibl.org

Abstract: Glass plays an important role in a vehicle, and it is essential to preserve the safety and security of passengers. It is called safety glass because it does not create sharp edges in the case of breaking, which can injure drivers and passengers, but scatters into many small pieces that are not dangerous (tempered glass) or remain held in place by means of an interlayer (laminated glass). During the operation of the vehicle, the glass is often damaged due to the impact of stones or other objects. The level of damage (size and character) and its location on glass determine the extent to which the safety of the use of glass is impaired. This paper presents the basic characteristics of safety glass installed in vehicles, the requirements which must fulfilled, as well as criterion for assessing its technical correctness in the technical inspection procedure. The significance of this research lies in the fact that state institutions can improve the legal regulations in the field of technical inspections of vehicles, while transportation companies can achieve appropriate financial savings in vehicle maintenance.

1. Introduction

Glass plays an important role as active and passive element of a vehicle safety. That is why the automotive industry uses the so-called safety glass. It is called safety glass because when damaged, it does not create sharp edges that can cause injuries to drivers and passengers, but scatters into many small pieces that are not dangerous (tempered glass) or pieces remain attached to the central foil (laminated glass), figure 1.

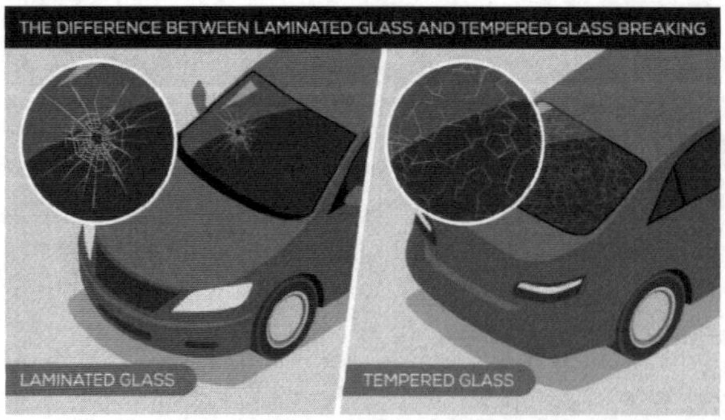

Figure 1. Broken glass [1].

IOP Conf. Series: Materials Science and Engineering 1271 (2022) 012020 doi:10.1088/1757-899X/1271/1/012020

Windshield glass is necessarily made as a laminated, while side car windows are mostly tempered glass. An important characteristic of laminated glass is that it can be repaired after damage, while tempered glass must be replaced after damage. The windshield affects the strength of the vehicle's structure by up to 30%, and also the effectiveness of the airbags which resist to the windshield during inflation.

Damage of the windshield is a common occurrence in a traffic. A cracked windshield can break at any time, which can seriously jeopardize driving safety. According to the fact that smaller damages can be located at such location on the windshield so they do not interfere with the clear visibility of the driver's field of vision, there is a dilemma - what level of damage of the windshield is acceptable without jeopardizing the safety of the vehicle.

Some countries have stipulated in their regulations the level of acceptable damage of windshields from the safety aspect, as well as procedures for testing the level of damage. For example, in Great Britain, the standard related to safety glass of vehicles (BS AU 178/ BS 857-2 / BS 5282) prescribes the permissible limits of windshield damage, which are not taken as a reason for the vehicle not passing the technical inspection. The British Ministry of Transport (MOT) has issued the Guide for carrying out a technical inspection of vehicles, which contains the procedure for checking vehicle windshield from the aspect of safety, [2]. Also, the Ministry of Transport of Ireland has issued Guide for the technical inspection of heavy commercial vehicles (category M2, M3, N2, N3, O3 and O4), [3]. This Guide, among other things, prescribes the procedure for checking the glazing of the vehicle.

Using available literature data, regulations and standards applied in other countries, the goal of our research was to set the limits of permissible damage to windshields that do not endanger the safety of the vehicle and do not affect the evaluation of the vehicle's roadworthiness during its technical inspection.

2. Characteristics of safety glass

Rulebook of the European Commission R43 - Uniform provisions concerning the approval of safety glazing materials and their installation on vehicles, provides definitions of materials used in the manufacture of glazing for vehicles, definitions of glazed surfaces and their geometric characteristics, characteristics that must be met through prescribed tests, marking methods, [4]. According to the Rulebook R43, safety glasses are made as tempered (mainly side car windows) and as laminated (mandatory windshield).

Tempered glass is 4 to 5 times more impact resistant than ordinary glass. The minimum permissible surface pressure that tempered glass must withstand is 120 N/mm^2, and the typical breaking pressure is about 165 N/mm^2, while ordinary glass breaks at 40 N/mm^2. In addition to these characteristics, tempered glass also has increased resistance to temperature changes (ΔT=150 K).

Laminated glass is made as a "sandwich" of two panes, with a thin but strong plastic layer between them, made of PVB (polyvinyl butyral), figure 2.

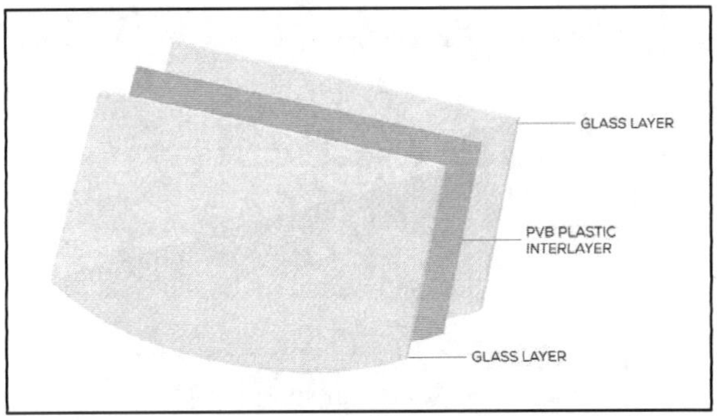

Figure 2. Laminated glass [1].

If the glass is subjected to an impact, the panes can split depending on the force of the impact, but the plastic interlayer must be preserved. In the case of an accident, this safety feature allows the glass to absorb the energy of the impact, helping to prevent the driver and passengers being ejected out the vehicle, and on the other hand, the penetration of objects into the vehicle. In addition, after the collision, the pieces of broken glass remain stuck to the plastic intermediate layer, which reduces the risk the driver and passengers being impacted by sharp glass fragments. Also, laminated glass improves the performance of side airbags due to the fact that window made of laminated glass tend to stay in its frame. Laminated glass provides a significant security advantage against burglary (in average, it takes 10 times longer to break a laminated window glass than it would be in the case of tempered window glass).

In addition to this basic characteristic of glass, which is safety for the driver and passengers, modern glass manufacturing technologies provide additional important characteristics: visibility and visual comfort, thermal comfort, acoustic comfort, style and visual impression, and contribute to the car's energy efficiency.

2.1. Visibility and visual comfort

Good visibility conditions are crucial for active vehicle safety, and visual comfort (for example, light reflection) is part of visibility. Sunlight cause glaring, raindrops could stain the windshield, condensation could fog up, and snow and ice almost block visibility.

Reflections occuring at the inner side of the windshield can disrupt the driver's visual comfort and concentration. This is also connected with the current design trends of installing control panels (instrument-panels) of light colors, which cause stronger reflections to the windshield. Anti-reflective glass, treated with special coatings, can reduce these reflections by up to 40% at a reflection angle of 60 degrees.

Hydrophobic glass are coated by a polymer layer that repels water, which causes rain or other water to leave the glass in the form of droplets, and also improves the efficiency of the wipers to completely remove the water from the windshield. Thanks to this characteristic, a visibility in a rainy conditions is significantly improved.

In heated windows, invisible wires or invisible metal coatings that are inserted into the interlayer of the laminated windshield react to the electrical supply by raising the temperature of the glass, in cold weather, and thus accelerate the removal of the resulting condensation and ice, preventing their further accumulation.

2.2 Thermal comfort

In recent years, the industry has developed a solar control glass, figure 3. This type of glass is specially designed to reduce the temperature in the vehicle cabin by up to 7°C or 8°C and reduce the penetration of heat into vehicle, which is exposed to solar radiation, by more than 25 %.

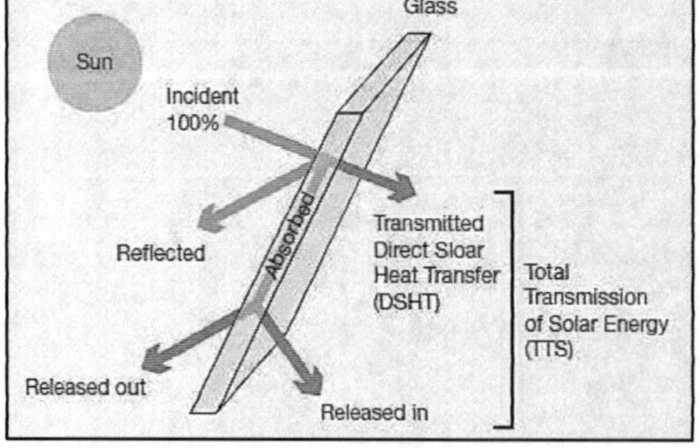

Figure 3. Solar control glass [5].

In this way, a significant amount of energy is saved instead of being used to power the air conditioning device, and passenger comfort is preserved. As part of the heat-reflective technologies, solar control glazing has a coating that allows it to reduce the transmission of solar heat from the external environment while still permitting the optimal light transmission essential to ideal driving condition [5].

2.3. Acoustic comfort

The technology of acoustically laminated glass was developed with the aim of reducing the noise transmitted into the vehicle cabin. The lower level of noise in the vehicle cabin ensures less driver fatigue and enables better communication inside the vehicle, thus contributing to general safety. When using certain acoustically laminated glasses, sound insulation can be achieved by an additional 3 - 5 dB [5], which is noticeable to the human hearing.

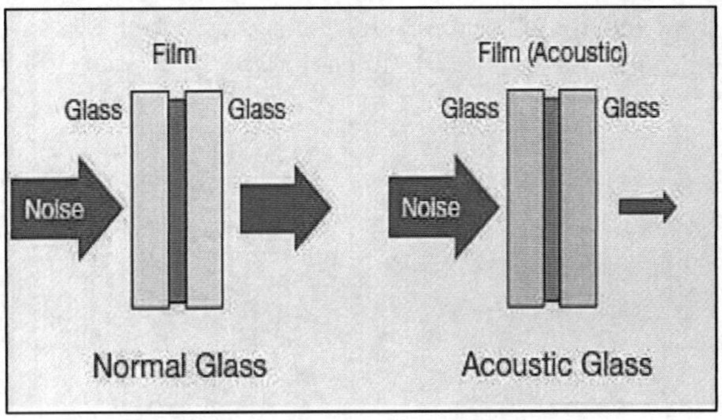

Figure 4. Acoustic glazing [5].

Acoustic glazing consists of two or more layers of glass, connected by one or more acoustic interlayers, figure 4. The interlayers act as a damping core, attenuating sound as it travels through the glass. Acoustically laminated glass also has all the advantages in terms of safety and security that standard laminated glass has. Moreover, acoustic windshield reduces the need for heavier glass, allowing car manufacturers to further reduce vehicle weight.

2.4. Style and visual impression

Glass represents an important visual and aesthetic component of the vehicle. The light and aerodynamic design of the car requires the use of glass of increasingly complex geometry. Therefore, manufacturers are continuously developing advanced glass forming methods that enable the bending and shaping of glass, allowing car manufacturers the freedom to achieve a desired design concept and style. Glass manufacturers also offer a number of different glass colors (e.g. green glass, dark tinted, etc.) that contribute to the harmony of the overall color of the car. Dark tinted windows keep the interior of the vehicle fresh by absorbing solar energy, while offering privacy for passengers at the back seat of the car.

With so-called "adaptive" car glass, the driver or passengers can control the amount of light entering the car - blocking it from 4% to 40%. The heat coming from the sun can be reduced from 2% to 20%. It is possible to choose a lighter shade of color during cloudy days and then darken it when the sun reappears, or simply change the lighting according to one's mood or create an atmosphere that suits an occasion.

2.5. Automotive glass and fuel consumption - energy efficiency

The automotive glass industry also, through its products, actively contributes to the reduction of fuel consumption and lower CO2 emissions. In this context, certain political initiatives of the European Union contributed to the use of energy-saving glass. European Commission launched the Mobile Air-Conditioning initiative (MAC initiative) in 2012, in order to reduce CO2 emissions due to the use of air conditioners [6], and thus encouraged technologies that would reduce the need to use air conditioners in vehicles or at least reduce the load on air conditioners. One of these technologies is automotive glass with solar control. Automotive glass with solar control enable a reduction of energy consumption needed for the operation of the air conditioner by an average of 10%, and can improve the overall efficiency of the vehicle by up to 2-4%, depending on the specifics of the vehicle.

In addition to significant reduction of the heat load inside the vehicle (which also leads to a reduction in fuel consumption), various types of advanced glazing are also designed to have less weight. In this way, it contributes to the reduction of the total weight of the vehicle, enabling further fuel savings.

Over the past 30 years, the glass industry has already achieved impressive results in reducing the weight of vehicle glass. Today, glass represents only 3% of the total weight of the vehicle. Also, vehicle glass is integrated as a structural part of the vehicle, reducing the necessity to use heavier materials.

3. Methodology of investigation of windshield glass damage level

Mechanical damages of the windshield, such as stone chips or small cracks, occur in different crack types: a star, a target (or bull's eye), a half moon, linear cracks, a combination type, figure 5.

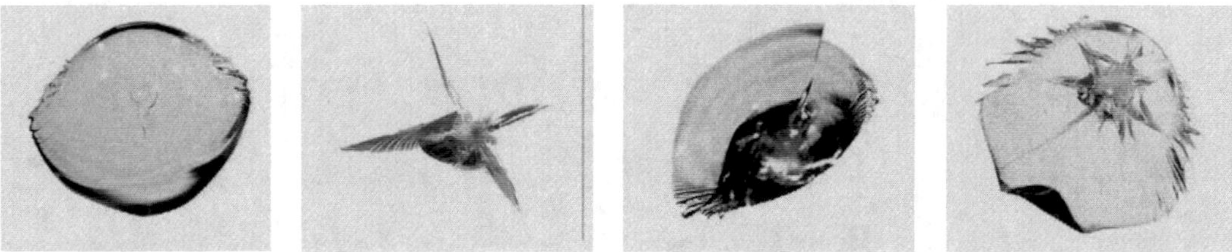

bull's eye *star* *half moon* *combination*

Figure 5. Different forms of windshield damages.

Regardless of the damage type, the damage must not disturb normal visibility through the windshield and must not have a tendency to spread further. From the security point of view, it is also necessary to check the transparency of the vehicle glass, the deformability of the image, the functionality of the movable windows.

Taking into account that smaller windshield damages can be repaired, or their locations do not interfere with the clear visibility of the driver's field of vision, follow questions can be asked:

- What level of windshield damage is allowed without jeopardizing the safety of the vehicle?
- What level of windshield damage is the reason the vehicle will not pass the technical inspection?

In the past period, the criteria for assessment the level of the windshield damage in the process of technical inspection of vehicles in Bosnia and Herzegovina wasn't fully define. Because of that, inspectors who carry out technical inspections were faced the dilemma.

Studying the international standards related to safety glasses, using the available literature [2 – 4, 7 - 9] and analysing the experience of others countries, especially EU countries that have high standards in traffic safety, the testing methodology and the permissible level of damage of windshields were defined by the technical inspection professionals. This methodology was adopted by the state institutions (Ministry of Transport and Communications of Bosnia and Herzegovina) and published in Rulebook No. 83/20, (Official Gazette of BiH), [10].

4. Results and discussion

According to the defined methodology [10] for testing the windshield, the permissible limits of damage are determined depending on the category of the vehicle and the position of the damage on the windshield. In order to uniformly define the position of damage, the windshield is divided into three zones - A, B and C, figure 6:

- Zone "A" is the area of the windshield that represents the driver's direct field of vision. The boundaries of this zone are determined as the area of the windshield which could be swept by the wipers. This is 290 mm wide area for vehicles of categories M1 and N1, or 350 mm wide area for vehicles of categories M2, M3, N2 and N3, measured so that the central axis of the steering wheel also represents the central axis of this area,

- Zone "B" is the rest of the windshield surface swept by the wipers,
- Zone "C" is the rest of the windshield surface, which is out of zones "A" and "B".

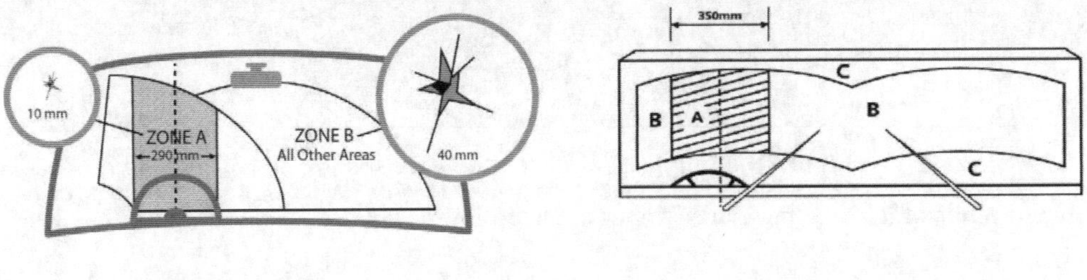

Vehicle categories M1 and N1 *Vehicle categories M2, M3 and N2, N3*

Figure 6. Windshield zones for different vehicle categories, [2, 3].

4.1 Template for determining the level of windshield damage

In order to assess the size of the windshield damage, a single tool - a template - has been defined. The template is in the form of a ruler with a length of 290 mm, for vehicles of categories M1 and N1, figure 7a, or 350 mm for vehicles of categories M2, M3, N2 and N3, figure 7b.

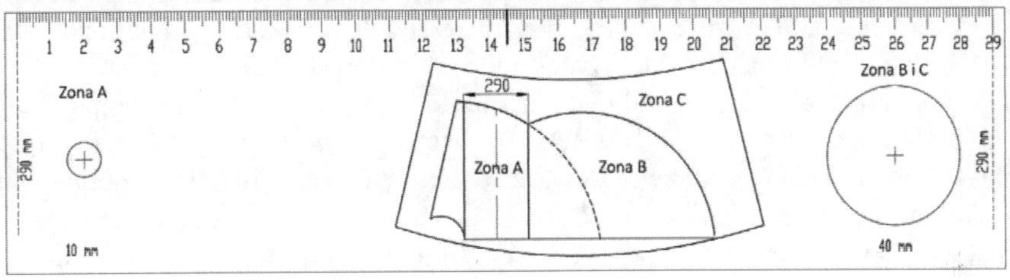

a) template for vehicle categories M1 and N1

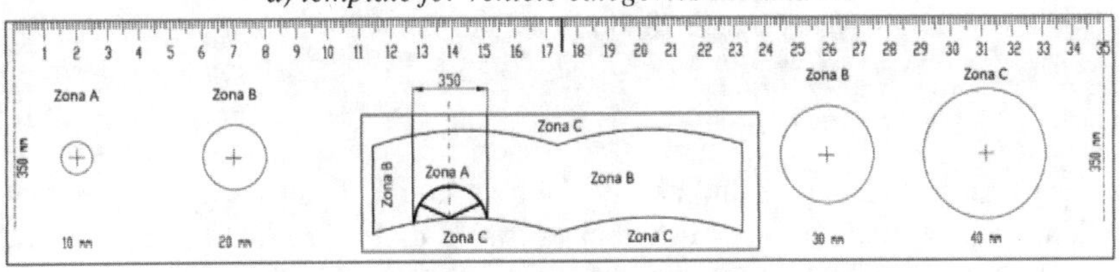

b) template for vehicle categories M2, M3, and N2, N3

Figure 7. Template for determining the level of windshield damage, [10].

Checking of windshield damage is done so that the inspector, sitting in the driver's seat, places the template on the windshield so that the center line of the template coincides with the center axis of the steering wheel. Then, with the wipers sweeping across windshield, the zones on the windshield are determined depending on the vehicle category (zone A, B or C), figure 8. Using the template having drawn circles (damage limits), it is checked whether the damage on the windshield exceeds the permissible limits prescribed for defined zones and specific vehicle categories.

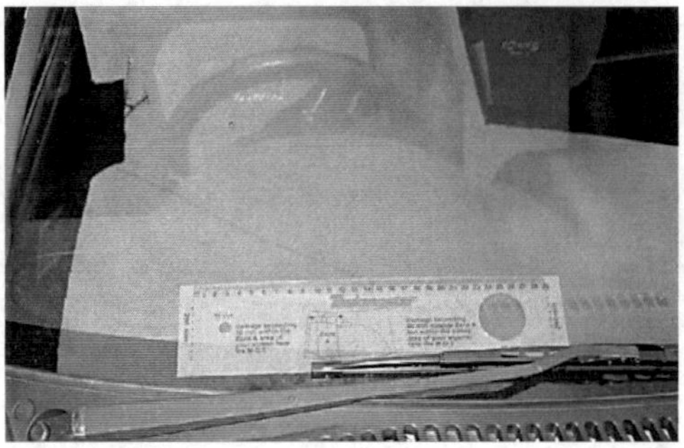

Figure 8. Template placed on the windshield [2].

Permitted limits of damage in defined zones are as follows:

- Zone "A": the damage in this area should be within a circle of 10 mm in diameter (a maximum of two damages are allowed, provided that they are at a distance of more than 100 mm),
- Zone "B": the damage in this area for vehicles of category M1 and N1 should be within a circle of 40 mm in diameter, and for other vehicles category it should be within a circle of 20 mm in diameter or it can be a linear crack 30 mm long (maximum of two damages are allowed, provided that they are at a distance of more than 100 mm),
- Zone "C": the damage in this area should be within a circle of 40 mm in diameter (a maximum of two damages are allowed, provided that they are at a distance of more than 100 mm).

In addition to the inspection of mechanical windshield damages, it is necessary to check that stickers do not obstruct the driver's field of vision, the sun visor on the driver's side is functional, and there are no other objects which could obstruct the driver's field of vision. Repaired windshields are checked according to the same criteria as unrepaired ones. It should be assessed whether the performed repairs impair the visibility of the windshield. An 'invisible' or barely detectable repair, finished flush with the surrounding glass, does not count as damage. Repaired damages that are "invisible" or barely noticeable, finished flush with the surrounding glass, do not count as damage, even if their dimensions exceed the damages prescribed as permissible. Scratches on the windshield are not considered as damage. However, the area where the scratches obscuring the view are concentrated, caused, for example, as a result of prolonged use of a wiper with a damaged blade, may be the reason for the vehicle not passing the inspection.

5. Conclusion

Glass plays an important role in active and passive vehicle safety. That is why the automotive industry uses the so-called safety glass. It is very frequent in traffic that windshield is damaged due to stone impacts. The safety of the vehicle depends on the size of the damage and the specific position on the surface of the windshield. The mere existence of damage to the windshield does not necessarily mean that it is technically defective. Failure due to windshield damage is only justified if the damage significantly affects the driver's view of the road. Especially critical area is the area of the driver's direct field of vision, so maximum damage below 10 mm in diameter is allowed in that zone. In other zones, greater damage is allowed depending on the type of vehicle. Also, if the damage is in the form of a crack, it is necessary to check its tendency to spread. However, even though the damage is not large, it is very important that the driver or the owner of the vehicle carry out the repair as soon as possible after the appearance of the damage, because getting dirt into the cracks reduces the quality of repair. Repaired glass will preserve its strength and provide the same level of safety for passengers as before the damage occurred. Successfully repaired damage (invisible and flush with the surrounding glass) is not a reason to vehicle fail the technical inspection.

References

[1] https://www.thewindscreenco.co.uk/help-advice/different-types-windscreens/ internet access 2th august 2022
[2] *The MOT Inspection manual* 2018 fourth edition, https://www.gov.uk/guidance/mot-inspection-manual-for-private-passenger-and-light-commercial-vehicles, internet access 2th august 2022, https://www.nationalarchives.gov.uk/doc/open-government-licence/version/3/
[3] *Heavy Commercial Vehicle testers manual* 2009 Government of Ireland, https://www.rsa.ie/docs/default-source/services/s4.8-cvrt/lcv-tester-manual-2020--5.1.pdf?sfvrsn=37681579_3
[4] *EC Regulation No. 43 - Uniform provisions concerning the approval of safety glazing materials and their installation on vehicles*, https://op.europa.eu/en/publication-detail/-/publication/6d819756-93c0-11e3-bf99-01aa75ed71a1
[5] *Glass - Contributing to innovative, green and safe transport solutions for Europe* www. glassforeurope.com
[6] *Position paper on the MAC Initiative – February 2011* Glass for Europe www. glassforeurope.com
[7] Green P and Burgess W 1981 *Windshield damage and driving safety*, *Technical Report UM-HSRI-81-35* Ann Arbour: The University of Michigan, Highway Safety Research Institute
[8] Hartley L, Lamble D and Penna F 1995 *Effects of reduced visibility caused by visually degraded windscreens on driver fatigue* Murdoch University Institute for Research into Safety & Transport
[9] Sayer J and Traube E 1994 *Factors influencing visibility through motor vehicle windshields and windows, A review of the literature* UMTRI-94-20 Ann Arbour: The University of Michigan, Transportation Research Institute
[10] *The Rulebook of technical inspections of vehicles* 2020 Official Gazette of BiH **83**/20, Ministry of Transport and Communications of Bosnia and Herzegovina, Pravilnik o tehničkim pregledima vozila, Službeni glasnik BiH 83/20

In-vehicle comfort assessment during fore-and-aft random vibrations based on artificial neural networks (ANN)

S Mačužić Saveljić[1], B Arsić[2], I Saveljić[3], J Lukić[1]

[1] University of Kragujevac, Faculty of Engineering, 6 Sestre Janjić Str., 34000 Kragujevac, Serbia
[2] Faculty of Science, Department of Mathematics and Informatics, Radoja Domanovića 12, Kragujevac, Serbia
[3] University of Kragujevac, Institute for Information technologies, Jovana Cvijića bb, Kragujevac, Serbia

E-mail: s.macuzic@kg.ac.rs

Abstract. Driving comfort is one of the important factors for vehicle users. There is a lot of research related to comfort or discomfort in a vehicle but there is still no well-defined way to assess it accurately. In this paper, an assessment of vehicle comfort during fore-and-aft random vibrations was made based on measured and predicted r.m.s. acceleration values. Measured values of r.m.s. accelerations were obtained by laboratory testing while the predicted values of r.m.s. accelerations obtained based on the ANN (Artifical Neural Network) model. 20 male subjects participated in the study. Their different anthropometric characteristics of the body were taken into account. Based on the measured r.m.s. values of acceleration formed ANN model which has ability to predict r.m.s. acceleration values based on measured values. The obtained results showed high accuracy of the model.

1. Introduction

During driving, certain forces are transmitted from the road to the vehicle in the form of vibrations. The appearance of these vibrations is felt by the driver and passengers and can lead to a feeling of discomfort. The definition of comfort is difficult to define, it changed during the research. Thus, for the definition of comfort based on the Cambridge dictionary [1], it can be said that it is a pleasant and satisfying feeling when a person is mentally and physically free from pain and suffering. Assessment of the oscillatory comfort of the human body can be achieved in two ways: subjective and objective methods. Subjective assessment of comfort is performed on the basis of experiments where subjective ratings are collected through a series of statements given in relation to a predefined scale. Standardized methods are used for the objective assessment of comfort. These include the ISO 2631 standards and the British standard BS 6841. In this work, the ISO 2631 standard was used, on the basis of which the comfort criteria were defined.

The role of vibration measurement is to record the movement of the body as a reaction to some kind of stimulus. The discomfort caused by vibrations varies depending on the frequencies present. In addition to the frequency, the duration of the vibration, the amplitude and the direction of the vibration (x, y and z axis) also have an impact on the appearance of discomfort. A review of the literature found that a significant number of studies dealt with the impact of vibrations, both on individual parts of the body and on the entire human body, in different environments. Exposure of the human body to the action of excitation in the direction of one axis is carried out in the fore-and-aft and vertical directions. Uniaxial vibrations along the vertical and fore-and-aft axis caused significant movements in the sagittal plane (vertically, fore-and-aft and laterally) of the upper body, suggesting a strong coupling effect [2], [3]. Whole body vibration is particularly significant in the frequency range from 1 Hz to 80 Hz. In this frequency range are also the main resonance points of certain organs and parts of the human body (e.g., head, eyes, stomach and spine) [4], [5]. Undesirable effects of these vibrations refer to the occurrence of a number of health problems, among which the most characteristic are back (lumbar region) and neck pains. Vibrations of the whole body in the area of extremely low frequencies (below 0.5 Hz) cause "seasickness" [6]. Table 1 shows the observations of the comfort of passengers of public urban transport for different intensities of vibrations according to the standard ISO 2631 (1997), based on laboratory tests. It is obvious that with an increase in the strength of vibrations, passenger comfort decreases [7].

Table 1. Passenger comfort perception according to ISO 2631 (1997).

Less than 0.315 m/s^2	Not uncomfortable
0.315 m/s^2 to 0.63 m/s^2	A little uncomfortable
0.5 m/s^2 to 1 m/s^2	Fairly uncomfortable
0.8 m/s^2 to 1.6 m/s^2	Uncomfortable
1.25 m/s^2 to 2.5 m/s^2	Very uncomfortable
Greater than 2 m/s^2	Extremely uncomfortable

A group of authors [8] showed that professional drivers are less sensitive to vibration levels. There are two reasons. First, drivers perceive obstacles on the road and are prepared for disturbances that reach their bodies (movements of internal organs are suppressed by increasing muscle tone). Second, the steering wheel allows the driver to moderate angular movements around the transverse axis and translational movements (fore-and-aft) of the body. The author [9] concludes that the perception of passenger comfort in real driving conditions corresponds to higher vibration intensities compared to the perception of the same level of comfort in laboratory tests. This is explained by the presence of numerous stimuli that affect the passenger's attention in real driving conditions.

Standard ISO 2631 (1997) [10] prescribes the total value of mean effective weighted accelerations, on the basis of which the impact on discomfort is assessed according to the formula:

$$a_v = \sqrt{\left(k_x \cdot \ddot{x}_{rms,w}\right)^2 + \left(k_y \cdot \ddot{y}_{rms,w}\right)^2 + \left(k_z \cdot \ddot{z}_{rms,w}\right)^2} \tag{1}$$

where:

k_x, k_y, k_z – are correction factors for r.m.s. values of weighted accelerations in the direction of the x, y and z axis,

$\ddot{x}_{rms,w}$, $\ddot{y}_{rms,w}$, $\ddot{z}_{rms,w}$ – are mean effective value of the weighted acceleration for the directions x, y and z, which are determined according to the formulas:

$$\ddot{x}_{rms,w} = \sqrt{\frac{1}{N}\sum_{i=1}^{N}\ddot{x}^2{}_{w_i}} \tag{2}$$

$$\ddot{y}_{rms,w} = \sqrt{\frac{1}{N}\sum_{i=1}^{N}\ddot{y}^2{}_{w_i}} \tag{3}$$

$$\ddot{z}_{rms,w} = \sqrt{\frac{1}{N}\sum_{i=1}^{N}\ddot{z}^2{}_{w_i}} \tag{4}$$

where:

$\ddot{x}_{w_i}$, $\ddot{y}_{w_i}$, $\ddot{z}_{w_i}$ – are *i*-th sample of weighted accelerations of the user for the directions of the x, y and z axis (m/s^2),

N – is number of signal samples of weighted accelerations.

Machine learning is the process of finding hidden laws and connections between data. It represents a new approach and modern problem solving in engineering. The task of machine learning is to focus on developing human-like learning algorithms. A set of such methods that mimic the work of the human brain in learning, and have bases in biology, are collectively called neural networks. This approach is adapted to well predict the nonlinear responses of the human body in a seated position. Artificial neural network schemes have been widely used in a wide range of applications [11-13] because they can model complex nonlinear systems by learning from the input and output signals of the system.

In this paper, based on the obtained experimental results, an ANN model was created in order to evaluate the oscillatory comfort of the human body.

2. Materials and methods

2.1. Laboratory experiment

Causing excitation of different amplitudes and frequencies was achieved using an electro-hydraulic pulsator HP-2007. This pulsator is designed to provide excitation in the frequency range 0.1-31 Hz and amplitude 0-50 mm. A portable vibration analyzer NetdB1 was used to measure the transmission of vibrations from the seat through the human body. AC102-1A acceleration transducers, weighing 90 g, sensitivity 100 mV/g, frequency range 0.5-15000 Hz, which can work in the temperature range of -50-121°C, were used to record the vibration response. Twenty male healthy subjects, with mean age 30.7±6.15 years, height 183.25±4.43 cm, weight 89.4±11.72 kg, BMI (BMI Body Mass Index) 26.57±2.96, seating height 88.35±4.79 cm, were exposed to random vibrations of the whole body for three excitation values 0.45, 0.8 and 1.1 m/s2 r.m.s., in the frequency range 0.1-20 Hz. The angle of inclination of the backrest was 90°. The duration of each experiment was 60 s, and the number of repetitions was 2. The parameters of the experiment are: sampling rate 51.2 kHz, block duration 80 ms, number of samples 4,096, where the number of averaging is 2, sampling step Δt = 0.0195 ms. Signal overlap is 75%. The frequency step Δf=0.390625 Hz, and the bandwidth is B = 39 Hz, where the Nyquist frequency criterion was satisfied, because the sample frequency is higher than twice the sampling rate. The measurement scheme is shown in figure 1.

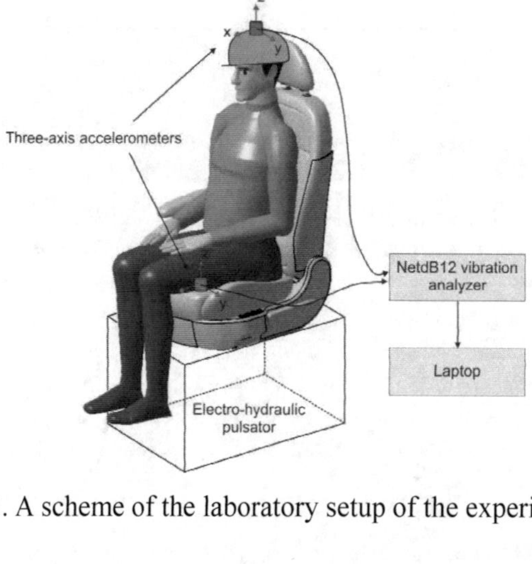

Figure 1. A scheme of the laboratory setup of the experiment.

2.2. Artificial neural network

Recurrent Neural Networks (RNN) were used in this work. Neural networks are formed in which the outputs of the neuronal elements of the following layers have synaptic connections with the neurons of the previous layers. This leads to the possibility of taking into account the results of information transformation by the neural network in the previous stage for processing the input vector in the next stage of the network operation. A recurrent network with a memory cell LSTM (Long-Short Term Memory) was used to determine the oscillatory comfort.

The data for training the neural network were data obtained from experimental measurements of 20 subjects exposed to random fore-and-aft vibration for three excitation amplitudes. Each subject was characterized by BMI, height, weight, sitting height and age. Figure 2 shows the scheme of the used neural network.

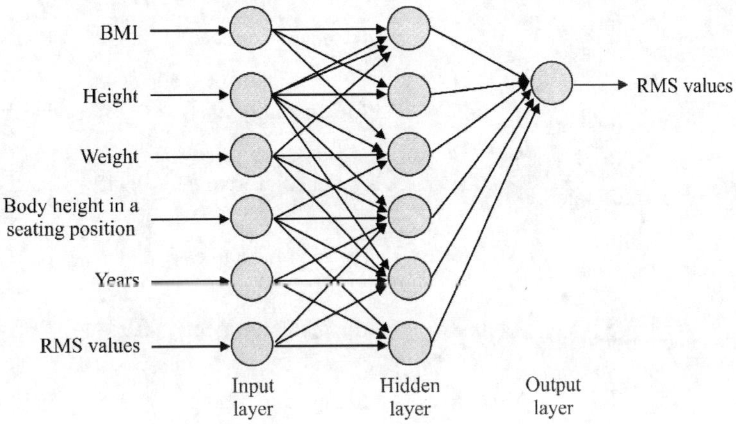

Figure 2. A scheme of a LSTM recurrent neural network.

The mean absolute error was used as the loss function in a training phase, while the effective Adam optimization algorithm was used to search for the minimum of the loss function.

3. Results

In this research, based on the ISO 2631 standard, an assessment of oscillatory comfort was performed based on the total mean effective value of the weighted acceleration. The ISO 2631 standard prescribes checking the value of the crest factor. The measured peak factor values were in the range of 1.391-2.784, which meets the requirements of the standard (peak factor < 9), so the weighting was done according to the ISO 2631 standard. Table 2 shows the total r.m.s. weighted acceleration values and oscillatory comfort ratings under the influence of fore-and-aft excitation for 20 male subjects. The lowest value of r.m.s. it was recorded in the subject number 3 (BMI = 34; the highest value) and is 0.213 m/s^2 for an excitation of 0.45 m/s^2 r.m.s. What characterizes the obtained results of all excitation directions is that with an increase in BMI, there is a decrease in the total r.m.s. acceleration values.

Table 2. R.m.s. weighted acceleration values and oscillatory comfort ratings under the influence of fore-and-aft excitation for twenty male subjects.

ID	0.45 m/s^2 r.m.s.		0.8 m/s^2 r.m.s.		1.1 m/s^2 r.m.s.	
1	0.381	A little uncomfortable	0.723	Fairly uncomfortable	0.984	Fairly uncomfortable
2	0.379	A little uncomfortable	0.714	Fairly uncomfortable	1.024	Uncomfortable

3	0.213	Not uncomfortable	0.603	A little uncomfortable	0.888	Fairly uncomfortable
4	0.227	Not uncomfortable	0.609	A little uncomfortable	0.896	Fairly uncomfortable
5	0.342	A little uncomfortable	0.679	Fairly uncomfortable	0.964	Fairly uncomfortable
6	0.355	A little uncomfortable	0.699	Fairly uncomfortable	0.988	Fairly uncomfortable
7	0.354	A little uncomfortable	0.714	Fairly uncomfortable	1.004	Uncomfortable
8	0.351	A little uncomfortable	0.683	Fairly uncomfortable	0.978	Fairly uncomfortable
9	0.282	Not uncomfortable	0.544	A little uncomfortable	0.928	Fairly uncomfortable
10	0.314	Not uncomfortable	0.606	A little uncomfortable	0.945	Fairly uncomfortable
11	0.322	A little uncomfortable	0.676	Fairly uncomfortable	0.964	Fairly uncomfortable
12	0.296	Not uncomfortable	0.653	Fairly uncomfortable	0.938	Fairly uncomfortable
13	0.236	Not uncomfortable	0.616	A little uncomfortable	0.908	Fairly uncomfortable
14	0.248	Not uncomfortable	0.624	A little uncomfortable	0.927	Fairly uncomfortable
15	0.382	A little uncomfortable	0.738	Fairly uncomfortable	1.045	Uncomfortable
16	0.374	A little uncomfortable	0.704	Fairly uncomfortable	0.996	Fairly uncomfortable
17	0.271	Not uncomfortable	0.634	Fairly uncomfortable	0.915	Fairly uncomfortable
18	0.334	A little uncomfortable	0.667	Fairly uncomfortable	0.954	Fairly uncomfortable
19	0.369	A little uncomfortable	0.701	Fairly uncomfortable	1.014	Uncomfortable
20	0.362	A little uncomfortable	0.705	Fairly uncomfortable	0.967	Fairly uncomfortable

Using an artificial neural network, it is possible to train the neural network based on the obtained experimental results and perform a comfort assessment based on ISO 2631 standard. The ANN model was trained with 100 training epochs, while the data series size was 20. The coefficient in training, validation and testing was over 98% for all types of stimuli. The root means square error of the predicted r.m.s. value was 0.024. R.m.s. the values of the original and the values after the process of training the network, for the fore-and-aft excitation, of the male subject under ID number 20, are shown in table 3.

Table 3. Prediction results of r.m.s. values of the male subject under serial number 20 for excitation in the fore-and-aft direction.

	0.45 m/s^2 r.m.s.		0.8 m/s^2 r.m.s.		1.1 m/s^2 r.m.s.	
	Original	Predicted	Original	Predicted	Original	Predicted
R.m.s.	0.362	0.347	0.705	0.682	0.967	0.974
Comfort rating (ISO 2631)	A little uncomfortable	A little uncomfortable	Fairly uncomfortable	Fairly uncomfortable	Uncomfortable	Uncomfortable

The developed model of artificial neural network shows that it has adequate precision of biodynamic modeling. The main feature of the ANN model is to consider height, weight, sitting height, BMI and age during whole body vibration exposure.

4. Conclusions

In this work, an assessment of oscillatory comfort was performed based on the ISO 2631 standard. Twenty male subjects were exposed to random uniaxial vibrations in the fore-and-aft direction. The R.m.s values obtained from these measurements were the input of the neural network, which had the task of evaluating the oscillatory comfort based on the input data. By using the weighted acceleration and the mentioned standard, it is possible to evaluate the comfort of a particular position of the subject. The neural network model was trained with 100 training epochs, while the size of the data series was 20. For all types of stimuli, the correlation factor in the training, validation and testing phase was over 98%. The root means square error of the predicted r.m.s. value of weighted acceleration was 0.024. The complexity of the model, which takes into account different anthropometric characteristics of the subjects, as well as different vibration amplitudes, is not a problem, but highlights the ability of the model to predict oscillatory comfort via r.m.s. values, in the time domain, defined by the ISO 2631 standard.

Future research will go in the direction of investigating the influence of the change in the angle of inclination of the seat back on the r.m.s. values, as well as determining the most influential factor on the response of the human body to the vibrations of the whole body.

Acknowledgments
This research was supported by the Ministry of Education, Science and Technological Development of the Republic of Serbia through Grant TR35041

References

[1] Cambridge dictionary https://dictionary.cambridge.org/dictionary/english/comfort
[2] Nawayseh N and Griffin M J 2004 Tri-axial forces at the seat and backrest during whole-body vertical vibration. *Journal of Sound and Vibration* **277** pp 309-326
[3] Rakheja S, Dong R G, Patra S, Boileau P-É, Marcotte P and Warren C. 2010 Biodynamics of the human body under whole-body vibration: Synthesis of the reported data, *International Journal of Industrial Ergonomics* **40**(6) pp 710-732
[4] Ahn S J 2010 Discomfort of vertical whole-body shock-type vibration in the frequency range of 0.5 to 16 Hz, *International Journal of Automotive Technology* **11**(6) pp 909-916
[5] Griffin M J 1996 *Handbook of Human Vibration*, Academic Press, UK p 988
[6] Hopcroft R and Skinner M 2005 C-130J Human Vibration *Defence Science and Technology Organisation*, Australia p 30
[7] Marjanen Y and Nevala K 2003 Whole body vibration analysis program for real measurements

and virtual model, Department of Mechanical Engineering University of Oulu Finland p 8
[8] Hassan R and McManus K 2001 Heavy Vehicle Ride and Driver Comfort *SAE World Congress, Session on Human Factors in Automotive Design* Detroit Michigan USA
[9] Granlund J 2008 Health Issues Raised by Poorly Maintained Road Networks *Swedish Road Administration Consulting Services* p 146
[10] ISO 2631-1 1997 *Evaluation of Human Exposure to Whole-Body Vibration. Part 1: General Requirements* International Organization for Standardization Geneva
[11] Won S H, Song L, Lee S Y and Park C H 2010 Identification of finite state automata with a class of recurrent neural networks *Neural Net. IEEE Trans* **21** pp 1408–1421
[12] May P, Zhou E and Lee C W 2012 Learning in fully recurrent neural networks by approaching tangent planes to constraint surfaces *Neural Net.* **34** pp 72–79
[13] Lukić J and Mačužić Saveljić S 2021 Ann driver model based on seat to head transmissibility *10th International Automotive Technologies Congress, OTEKON 2020* Bursa Türkiye

Optimization of the air distribution system in motor vehicle cabin using computational fluid dynamics

D Feher[1,2], D Ružić[1], S Galamboš[1], N Nikolić[1] and J Dorić[1]

[1]Faculty of Technical Sciences, University of Novi Sad, Trg Dositeja Obradovića 6, 21000 Novi Sad, Serbia

[2]Author to whom any correspondence should be addressed (daliborfeher@uns.ac.rs)

Abstract. The paper presents the application of CFD software for simulating the conditions that prevail inside the air distributor of the vehicle's ventilation system with the aim of increasing the efficiency of the distributor. The simulation results were validated based on the experimental data found in the literature. Based on the results of the simulation, the geometry of the distributor was optimized with the aim of reducing the pressure drop through the ventilation system. It was found that optimizing the distributor geometry to some extent can provide a more energy efficient ventilation system.

1. Introduction

The development of modern Computational Fluid Dynamics - CFD began with the advent of digital computers in the early fifties of the XX century. Computational fluid dynamics was popular in that period only in technologically superior engineering disciplines such as aeronautics and astronautics. Today, this method is widespread and is used to solve various engineering problems that require the application of fluid mechanics, thermodynamics or chemical engineering.

It is generally known that computer methods generally increase the efficiency of the process of designing and constructing completely new engineering systems, in addition, they are massively used during the process of improving already existing solutions, where the greatest contribution is provided by computer optimization. It is precisely for this reason that the application and development of computer methods are essential for the further development of technology. Numerous applications of CFD software for the analysis of various systems in motor vehicles are mostly encountered when analyzing vehicle aerodynamics, fluid flow and thermal properties of internal combustion engines, microclimatic conditions of the vehicle cabin, etc.

A large number of papers on this topic are devoted to different ways of testing and analyzing the effects that occur within the air conditioning, ventilation and windshield defrosting system. Mostly the papers are based on experimental testing or CFD simulations.

As part of the paper from Ležovič et al. [1], the air speed and the intensity of turbulence in the area of the ventilation opening, which is positioned on the side of the dashboard, was examined. The main goal was to examine whether, and to what extent, the direction of the nozzle that delivers air into the cabin corresponds to the direction of the valve guide. The test results showed that the axis of the nozzle is different from the axis of the guide on the valve and that there is no concrete correlation between their positions.

Fišer and Pokorný [2] studied the relationship between the amount of air supplied to the HVAC system of a moving vehicle and the speed of the vehicle. The data on the amount of air that is sucked in was obtained by measuring the air speed in front of the valve in the cabin. Data obtained from measurements during the test show that the air flow is highly dependent on the vehicle speed when the fan is set to the second level in combination with the heating flap position half open.

An experimental study of Fojtlin et al. [3] was carried out with the aim of developing a simple and relatively accurate method for measuring air flow. The goal was also to create results that can be used as a reliable source of boundary conditions for numerical simulations of vehicle cabin microclimates. The mean flow values at the point of air inlet and outlet of the system are approximately the same. The results showed that for the case of summer the HVAC unit provides larger amounts of air (57 l/s) to compensate for the higher heating of the cabin compared to the cases of winter and spring/autumn (42 l/s). In summer, air distribution is directed to the upper part of the passenger's body (76%) and the windshield (20%). In the case of winter and spring/autumn, the air distribution is directed towards the feet (40%), the rest of the air is used to heat the windows.

Pokorný et al. [4] performed measurements to determine the amount of air sucked into the system depending on the setting of the interior valves. The objective was also to find the relationship between fan rotation speed and volumetric air flow, as well as to determine the pressure difference which occurs at the fan position. The experiment also showed that for a specific vehicle there is a linear dependence between the voltage of the fan and the volume flow of air.

Another study [5] describes the optimization of the part of heating, ventilation and air conditioning (HVAC) system of an electric car. The geometry of the air duct connected to the heater unit on one side and the dashboard on the other. In order to optimize the initial geometry of the air duct, approach of using gradient methods was applied. According to the results of the study, detailed concept of optimizing the air duct of an electric car is presented. The optimized geometry was tested with the use of methods of computational aerodynamics.

During the analysis of the literature, it was found that very few papers deal with the air distributor that conducts out air into the cabin, and the application of software for simulating the conditions that prevail inside the air distributor of the vehicle's ventilation system is presented in the paper. Based on the results of the simulation, the geometry of the distributor is optimized with the aim of reducing the pressure drop through the ventilation system. Optimizing distributor's geometry can provide an energy efficient ventilation system. The need for energy-efficient systems is especially pronounced today in electric vehicles, where the low energy density of the electrical source is still their most debatable characteristic.

2. Methodology

Within the air distribution system there is a relatively large number of elements and obstacles of complex geometry. The result is local losses that occur with each change of the velocity vector. In the characteristic places where losses occur, in general, there is a local pressure drop. The total pressure drop is a very important parameter of the air distribution system, depending on it, a fan with appropriate characteristics is adopted that will be able to overcome all the losses that occur in the system. The total pressure drop represents the sum of local losses that occur between the point of air inlet into the ventilation system in the engine compartment and the point of air outlet at the cabin ventilation openings. The subject of study in this paper is one of the system segments, the air distributor.

2.1. 3D model creation

In order to approach the CFD analysis of the air flow through the distributor, it is necessary to create an appropriate 3D model of it. The model was formed on the basis of real dimensions. Some of the dimensions, such as the dimensions of the ventilation openings, their mutual distance and the dimensions of the dashboard, are shown in Figure 1.

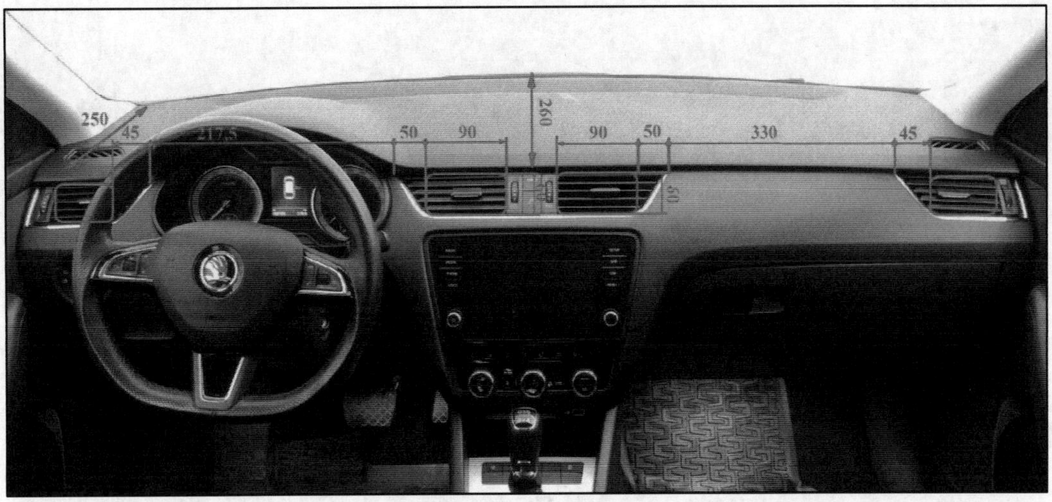

Figure 1. Characteristic dimensions of the air distribution (mm)

When creating the model, special attention was paid to three parts of the distributor, namely the pipelines through which the air flows to the ventilation openings (Figure 2 a), the adapter that divides the main stream into four corresponding ones (Figure 2 b), and the extension of the main stream whose role is to ensure that steady flow is achieved before the fluid reaches the bifurcation. Although the adapter and pipelines are physically separate elements, in this case they are modeled as a single element. The additional segment has two roles, the first is to ensure that the air flow is established before entering the distribution lines, and the second is to take into account the pressure drop caused by the bifurcation at the point of division of the lines.

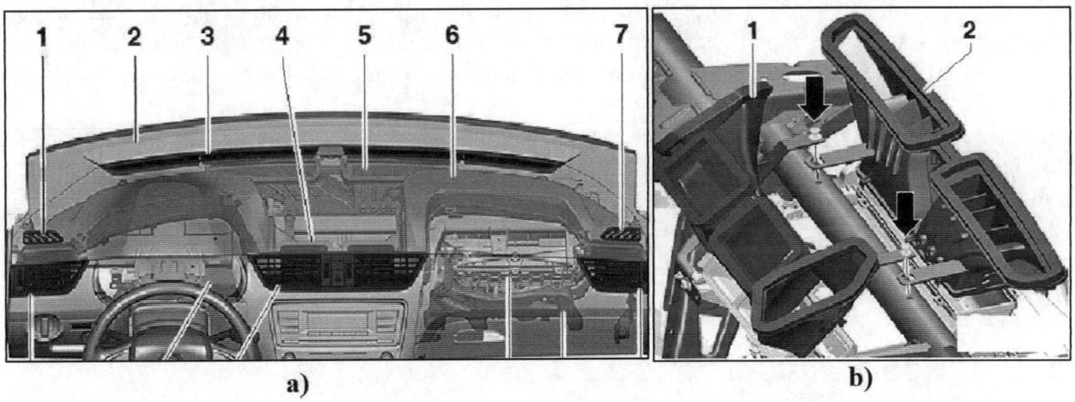

Figure 2. Display of the elements of the air distribution system on a specific vehicle [6]

The model formed on the basis of the real dimensions of the air distributor represents the air flow domain inside the distributor itself, and not its external shape. A view of the 3D CAD model is given in Figure 3.

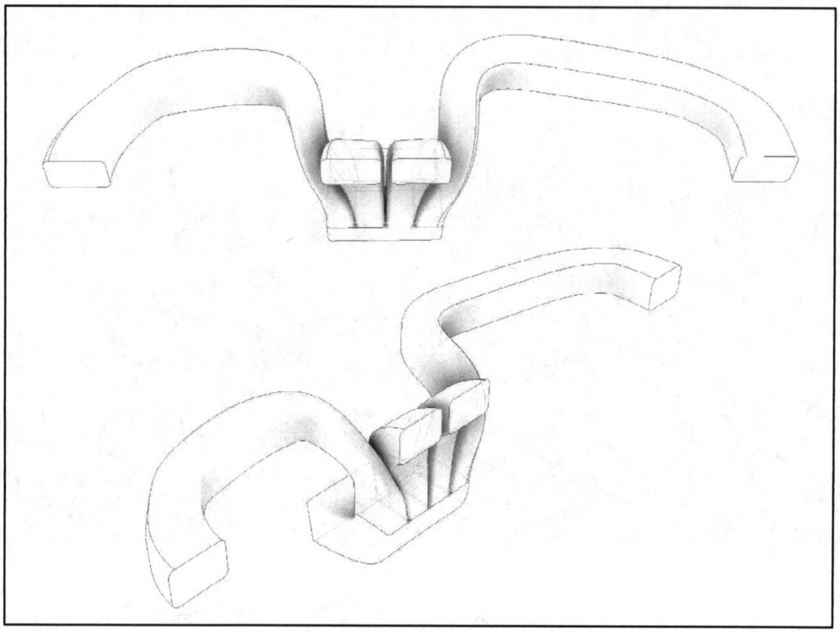

Figure 3. 3D model of the air distributor

2.2. Pre-processing

After the model is imported into the CFD software, type of boundary surfaces are defined. By analyzing the experimental data presented by Jan Pokorny and others [5], it was concluded that the most pronounced pressure drop occurs when the side vents are open, at maximum fan speed. Same case is analyzed here because in that regime the greatest improvement is expected in terms of the reduction of flow resistances that are a consequence of the geometry of the distributor. In order to be able to validate the simulation results with experimental data, the same properties of the boundary conditions are used as in the case of Jan Pokorny et al. Names of boundary surfaces, their types and characteristic values are presented in table 1.

For steady-state simulations, which is the case here, the converged solution should be independent of the initial conditions. However, the path to convergence, and the computational resources used to achieve convergence are not independent of the initial conditions. Some of the initial conditions that can be defined are: pressure, temperature, velocity components, turbulent quantities, etc. For the problem considered in this paper, a global initial condition of atmospheric pressure in the cabin of 102375 Pa is defined.

Table 1. Names of boundary surfaces and their types

Name of boundary surface	Boundary condition	Value
Inlet	*Velocity inlet*	*6,435 m/s*
Left outlet	*Pressure outlet*	*102 375 Pa*
Right outlet	*Pressure outlet*	*102 375 Pa*
Left-centred outlet	*Wall*	*No-slip*
Right-centred outlet	*Wall*	*No-slip*
Faces	*Wall*	*No-slip*

The last step of the preprocessing is the generation of the mesh. The properties selected when defining the mesh are as follows: surface remesher, polyhedral mesher, automatic surface repair, thin mesher, prism layer mesher. In order for the mesh to be optimal for a specific problem, it is necessary to define a large number of influential parameters. The parameters of importance that are varied until the initial mesh is found, which will give the expected results and the convergence of the residuals, are shown in Table 2. The mesh generated according to the specified parameters contains 509933 cells and is shown in Figure 4.

Table 2. Initial mesh parameters

Parameter	Value
Base Size	*20 mm*
Target Surface Size	*30 mm*
Minimum Surface Size	*5 mm*
Surface Growth Rate	*1,3*
Number of Prism Layers	*2*
Prism Layer Stretching	*1,2*
Prism Layer Total Thickness	*6 mm*
Volume Growth Rate	*1,2*

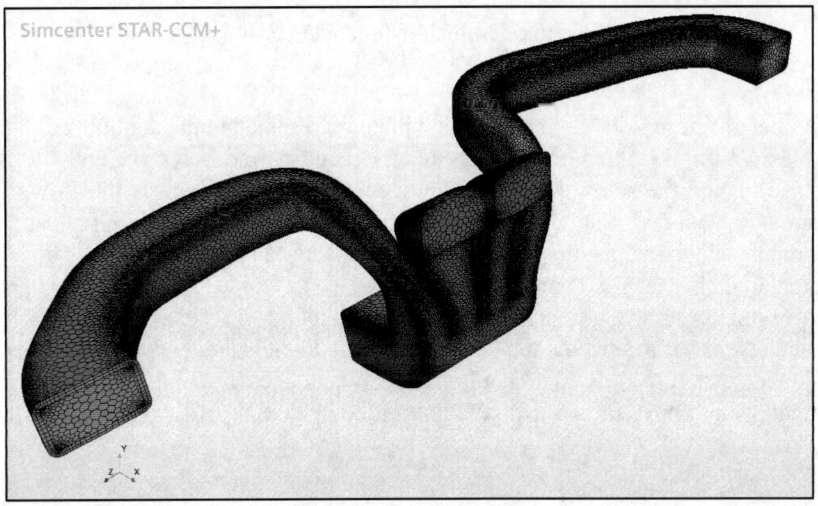

Figure 4. Initial mesh

2.2.1. Mesh independence study. In order to determine whether the formed mesh is fine enough, or too fine, the independence of the mesh was examined. Figure 5 presents a graph of the mesh independence study that was performed for the specific problem of pressure drop in the air distributor. The study was performed for fourteen different mesh settings with the coarsest setting forming a mesh of 73788 cells and the finest setting forming a mesh of 1222716. By observing the results, it can be noticed that the results do not deviate to a large extent, and it can be concluded that the problem being analyzed is relatively simple, however, the results are more stable when the mesh consists of 500,000 or more cells, and in that case the value of pressure drop varies within the range of 1Pa.

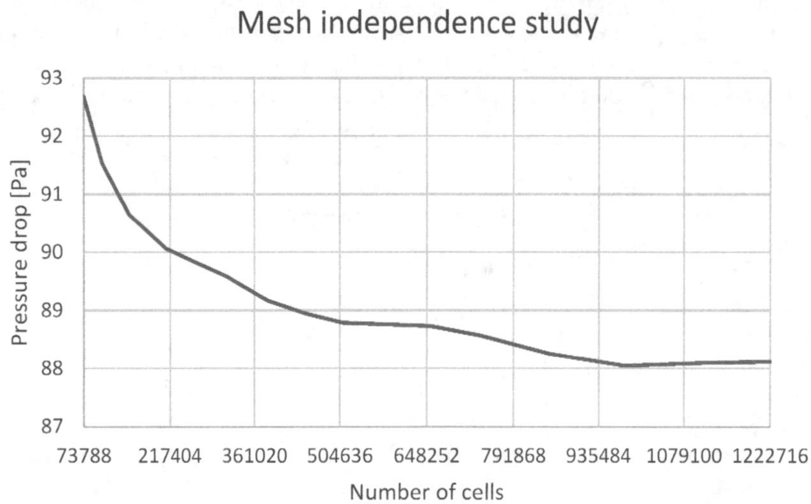

Figure 5. Mesh independence study

2.3. Problem solving

Physical models determine how some phenomena will be represented in the continuum. They define the primary variables within the simulation (pressure, temperature and velocity) and the mathematical formulations used to find the solutions. Since the problem being analyzed is the flow of air through pipes, it is a three-dimensional problem, air is a gas that moves at a relatively low speed, so it can be considered incompressible and its flow is steady. Given that it is a turbulent air flow regime, two different turbulent models were used during the simulations: Realizable K-Epsilon model and K-Omega. The model that in this case provides a more stable solution and faster convergence is Realizable K-Epsilon, so in the rest of the paper only the simulation in which it is active will be shown. By selecting the aforementioned models, STAR-CCM+ automatically activates Two-Layer All y+ Wall treatment and Wall distance, which have the task of calculating the behaviour of the fluid along the distributor walls.

2.4. Evaluation of convergence and analysis of results before optimization

The value of the residuals and the decrease in their value can vary depending on the simulation. For one problem a drop in the change of the residual values below the third decimal place may be acceptable, and for another it may be an insufficient drop. For this reason, it is always recommended that along with the residuals, a physical or engineering quantity of interest, such as force, pressure change, mass flow, etc., is also monitored. In this paper, residuals (Figure 6 a) and pressure drop (Figure 6 b) are monitored at the same time. Based on the images, it can be noticed that already after 285 iterations, the solution changes only at the third decimal point, which in this case is quite sufficient precision, bearing in mind that the observed quantity is the pressure drop, which is expressed in Pa, and that the pascal is a unit that is relatively small. It takes between 600 and 650 iterations, i.e. 31 minutes, for residuals to converge.

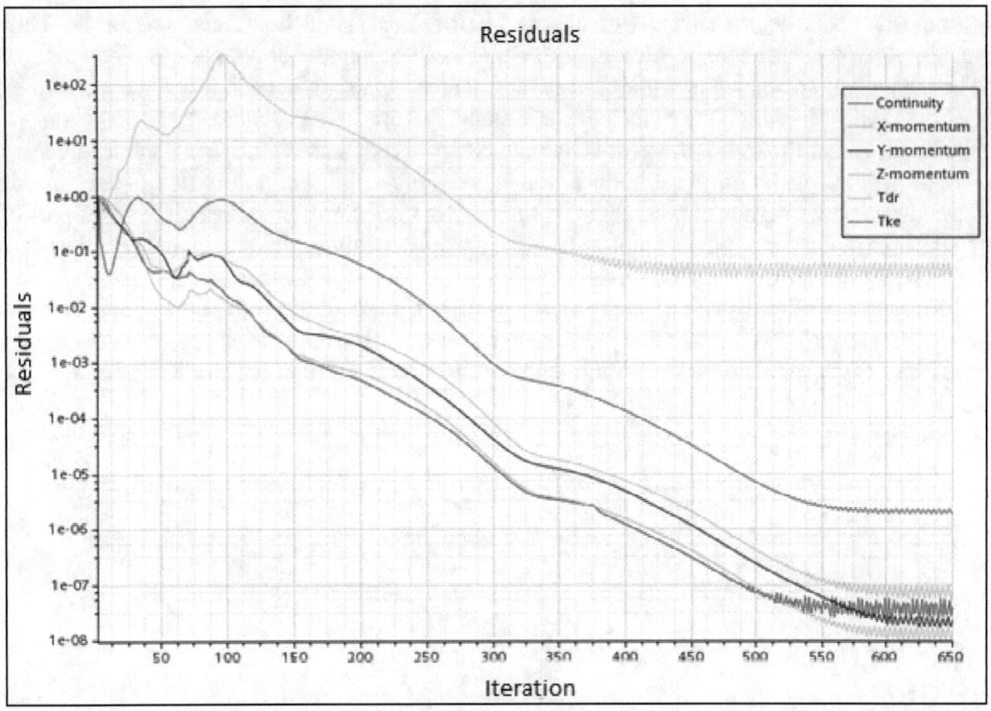

a)

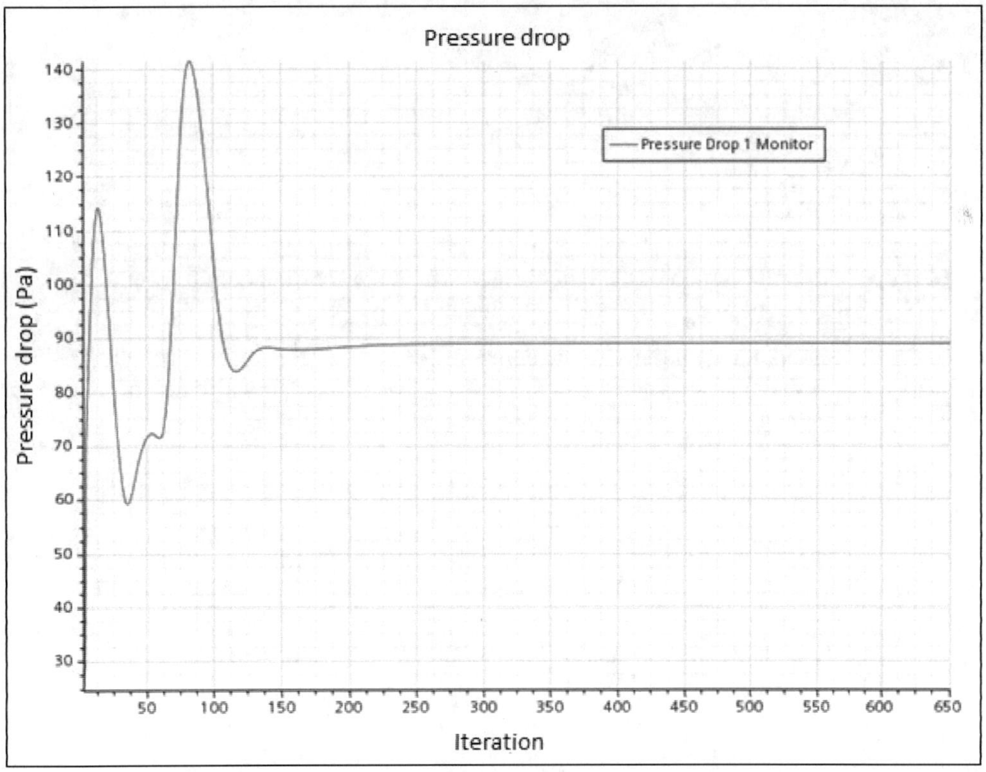

b)

Figure 6. Display of residuals (a) and pressure drop (b)

The first important parameter that must be checked before analyzing the others is Wall y+. This is a parameter that is a key indicator of the quality of the mesh along the domain wall. The value of the distance of the centroid of the cells to the wall is represented by a scalar scene (Figure 7) in which only the area of the domain in which the values of the distance y+ are in the range 1 < y+ < 30 is displayed. During the formation of the mesh, the goal is that the centroids of the boundary cells are located as little as possible in the area of the transient boundary sublayer. The larger the part of the domain that is not in the defined range, the higher the precision of the results. In Figure 7, it can be seen that part of the domain in the area of the central lines belongs to an inappropriate range, and that in that area the accuracy of the results may be reduced. In the area of the central lines, the fluid flow rate is relatively low compared to the rest of the distributor due to the closure of the central ventilation openings. A lower flow rate significantly affects the fact that the current losses are also small, so any deviation of the results in this area can be ignored. Based on the above, it can be concluded that the mesh is good enough for the results of the other sizes to be considered accurate.

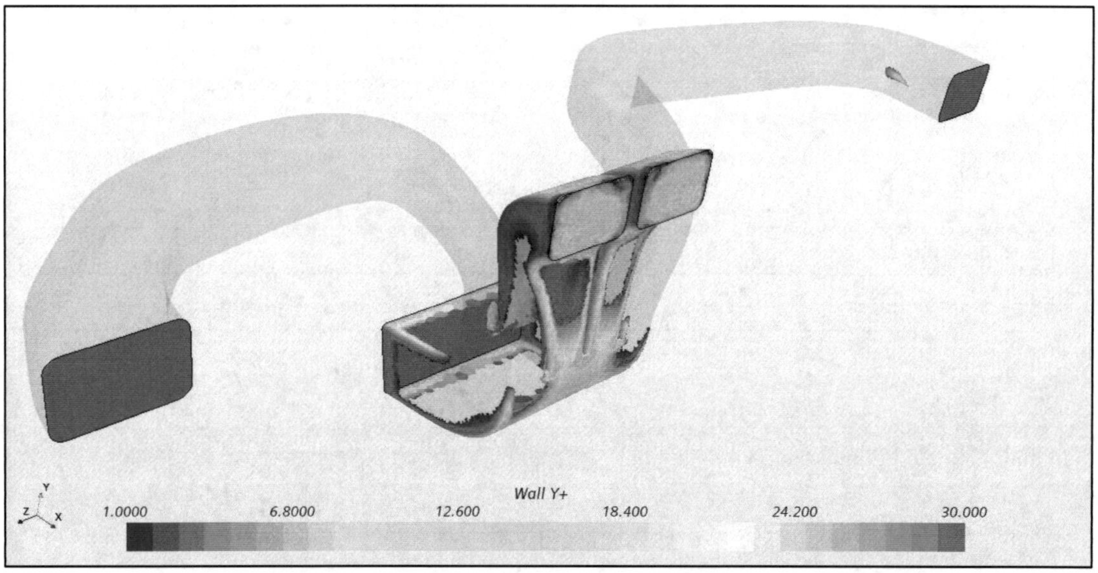

Figure 7. Representation of the domain area where the centroids of the boundary cells are located in the transition boundary sublayer (STAR-CCM+)

As the boundary condition at the entrance to the domain is defined by the air flow velocity (6.435 m/s), it is useful to observe the field of velocity intensity over the computational domain, as well as the velocity of fluid discharge at the exits from the domain, averaged over the respective boundary surfaces.

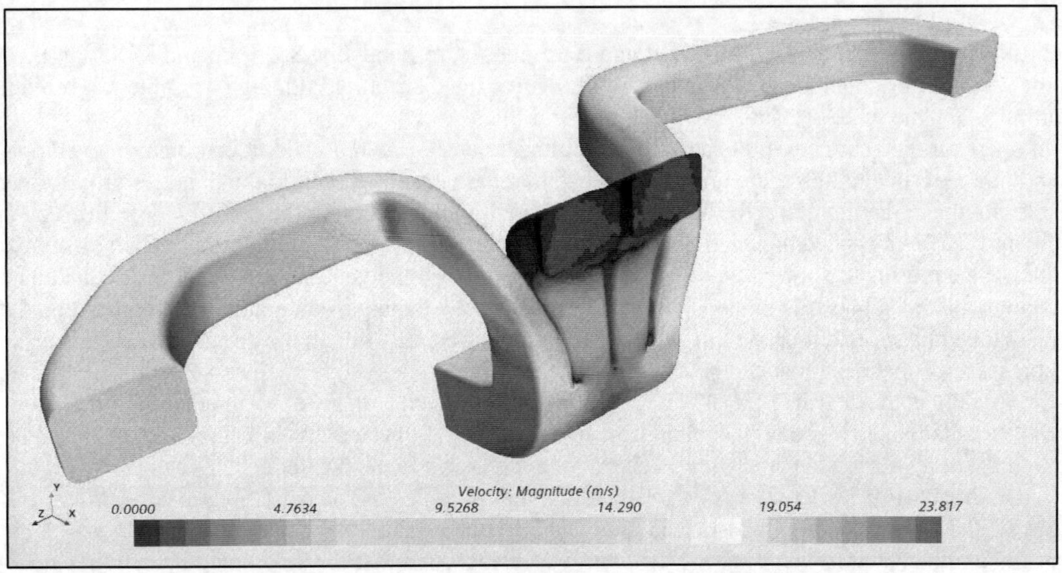

Figure 8. Air flow rate through the distributor (STAR-CCM+)

In Figure 8, it can be seen that the maximum velocity of the fluid (23.8 m/s) occurs at the knee points and on the inner part of the curve. The speed of the air flow at the outlet from the ducts is: left duct - 9.3 m/s, right duct - 9.7 m/s.

Another important quantity to monitor is the distributor pressure. Based on the pressure difference at the inlet to the distributor and the outlet from the corresponding line, data on the pressure drop for a specific case can be obtained. When defining the boundary conditions, the static pressure that prevails in the cabin is defined at the entrance as 102375 Pa, however, the total pressure is the one that is valid when determining the pressure drop. The total average pressure at the entrance is 102521 Pa, and at the exits from the lines it is: left - 102432 Pa, right - 102433 Pa. In Figure 9, it can be observed that the lowest pressure occurs in the bifurcation zone as a result of flow separation, as well as that the pressure at the central openings is higher than at the side openings, which is a consequence of the central ventilation openings being closed.

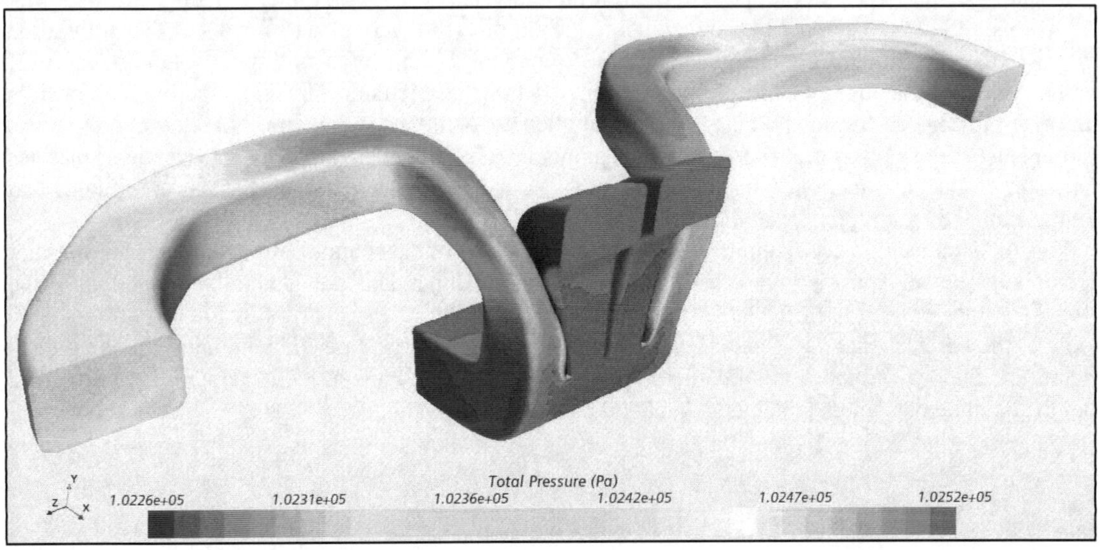

Figure 9. Pressure drop through air distributor (STAR-CCM+)

The value of the pressure drop for the left and right lines is, respectively, 88.79 Pa and 87.89 Pa. Based on the results, it is concluded that the current design of the distributor, although asymmetric, provides approximately the same pressure drop for the lateral lines.

In order for the obtained results and model settings to be considered valid, it is necessary to compare them with experimental pressure drop data. Since there is very little data on the pressure drop through the distributor in the literature, the results are compared with the data on the pressure drop through the entire air distribution system, which, for the same vehicle, was presented in the paper of Jan Pokorny et al. [4]. The experiment showed that the pressure drop through the ventilation system is 368.5 Pa at an air flow of 97.5 l/s [4]. It is evident that it is not possible to compare the value of the pressure drop through the entire distribution system, with the pressure drop only through the distributor, for this reason the pressure drop in the rest of the system is approximated, and then the pressure drop before the distributor and in the distributor itself are added up and compared with the experimental data. It is known that in the main air flow, when the cabin heating is turned off, there is a cabin filter and an evaporator that represent resistance to the air flow. The pressure drop on the cabin filter can be read from the graph shown in the paper [7] where it was established that for combined traffic conditions after approximately one year, the filter provides a resistance of 55 Pa [7]. The pressure drop on the evaporator can be read from the graph shown in the paper [8], where it was established that for a volumetric air flow value of 5.83 m^3/min (97.5 l/s), a pressure drop of 6.5 mmH_2O occurs, i.e. 65 Pa [8]. When those values are added to the mean pressure drop obtained by simulation (88 Pa), a pressure drop of 208 Pa is reached. Based on the value of the pressure drop through the distributor, it can be assumed that the difference in the pressure drop, between the experimental data from the literature and previously obtained by simulation, occurs due to the resistance that exists between the place of air inlet and the cabin filter, fan and evaporator, that is, the evaporator and distributor. As the mentioned resistances data are very difficult to find in the existing literature, based on the exposed experimental data, the simulation results are considered acceptable.

2.5. *Adjoint geometry optimization*

Adjoint analysis is an effective way to predict the influence of structural characteristics and input physical parameters on some engineering quantity of interest for a specific problem. In other words, the adjoint method calculates the sensitivity of some output quantity (eg pressure drop) in relation to some constructive properties (eg domain geometry). This method is applied to determine how much influence distributor geometry has on pressure drop. Adjoint analysis is carried out in two main steps. The first step is to simulate the established problem using the coupled flow model. The results of that simulation are called the primal solution. The second step is to set up the adjoint solver and start the adjoint analysis, which builds on the already formed initial solution. After the calculation using the adjoint solver, it is necessary to choose the method that will be applied to optimize the geometry. The geometry was optimized with use of control points. Control points represent elements that user arbitrarily generates using the Point Set function. Once created, set of points is connected to a physical continuum that contains an active mesh deformation model.

Figure 10. presents a set of points created for the purposes of distributor optimization. The points are positioned centrally through the cross-sections of the pipelines and are generally divided into two groups, in the left and right lines. As the goal is optimization of the distributor for use on the same vehicle, the surfaces that will retain their initial geometric properties are the ventilation openings, i.e. the outlet boundary surfaces and the inlet boundary surface. The walls of the distributor are defined so that they can be deformed.

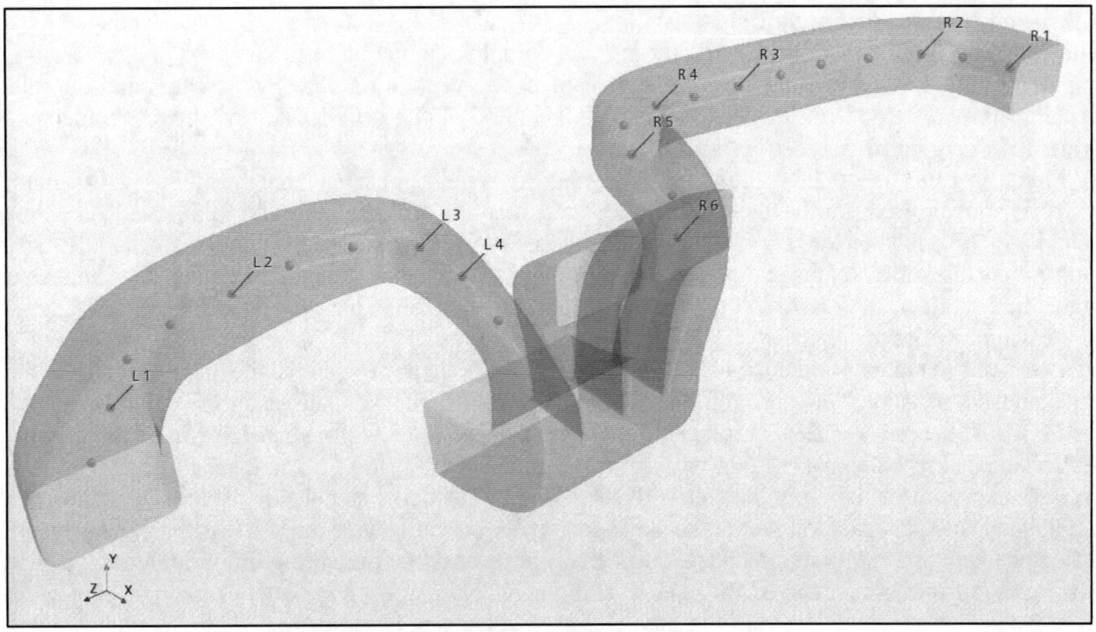

Figure 10. Set of control points (STAR-CCM+)

The next step is to calculate the sensitivity of the mesh. The calculation of the sensitivity is based on the previously realized adjoint solution. Mesh sensitivity represents the connection between the sensitivity of the cost function and mesh deformation. The results of the calculation can be presented in the form of a vector displayer on the geometry scene, where for each control point a vector is displayed Displayer shows what kind of deformation would lead to a reduction in pressure drop (Figure 11).

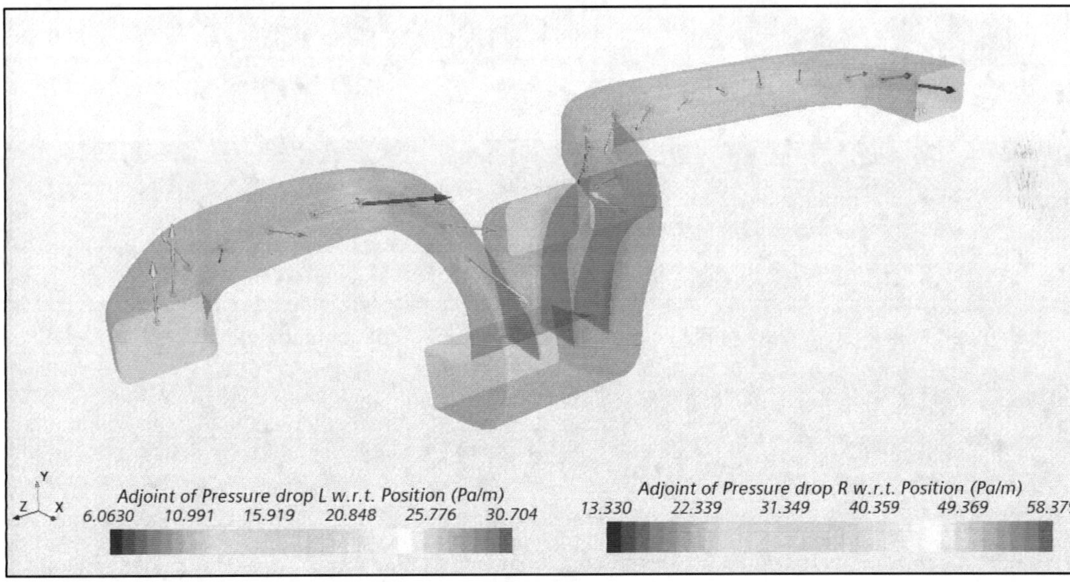

Figure 11. Display of mesh sensitivity for the left and right lines of the distributor (STAR-CCM+)

By applying the mesh deformation solver, the control points are moved and, together with them, the distributor geometry is modified. The time necessary to perform one optimization cycle is approximately 80 min, and the total optimization time, including the calculation of the initial solution, amounts to less than 12 hours of active processor work.

3. Optimization results and discussion

The optimized geometry of the distributor is more complex compared to the original one, where it is important to emphasize that the position and shape of the ventilation openings in cabin and the inlet opening have remained unchanged. A comparative view of the original and modified geometry is presented in Figure 12.

From Figure 13, where the air flow velocity is shown, it can be observed that the modified distributor provides a lower maximum velocity value (21.7 m/s) than the original distributor (23.8 m/s). The flow velocity in the region of the inner part of the bend is also lower, and the velocity field is generally more uniform. If the outlet velocities are compared, it can be concluded that the modified distributor has higher outlet velocities (9.8 m/s - left; 10.7 m/s - right) due to a lower pressure drop, which was the goal of the geometry optimization.

The quantity in relation to which the optimization was performed is the pressure inside the distributor. That is the pressure drop that occurs as a result of losses caused by the geometry of the distributor. If the pressure field is compared to the field of the original distributor, it will be noticed that the modified distributor has a more uniform distribution of pressure, i.e. there are no sharp pressure changes except for the zone in the area of the burification, which was not the subject of optimization within this paper. In order to monitor the pressure drop, two reports were created that show the pressure difference between the inlet and outlet. Through optimization, the pressure drop through the left pipe was reduced from 88.79 Pa to 62.89 Pa (29%), and between the inlet and outlet of the right pipe from 87.89 Pa to 73 Pa (17%). Figure 15 presents a graph of the mean pressure drop (for both outlets) where are step changes in pressure due to each modification of the distributor's geometry.

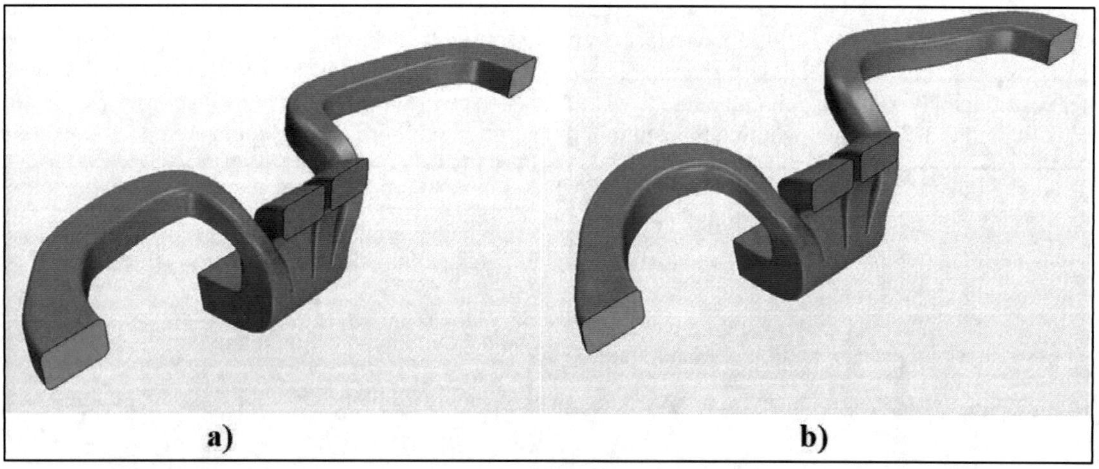

Figure 12. Isometric view of the original (a) and modified (b) geometry (STAR-CCM+)

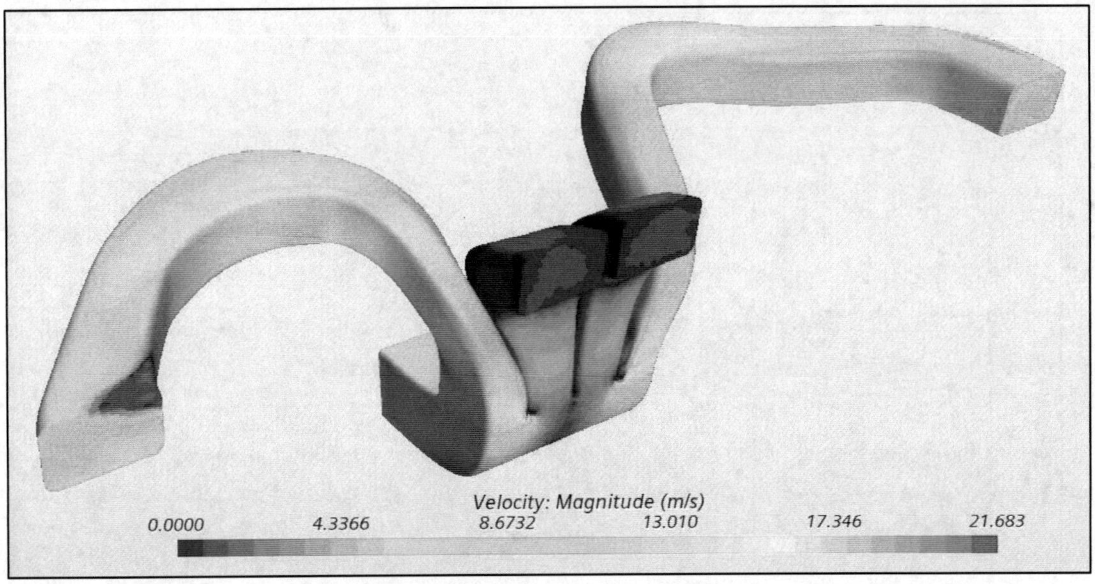

Figure 13. Speed of air flow through the modified distributor (STAR-CCM+)

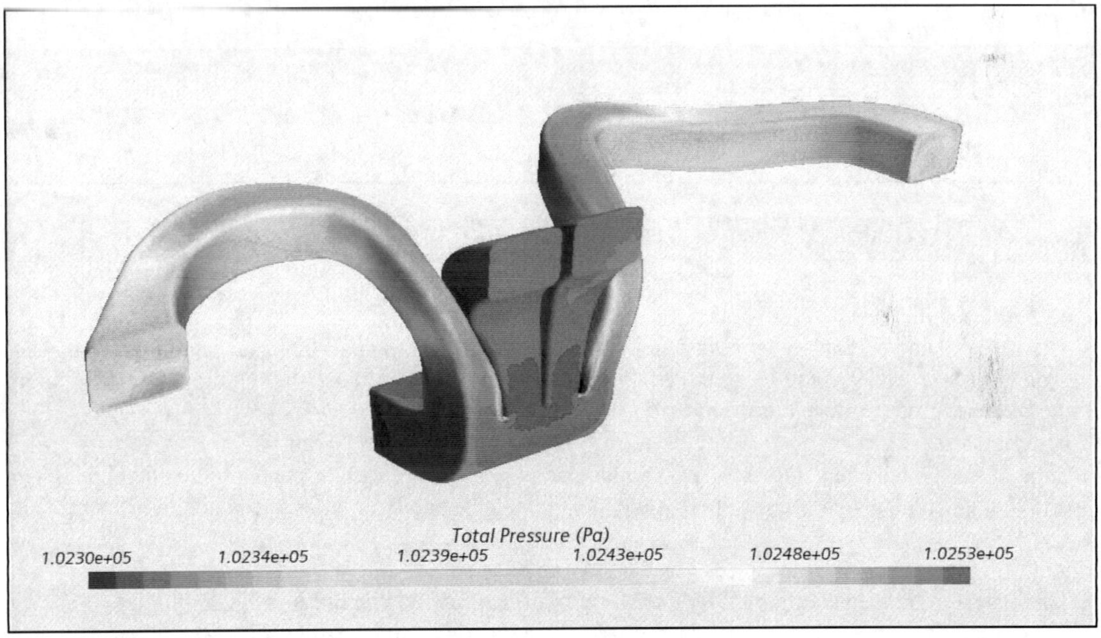

Figure 14. Pressure distribution inside the modified air manifold (STAR-CCM+)

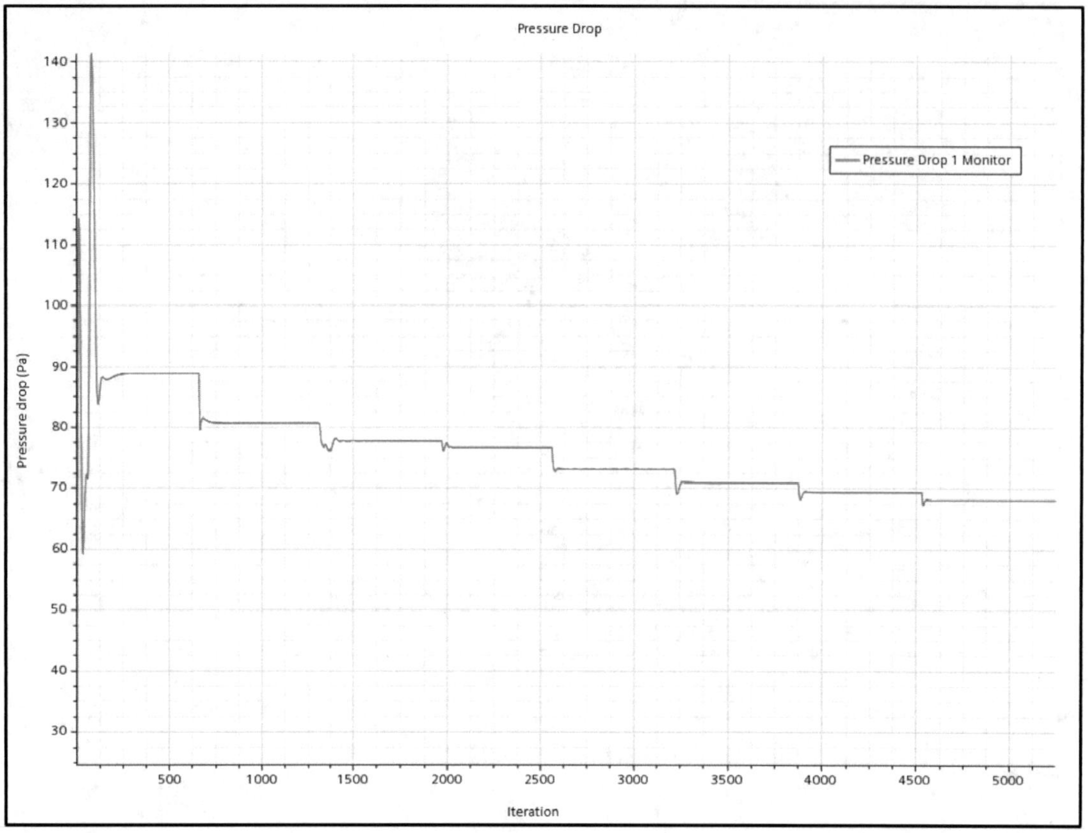

Figure 15. Display of the step change of pressure depending on modification of the distributor geometry (STAR-CCM+)

4. Conclusion

The results showed that by applying adjoint optimization in the specific case, a reduction in pressure drop from 17% to 29% can be achieved. Although in the analyzed case the mentioned improvement of the geometry of the distributor have a relatively small effect on the efficiency of the air distribution system, if the described procedure were applied to the rest of the ventilation system, a significant improvement could be achieved. Also, considering the large number of mass-produced vehicles whose efficiency can be increased, the energy savings that can be achieved in this way are significant.

Acknowlegments

This research has been supported by the Ministry of Education, Science and Technological Development through project no. 451-03-9/2021-14/ 200156: "Innovative scientific and artistic research from the FTS activity domain".

References

[1] Ležovič T, Lizal F, Jedelsky J and Jicha M 2013 HVAC automotive vents evaluation and their performance, *HVAC&R Research* **19** 1073-1082

[2] Fišer J and Pokorný J 2014 Effect of car speed on amount of air supplied by ventilation system to the space of car cabin, *EPJ Web of Conferences* **67** 02027

[3] Fojtlin M, Planka M, Fišer J, Pokorný J and Jícha M 2016 Airflow Measurement of the Car HVAC Unit Using Hot-wire Anemometry, *EPJ Web of Conferences* **114** 02023

[4] Pokorný J, Poláče F, Fojtlin M, Fišer J and Jícha M 2016 Measurement of airflow and pressure characteristics of a fan built in a car ventilation system, *EPJ Web of Conferences* **114** 02097

[5] More references Maksimov A, Igoshina D, Petrov R, Klyavin O 2020 Optimization of the electric vehicle HVAC duct system based on gradient method, *International Journal of Mechanics* **14** 135-140

[6] ŠKODA Workshop Manual: Heating, Air conditioning, Edition 07.2013.

[7] Xu B, Liu S, Liu J and Zhu Y 2011 Effects of Vehicle Cabin Filter Efficiency on Ultrafine Partcle Concetration Rations Measured in-Cabin and On-Roadway, *Aerosol Science and Technology* **45** 234-243

[8] Hyun L K and Jong-Pil W 1999 Thermal Design Study of A High Performance Evaporator for the Automotive Air Conditioner, *SAE Transactions* **108** 2216–2223

IOP Conf. Series: Materials Science and Engineering 1271 (2022) 012023 doi:10.1088/1757-899X/1271/1/012023

Coordinated control of ESC with active front steering and active suspension normal force control for better vehicle's dynamic response

V Jordanoska[1], V Changoski[2], D Danev[3]

[1] Assistant Professor, UKIM Faculty of Mechanical Engineering – Skopje, MK
[2] Teaching Assistant, UKIM Faculty of Mechanical Engineering – Skopje, MK
[3] Full-Time Professor, UKIM Faculty of Mechanical Engineering – Skopje, MK

E-mail: vase.jordanoska@mf.edu.mk

Abstract. Advances in vehicle technologies brought changes into architectures of integration of chassis control systems. Alongside other improvements, integration can provide better vehicle handling, stability and safety. Different chassis control systems can use different control methods. Considering while in motion vehicle has movements in longitudinal, lateral and vertical direction, different control systems target different motion. If there is no coordination among active control systems, interaction and performance conflict can arise. This paper presents coordination of three control systems: electronic stability control (ESC), active front steering control (AFSC) and active suspension normal force control (ASNF). ESC is actually direct yaw and anti-roll control using selective wheel braking. ASNF is considered only on front axle. All systems use fuzzy-logic as a control method. Vehicle is presented as 14-DOF nonlinear full vehicle model and for the control purposes 3-DOF reference model was introduced. Emphasis was given on vehicle parameters characterizing dynamic behavior in longitudinal and lateral direction. Benefits of coordinated action of the three systems can be seen from the results gained through simulation of cornering event and single lane change in Matlab/Simulink. Coordinated control adds to the action of ESC itself, resulting in improved stability and handling, and thus safety of the vehicle.

1. Introduction

Every main vehicle's system nowadays is updated with automated electronic system. The purpose of these advancements is to facilitate driver's operation and lower his influence during the driving process. It is due to improve safety, protect environment and to increase cost-effectiveness. Current focus is on making vehicles more "open" for communication with other vehicles and infrastructure. Part of this focus is closely related to vehicle safety systems, which affect vehicles dynamic behavior.

It is considered that active vehicle systems began with the emergence of Anti-lock Braking System (ABS) in 1950s. First it has been available on luxurious vehicle's rear wheels in late 1960s, but demand increased dramatically in the early 1990s when the benefits of vehicle dynamic control were recognized and widely accepted. In 1987, Bosch combined the Traction Control System (TCS) function with the ABS system. The further development of active vehicle's systems leads to electronic dynamic stability control system in 1995. Depending on developer, this kind of system has different names: Vehicle Dynamics Control (VDC), Vehicle Stability Control (VSC), Electronic Stability Control (ESC) or

Electronic Stability Program (ESP), Vehicle Stability Assist (VSA), Dynamic Stability Control (DSC). In this paper, the abbreviation ESC will be used. In today's vehicles ESC control usually consists of two modes of operation: direct yaw and anti-roll control. The former helps the driver to maintain the desired direction of the vehicle, even when the vehicle is at the edge of adhesion. The second reduces the possibility of the vehicle rollover. Sports utility vehicles (SUVs) and other vehicles with higher center of gravity have a higher tendency to roll over. So, a rollover stability control system is particularly useful. The electronic stability control system helps the vehicle maintain stability and thus safety, by individual braking of wheels [1, 2, 3, 4, 5].

The function of active control in a vehicle is done through engine control or through control of main systems (steering, braking, suspension), but also through additional electronic systems. Control logic used in these systems can be logic threshold, PID, sliding mode control, fuzzy-logic, MPC, LQR or neural networks. In order to improve overall vehicle performance, the method of integrated or coordinated cooperative control is introduced. Numerous research and papers are produced in the area, which offer different solutions for the cooperative function of the yaw moment control, through coordinated operation of differential braking and other systems i.e. active steering or suspension. Scientific papers offer different approach and control method to achieve coordinated cooperative operation of systems to get a better vehicle response and safety. Detailed explanation of the state-of-art can be found in [6].

Paper [7] gives the efficiency of the ESC system when fuzzy-logic is used as a control method. It also uses β-$\dot{\beta}$ phase plane method, to avoid ESC to control the vehicle frequently. Results show lower variation range of the yaw rate and sideslip angle and the needed time to reach steady state are smaller. The control manages to reduce the vehicle oversteer trend and brings it closer to neutral steering. Integrated control system of active front steering and normal force control using fuzzy-logic to enhance the vehicle-handling performance was developed by [8]. It is proven by simulation results that the integrated chassis control scheme, utilizing the steering and suspension controllers, is more effective in attaining the desired performance that would not be attained by systems acting individually. [9] using fuzzy-logic control coordinates systems given in [7] and [8], i.e. ESC, active steering and active suspension control. It evaluates how much further improvement can be done in the dynamic behavior control of the vehicle during sudden maneuvers, compared to coordinated control of just pairs of those systems.

2. Vehicle modelling

Through the development vehicle dynamic models go from the traditional lumped parameter model with 2 DOF (degrees of freedom) to 27 DOF vehicle model in the software CarSim [10]. For the purpose of this research 14-DOF full vehicle model was introduced (Figure 1), used by [11, 12]. The model is validated using CarSim and ADAMS/Car [13]. The 14-DOF nonlinear model was implemented in Matlab/Simulink to describe the dynamic behavior of the vehicle. Vehicle mass, which represents sprung mass, is considered a lumped mass at the center of gravity and has six degrees of freedom, along and around X, Y and Z axis. The model has another four lumped masses of wheels i.e. unsprung mass, each one having two degrees of freedom: vertical travel and wheel rotation. Assumptions made for this model are: vehicle body is being modelled as rigid, steer angles of the outer and inner wheel are same, tires have contact with the ground all the time and the aerodynamic force is neglected. Yet the 14-DOF vehicle model is sufficient to study vehicle dynamics in longitudinal, lateral, and vertical direction.

The sprung mass vertical motion is described by equation1, pitch motion by equation 2 and roll motion by equation 3:

$$m_{pm} \cdot \ddot{z}_{pm} = F_{pr\_pl} + F_{am\_pl} + F_{pr\_pd} + F_{am\_pd} + F_{pr\_zl} + F_{am\_zl} + F_{pr\_zd} + F_{am\_zd} \tag{1}$$

$$I_y \cdot \dot{\omega}_y = \left(F_{pr\_zl} + F_{am\_zl} + F_{pr\_zd} + F_{am\_zd}\right) \cdot l_z - \left(F_{pr\_pl} + F_{am\_pl} + F_{pr\_pd} + F_{am\_pd}\right) \cdot l_p \tag{2}$$

$$I_x \cdot \dot{\omega}_x = \left(F_{pr\_pl} + F_{am\_pl} + F_{pr\_zl} + F_{am\_zl}\right) \cdot \frac{b}{2} - \left(F_{pr\_pd} + F_{am\_pd} + F_{pr\_zd} + F_{am\_zd}\right) \cdot \frac{b}{2} \tag{3}$$

Each wheel vertical motion is described by equation 4:

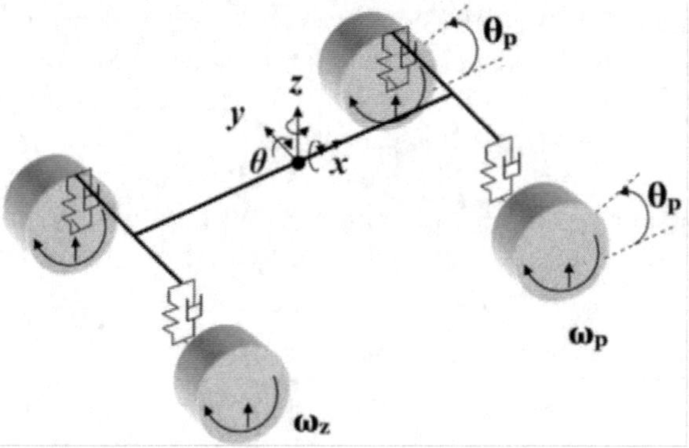

Figure 1. Non-linear full vehicle model with 14 DOF.

$$m_{nm\_ij} \cdot \ddot{z}_{nm\_ij} = F_{t\_ij} - F_{pr\_ij} - F_{am\_ij} \tag{4}$$

where index *ij* stands for the wheel position (i - front, rear; j - left, right).

The handling model has longitudinal, lateral and yaw motion of the sprung mass. The equation of longitudinal motion is given by equation 5, lateral motion by equation 6 and yaw motion by 7:

$$m \cdot a_x = F_{xpl} \cos\theta_p - F_{ypl} \sin\theta_p + F_{xpd} \cos\theta_p - F_{ypd} \sin\theta_p + F_{xzl} + F_{xzd} \tag{5}$$

$$m \cdot a_y = F_{ypl} \cos\theta_p + F_{xpl} \sin\theta_p + F_{ypd} \cos\theta_p + F_{xpd} \sin\theta_p + F_{yzl} + F_{yzd} \tag{6}$$

$$I_z \cdot \dot{\omega}_z = -\frac{b}{2} F_{xpl} \cos\theta_p + \frac{b}{2} F_{xpd} \cos\theta_p - \frac{b}{2} F_{xzl} + \frac{b}{2} F_{xzd} + \frac{b}{2} F_{ypl} \sin\theta_p - \frac{b}{2} F_{ypd} \sin\theta_p + l_f F_{xpl} \sin\theta_p + l_f F_{ypl} \cos\theta_p + l_f F_{xpd} \sin\theta_p + l_f F_{ypd} \cos\theta_p - l_z F_{yzl} - l_z F_{yzd} + M_{zpl} + M_{zpd} + M_{zzl} + M_{zzd} \tag{7}$$

The equation of motion for each wheel rotation is:

$$I_t \cdot \dot{\omega}_{ij} = T_{v_{ij}} - T_{k_{ij}} - F_{xij} \cdot r_t \tag{8}$$

Vehicle parameters used in the model are given in Table 1 and correspond to vehicle body such as sport utility vehicles or crossover vehicles.

Table 1. Vehicle parameters.

Parameter	Value
Sprung mass m_{pm} (kg)	1440
Front track b_p (m)	1.5
Rear track b_z (m)	1.5
Haight of CM h (m)	0.75
Total vehicle mass m (kg)	1600
CM distance to front axle l_p (m)	1.016
CM distance to rear axle l_z (m)	1.524
Pitch moment of inertia I_y (kgm^2)	2000
Roll moment of inertia I_x (kgm^2)	900
Yaw moment of inertia I_z (kgm^2)	2000
Unsprung mass per wheel m_{nm} (kg)	40
Rolling radius r_t (m)	0.334
Wheel rotational inertia I_t (kgm^2)	3
Front spring stiffness c_p (N/m)	30000
Rear spring stiffness c_z `(N/m)	20000
Front axle damping coefficient k_p (Ns/m)	1897
Rear axle damping coefficient k_z (Ns/m)	1265

Tire stiffness c_t (N/m)	200000
Cornering tire stiffness K_δ (N/rad)	60000

2.1. Tire model
To present the non-linear behavior of tires, Pacejka's magic formula was also implemented in Matlab/Simulink, according to [14].

2.2. Reference model
3 DOF model was used as a control reference, to obtain the desired values upon which the control determines the error. Reference model includes lateral acceleration, yaw velocity and roll angle and gives values for desired yaw velocity (ω_{Zpos}), side-slip angle (β_{Cpos}) and roll angle (ϕ_{pos}) according to the inputs [15]. Inputs in the model are the steering angle and the vehicle speed. Equations describing the reference model are:

$$\omega_{z\_pos} = \frac{v_x}{l + \left(\frac{m \cdot (l_z \cdot K_{\delta z} - l_p \cdot K_{\delta p})}{2 \cdot K_{\delta p} \cdot K_{\delta z} \cdot l}\right) \cdot v_x^2} \cdot \theta_p \tag{9}$$

$$\beta_{c\_pos} = \frac{l_z - \left(\frac{l_p \cdot m}{2 \cdot K_{\delta z} \cdot l}\right) \cdot v_x^2}{l + \left(\frac{m \cdot (l_z \cdot K_{\delta z} - l_p \cdot K_{\delta p})}{2 \cdot K_{\delta p} \cdot K_{\delta z} \cdot l}\right) \cdot v_x^2} \cdot \theta_p \tag{10}$$

$$\phi_{pos} = 0 \tag{11}$$

It should be noted that due to limitations of the simplicity of the reference model, maximum values of yaw velocity and side-slip angle are limited by the adhesion coefficient.

3. Control model
Electronic Stability Control (ESC) is actually direct yaw and anti-roll control using selective wheel braking. Therefore, control actions of ESC are performed by calculation of yaw moment M_{yaw} and roll moment M_ϕ (equation 12, 13), where M_ϕ uses feedforward and feedback components. Brake distribution is considered to be on the wheels of the same side of the vehicle. Based on this adopted principle, the direction of the compensation moment can be determined through the steering wheel angle and the difference between the actual and the reference yaw velocity [7].

$$M_{yaw} = F_{kpi} \cdot \frac{b_p}{2} + F_{kzi} \cdot \frac{b_z}{2} \tag{12}$$

$$M_\phi = M_{\phi ff} + M_{\phi fb} \tag{13}$$

To prevent frequent activation of the electronic stability control system, β-$\dot{\beta}$ phase plane method is used, explained in [16, 17]. Fuzzy-logic method is used as control strategy for all outputs of the ESP and control of active front steering (AFS) and active suspension normal force (ASNF). All membership functions and linguistic rules can be found in [6].

In the next step, after ESC control was defined, ASNF control was added. Studies by [8], as well as [18] have shown the benefits of adapting the normal force through active suspension system control during turning maneuvers. It has been observed that the application of active force on the front angle, opposite to the turn direction, gives the best results, in terms of the lateral force impact during cornering. It also gives the lowest roll movement, according to the combinations examined. So, the force generated by the active and passive components can be determined by equation 14.

$$F_{sus} = c(z_{nm} - z_{pm}) + F_a + k(\dot{z}_{nm} - \dot{z}_{pm}) \tag{14}$$

Lastly, active front steering system control was modeled, where the actual turning angle θ_p is a difference between the input angle of the steering wheel θ_{ut} and the corrective angle generated by the fuzzy-controller θ_{corr}. The steering actions are performed by calculations of the correct turning angle of the front wheels according to equation 15.

$$\theta_{corr} = \theta_{ut} - \theta_p \tag{15}$$

Coordination of the ESC with AFS and ASNF control was done using scaling coefficients, so every system contributes with a certain share in dynamic behavior corrective actions. The coefficients have been obtained by tryouts, so that the control requires lower forces that should be provided by active components and at the same time it effectively stabilizes the vehicle. Schematic illustration of the coordinated control of systems is given on Figure 2.

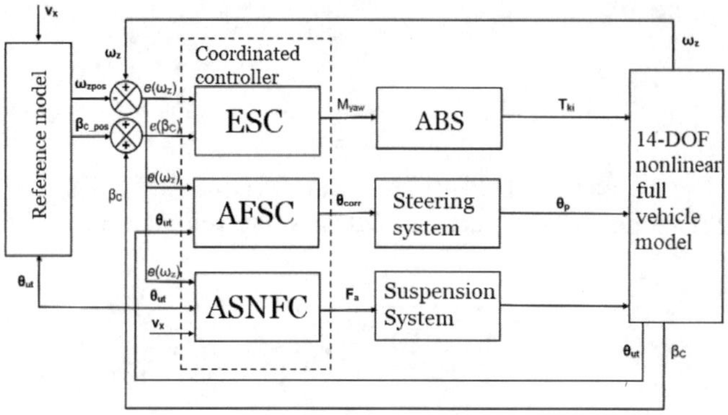

Figure 2. Structure of active control.

4. Simulation and results

According to ISO 7401 there are several open-loop test methods for determining the transient response behavior of road vehicles, in time and in frequency domain. Test maneuvers in time domain are cornering event (step input) and single lane change (one period of sinusoidal input) are used in this paper. Both maneuvers are performed with initial vehicle speed of 130 (km/h). The speed is chosen according to the traffic regulations in the majority of European countries. It is a maximum speed at which the road infrastructure allows safe driving.

The step input in the first maneuver is 90° on the steering wheel which corresponds to wheels turning of 7.5°, during the time of 0.1 (s) and after 1 (s) of moving straightforward. One period of sinusoidal input with 90° amplitude (7.5° on wheels) is used for second maneuver, during the period of 2 (s) and after 1 (s) of moving straightforward. Road is assumed to be dry, with adhesion coefficient of 0.9. Results that characterize vehicle transient response are given below. The legend of control combinations is presented on figure 3. Results on figure 4 to 7 in the left column are related to cornering event (a) and in the right to the single lane change (b).

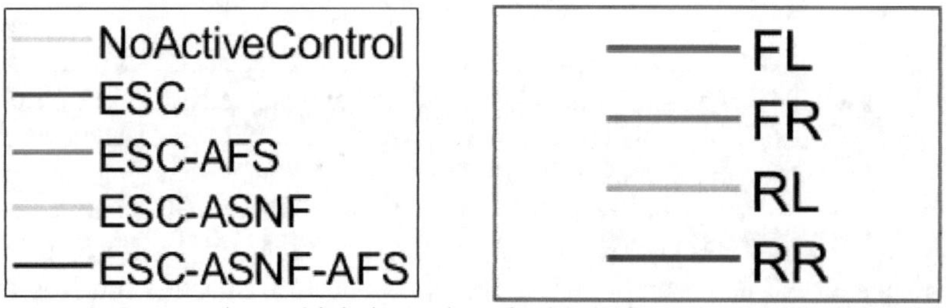

Figure 3. System that vehicle is equipped with (left) and wheel position (right).

IOP Conf. Series: Materials Science and Engineering 1271 (2022) 012023 doi:10.1088/1757-899X/1271/1/012023

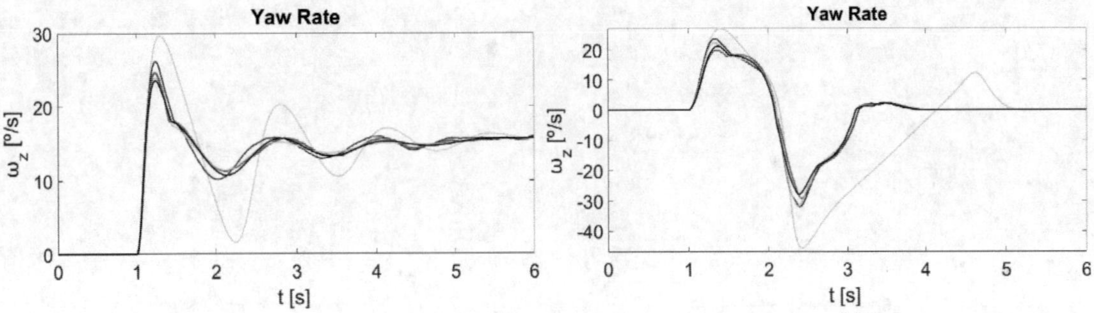

a) b)

Figure 4. Yaw velocity.

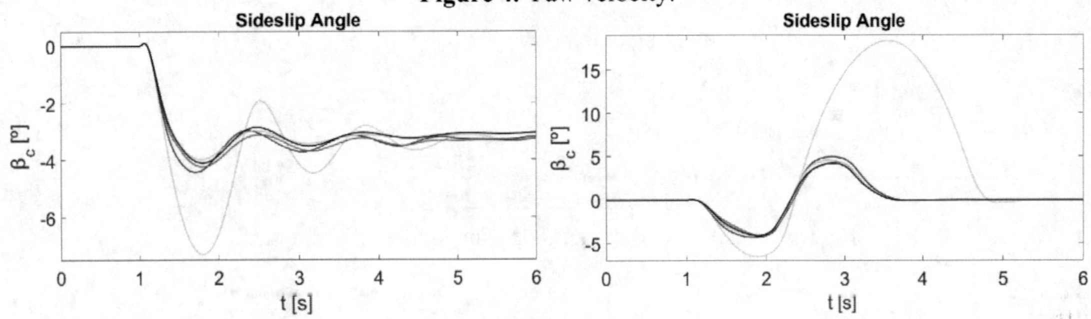

a) b)

Figure 5. Side-slip angle.

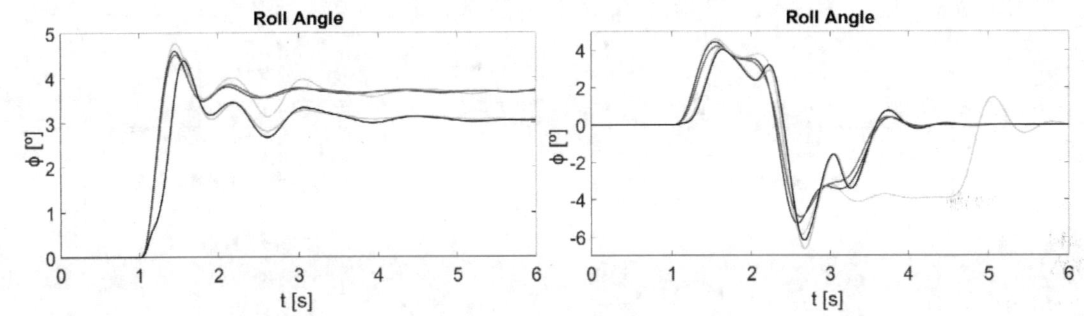

a) b)

Figure 6. Roll angle.

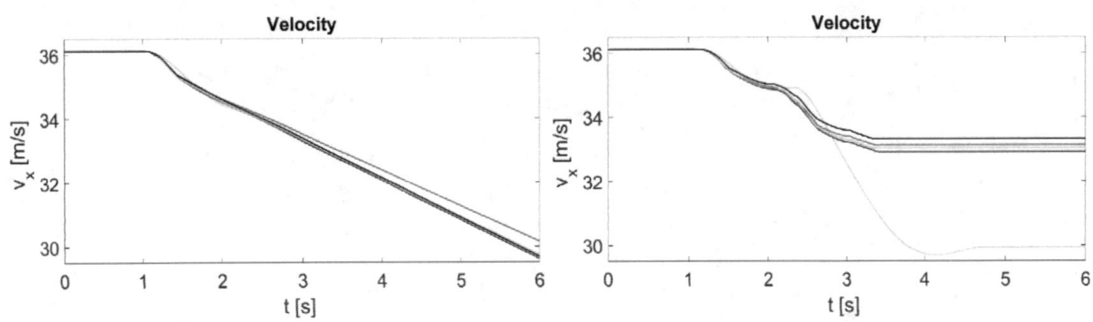

a) b)

Figure 7. Speed change.

Next results are wheel related parameters, presented only for single lane change maneuver.

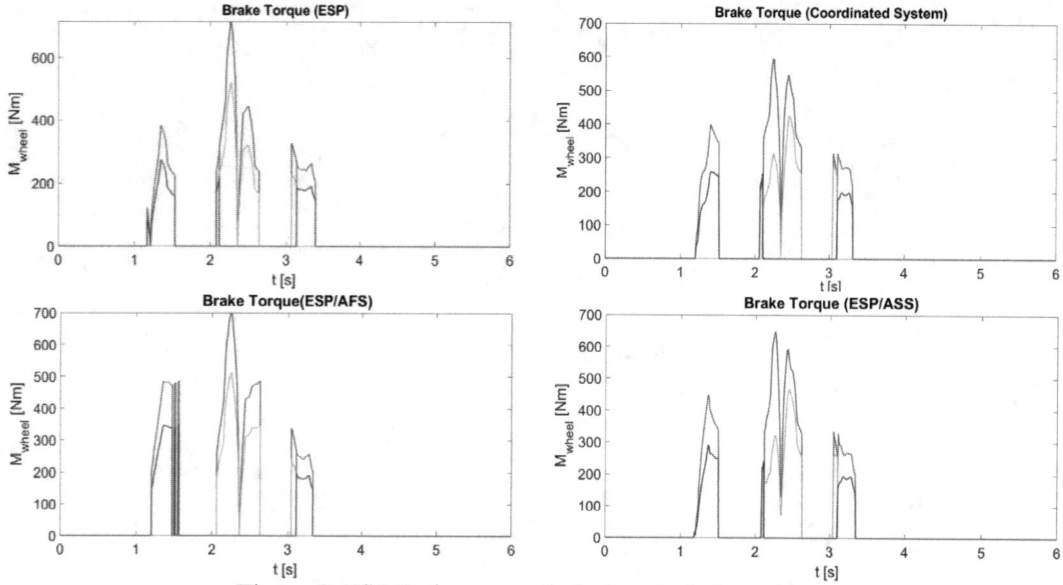

Figure 8. ESP Brake moments during single lane change.

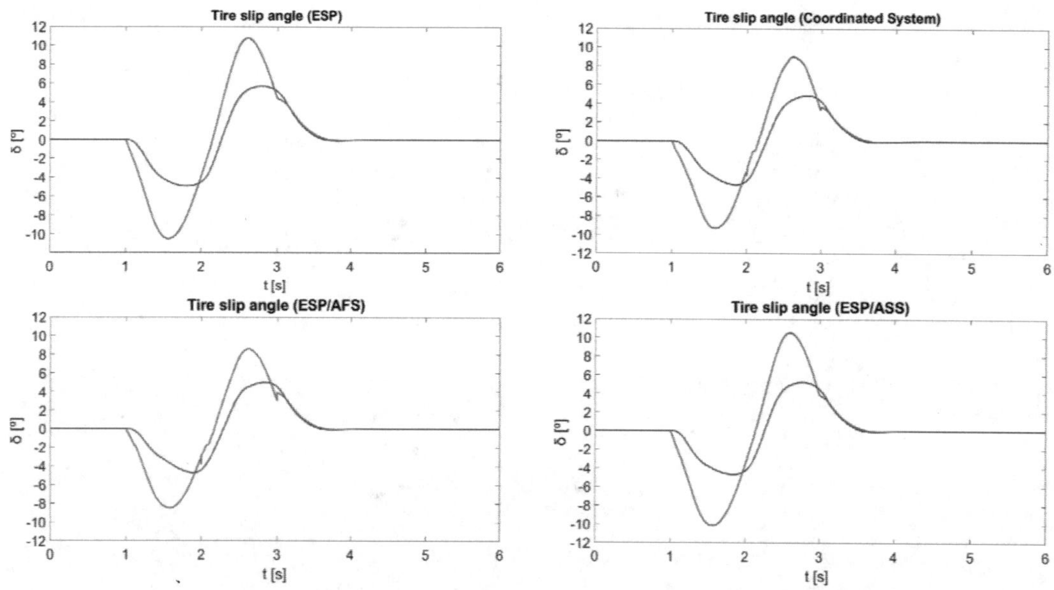

Figure 9. Wheels slip angles during single lane change.

5. Discussions and conclusions

The excellent results that ESC system have shown through the years, made it mandatory for all vehicles. The main disadvantage of the system is lowering the average vehicle speed, since it acts through the ABS system through brakes. It also impacts the rate of braking pads wearing out. As can be seen on Figure 7 ESC coordination with other systems influence on the speed change, followed by Figure 8 which shows lower brake torque for the same maneuver, because of the simultaneous action of other systems. Wheels slip angles in Figure 9 also show lower values for the coordinated system. It can be seen that only ESP-AFS coordinated control also gives good results, similar to the coordination of ESP with ASNF and AFS, for the yaw velocity, side-slip angle and vehicle speed change, but not when it comes to lowering braking torque and also for rollover movement of the vehicle.

Simulation results showed that during sudden maneuvers and at higher speeds, the ESC system has the highest efficiency. Acting alone, the impact of the AFS system is insignificant, but at lower speeds (city speeds) it improves the driver's sense of maneuverability. The normal force control alone, gives better results than the AFS system, but under the same driving conditions its action is not enough to keep the vehicle in the stable region. So, the coordination of all three systems gives better dynamic behavior and stability in every direction of movement of the vehicle.

In this research, further steps can be made by modeling and upgrading the coordinated system with traction control system (TCS) and potentially to make a sequence in which systems would activate in case of need.

NOMENCLATURE

a_x, a_y – Longitudinal, lateral acceleration
b_p, b_z – Front, rear track
c – Spring stiffness
F_a – Normal active force
F_{am} – Damping force
F_k – Brake force
F_{pr} – Spring force
F_t – Vertical tire force
F_x, F_y - Longitudinal, lateral tire contact force
I_t – Wheel rotational inertia
I_x, I_y, I_z – Roll, pitch, yaw moment of inertia
k – Damping coefficient
K_δ - Cornering tire stiffness
l_p, l_z – CM distance to front, rear axle
m, m_{nm}, m_{pm} – Total vehicle mass, unsprung, sprung mass
M_z – Stabilization moment
r_t – Rolling radius
T_v, T_k – Traction, braking torque
v_x – Longitudinal velocity
z_{nm}, z_{pm} – Vertical travel of unsprung, sprung mass
β_c – Side-slip angle
θ_p – Front wheels angle
φ – Roll angle
ω_x, ω_y, ω_z – Roll, pitch, yaw rate

References

[1] Van Zanten A T, Erhardt R and Pfaff G 1995 VDC, the vehicle dynamics control system of Bosch *SAE transactions* 1419-1436
[2] Van Zanten A T, Erhardt R, Landesfeind K and Pfaff G 1998 VDC systems development and perspective *SAE transactions* 424-444
[3] Van Zanten A T 2000 Bosch ESP systems: 5 years of experience *SAE transactions* pp.428-436
[4] Liebemann E K, Meder K, Schuh J and Nenninger G 2004 Safety and performance enhancement: The Bosch electronic stability control (ESP) *SAE Paper* ***20004*** 21-0060
[5] Kin K 2004 Enhanced vehicle stability and steerability with VSA *A VEC'04* 531-536
[6] Jordanoska V 2020 Coordinated cooperative control of active automotive systems for improved vehicles' stability and safety *PhD Thesis* University "Ss. Cyril and Methodius" Faculty of

Mechanical Engineering, Skopje

[7] Jin L, Xie X, Shen C, Wang F, Wang F, Ji S, Guan X and Xu J 2017 Study on electronic stability program control strategy based on the fuzzy logical and genetic optimization method *Advances in Mechanical Engineering* **9(5)** 1687814017699351

[8] March C and Shim T 2007 Integrated control of suspension and front steering to enhance vehicle handling *Proceedings of the Institution of Mechanical Engineers, Part D: Journal of Automobile Engineering* **221(4)** 377-391

[9] V Jordanoska et al 2021 Evaluating Coordinated Cooperative Control of Three Active Car Systems Using Fuzzy-Logic *IOP Conference Series: Materials Science and Engineering* **1190** 012029

[10] Yang S, Lu Y and Li S 2013 An overview on vehicle dynamics *International Journal of Dynamics and Control* **1(4)** 385-395

[11] Shim T and Toomey D 2004 Investigation of active steering/wheel torque control at the rollover limit maneuver *SAE Transactions* **(2004-01-2097)** 1133-1140

[12] Safarudin M, Setiawan J D and Sihombing H 2010 Vehicle active roll control using roll moment rejection algorithm *Strojnícky časopis (Journal of Mechanical Engineering)* **61 (5-6)** 301-322

[13] Setiawan J D, Safarudin M and Singh A 2009 Modeling, simulation and validation of 14 DOF full vehicle model *International Conference on Instrumentation, Communication, Information Technology, and Biomedical Engineering* 1-6

[14] Bakker E, Pacejka H B and Lidner L 1989 A new tire model with an application in vehicle dynamics studies *SAE transactions* 101-113

[15] Rajamani R 2011 *Vehicle Dynamics and Control* (New York: Springer Science & Business Media)

[16] Inagaki S, Kushiro I and Yamamoto M 1995 Analysis on vehicle stability in critical cornering using phase-plane method *JSAE Review* **2(16)** 216

[17] Koibuchi K, Yamamoto M, Fukada Y and Inagaki S 1996 Vehicle stability control in limit cornering by active brake *SAE technical paper* 960487

[18] Blank M and Margolis D 1996 The effect of normal force variation on the lateral dynamics of automobiles *SAE Technical Paper* 960484

IOP Conf. Series: Materials Science and Engineering 1271 (2022) 012024 doi:10.1088/1757-899X/1271/1/012024

The determination of the disc brake thermal stresses for different vehicle speeds

N Stojanovic[1], I Grujic[1], OI Abdullah[2,3,4], A Belhocine[5], J Glisovic[1]

[1] University of Kragujevac, Faculty of Engineering, 6 Sestre Janjić Str., 34000 Kragujevac, Serbia

[2] Department of Energy Engineering, College of Engineering, University of Baghdad, Baghdad, Iraq

[3] Department of Mechanics, Al-Farabi Kazakh National University, Almaty 050038, Kazakhstan

[4] System Technologies and Engineering Design Methodology, Hamburg University of Technology, Hamburg, Germany

[5] Department of Mechanical Engineering, University of Sciences and the Technology of Oran, L.P 1505 El -MNAOUER, USTO 31000 ORAN (Algeria)

E-mail: nadica.stojanovic@kg.ac.rs

Abstract. From the aspect of traffic safety, the brake system is one of the most important systems. For all exploitation conditions, for which vehicles are projected, and for loads in prescribed boundaries, no matter to the vehicle speed, it is important, for the brake system to be reliable and to provide a very fast stop, or speed adjustment to road conditions. During the stop or during the speed adjustment it comes to the disc brakes temperature increment. So, because of this, the experimental investigations were conducted, in order to determine final values of the temperature for multiple braking processes until stop. In this paper, were conducted two experimental tests for the same vehicle, but for different speeds, which were 60 km/h and 80 km/h, while the other parameters (initial temperature, the mass of the simulated vehicle, number of repetitions) were the same for both tests. The conclusion is that by increment of the speed, rises the necessary time to stop the simulated vehicle, and by this, rises and the stopping distance. Also, rises and the temperature of the disc brake, with the increment of the vehicle speed.

1. Introduction

Besides the steering system, the brake system is one of crucial systems on the vehicle, from the aspect of the traffic safety. The brake system should provide deceleration, that is, the speed adjustment to road conditions. Also, it should provide the vehicle stop, as well as the vehicle maintain in steady state, in the case when the vehicle is parked. In all situations, the brake system should be reliable and to not endanger the safety of the driver, as well as the safety of other traffic participants.

The two main types of brakes which are widely used are disk brakes and drum brakes. In the case of disk brakes, the disk can be solid disc, or a vented disc. However, today the widest application have vented discs [1], and because of this, this type of disc will be analysed in the paper. The vented discs are much more convenient for use, because their ribs provide a much better cooling and heat release into the environment [2]. Thanks to this, it is not endangered the reliability of brake elements during the extreme conditions of exploitation, compare to the solid discs [3]. The only thing, about which should take care, when use the vented disc, is the geometry of ribs [4]. Because, only by correct choice of the shape of ribs, that is, by their modification, the heat release will efficient [5]. The Limpert [6] in his research shown that 60% of heat releases from ribs, while the rest amount of heat, releases into the environment from the contact surface of the brake disc.

During the braking process comes to the increment of the temperature of the brake disk and brake pads [7]. Also, this can cause the reduction of the friction coefficient [8], which further leads to the longer stopping distance, as well as to the longer time, which is necessary to stop the vehicle. In the case, if the heat does not release into the environment, or the heat release is very slow, a hot spots can appear, or even can come to cracks [9, 10]. The temperature field is completely connected with the field of stresses which appear on the contact surface of the brake disc [11], which further results with the damage of the brake disc. Depend from the type of braking, will depend and type of heat stress. During the normal braking, a great heat stresses appear, however, due to the extreme braking, an enormous heat stresses appear [12].

One of main questions is, will a brake system provide safe vehicle stop, no matter to the temperature of the brake disc and brake pads. Also, how much will heat up the brake disc and brake pads, in respect to the vehicle speed. This paper shows, how much are the temperatures for different vehicle speeds, as well as how the speed influences on the temperature increment. Besides these answers, the paper provides and thermographic presentation of temperature distribution on the contact surface of the brake disc.

2. The experimental installation

The test rig for the investigation of disk brakes thermal stresses – BRAKE DYNO 2020, is shown on figures 1 and 2, while the scheme of the test rig is presented by figure 3. The connection between the shaft of the electric motor (1) and flywheel mass (5) is provided by the electromagnetic clutch (2), which turns on and turns off, by brush (14). The position of electric motor by height can be adjusted, by adjusters (23). The control of the speed of the electric motor is achieved by frequency regulator (13). In order to have constant information about the angular speed of the flywheel mass, on the driven part of the clutch is mounted wheel with teeth's (13), where the inductive sensor (17) serve to measure the rotation speed. The rotation speed can be followed during the acceleration, as well as during the braking process. The shaft of the flywheel mass, on both sides lays on two bearings (4) on the support construction of the test rig (22). On the opposite side from the electric motor is the brake disc (6), which by screws is connected with the shaft of the flywheel mass. The brake calliper (27) is connected with the base by carrier (28) and force sensor (19). Inside each brake pad are mounted two temperature sensors (18), for temperature measure on the entering and exiting side of the brake pad, while for the measure of the temperature of brake disk thermal camera is used (8). The brake activates by hydraulic braking installation (12 and 26), while the hydraulic brake installation activates by pneumatic installation (9), which use air from compressor (10). Firstly, rises the air pressure in the pneumatic installation, where the air from the compressor to the pneumatic part comes through the pipe (25). After that, the pneumatic part activates the brake cylinder (12), and by further rise of oil pressure in the hydraulic installation (26), brake pads press the brake disc, where value of the brake pressure rises until the value of the pressure equalizes with the value defined before the measurement. During the measurement, are followed the values of the pressure in the pneumatic installation (21), and in hydraulic installation (20). The control of the test rig is automatized, that is, the commands to the test rig are sending from the computer (11), which receives and sends signals trough A/D converter 6353. During the measurement, all data are collecting and save on the computer memory.

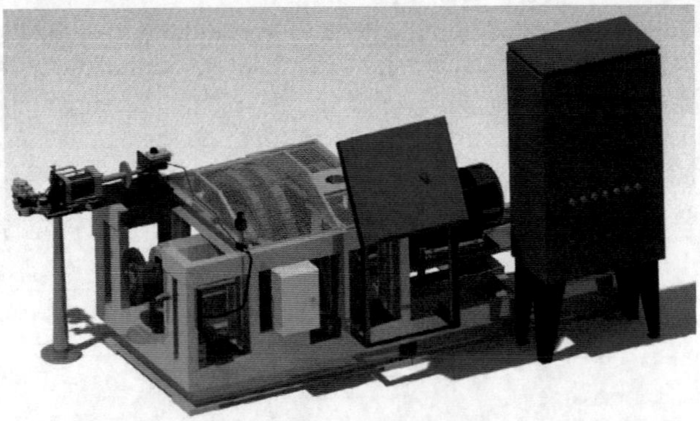

Figure 1. 3D model of the test rig.

Figure 2. The test rig – BRAKE DYNO 2020.

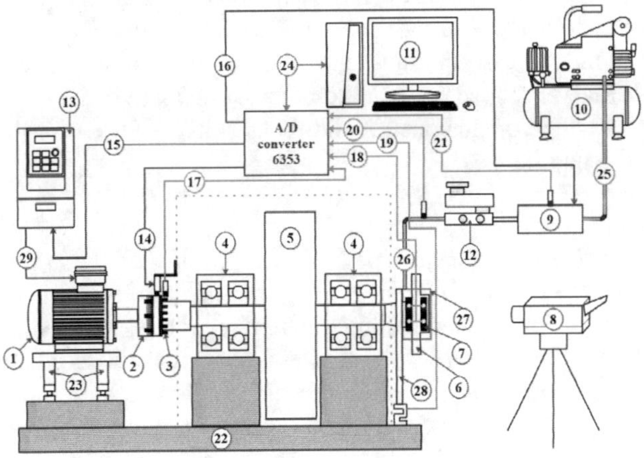

Figure 3. The scheme of the test rig BRAKE DYNO 2020.

3. The plan of the experiment

With aim to determine how the vehicle speed influences on the value of the temperature of the disc brake, it was defined the measurement procedure, figure 4. The first step, includes the defining of the investigation conditions, such are the quarter of the vehicle mass (the test rig simulates the quarter of the vehicle mass ($m_{1/4} = 300$ kg)), brake pressure and the desired speed of the simulated vehicle. The next step is to define parameters which should be followed during the investigation process – the output parameters. The output parameters are temperature of brake pads and of the brake disc, the angular speed, the friction coefficient, the brake moment and change of the pressure during the braking process. Besides the defined investigation conditions, on the heat up of the brake disc and brake pads, influence have and the material characteristics, table 1.

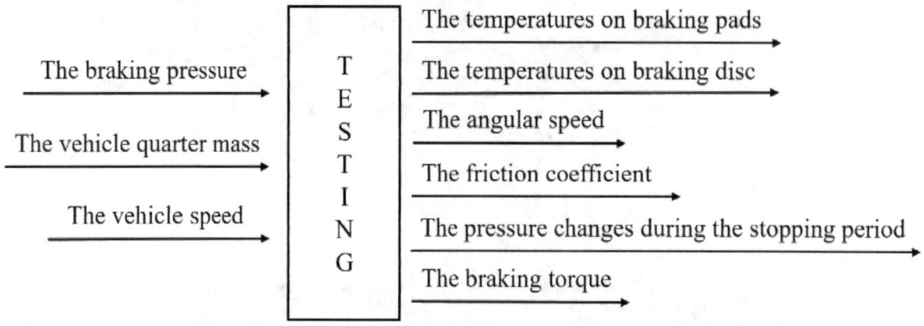

Figure 4. The measurement procedure.

Table 1. The material characteristics of the brake disc and brake pads.

	Disc	Pad
Density, kg/m^3	7100	2300
Elastic modulus, GPa	118	20
Poisson ratio, -	0.32	0.3
Thermal conductivity, W/m·°C	53.3	3
Specific heat, J/kg·°C	490	1200
Thermal expansions, °C^{-1}	10.85×10^{-6}	10×10^{-6}

The next step the is the experimental investigation in the laboratory, that is, start of the test rig, and after the achievement of the desired speed for the defined vehicle, the brake activates, simulated vehicle stops, and during the braking process, the input data and output data are collecting and saving on the computer memory. After that, all collected data are processing, analysing and show by diagrams, and on the basis of them, come the conclusions. While the results captured by the thermal camera, are showing in the shape of thermal images.

4. Results and discussion

The change of the temperature on the brake pads, has the shape of the stairs, that is, has constant increment, figures 5 and 6. The greatest increment of the temperature happens after first braking cycle, while with each next braking cycle, the temperature increases more slowly, and to the same conclusion has come and Grzes [13]. By observation of the figure 5, in the case when the speed of the simulated vehicle was 60 km/h, during the first acceleration process, the temperature on brake pads is constant. During the braking it comes to the increment of the temperature, while the greatest increment appears when doesn't exist rotational movement, that is, when the brake disc is in the steady state. This happens, because, it is necessary time for the heat generated in the contact between the brake disc and brake pads, to conduct inside the brake pads, where are positioned temperature sensors. After that, it follows the next cycle of acceleration, and during this cycle, the temperature on the brake pads has a slight increment. The greatest increment appears, as in the first case, when doesn't exist rotational movement of the brake disc. Which means, that is necessary to pass some time, in order for the heat generated in the contact of brake disc and brake pads, to conduct by depth of brake pads, all to the temperature sensors, how previously was said. Of course, the temperature of brake pads increases and by depth which is behind the temperature sensors. This phenomenon, repeats until the last testing cycle, with that, that the measured temperatures are higher and higher after each braking cycle.

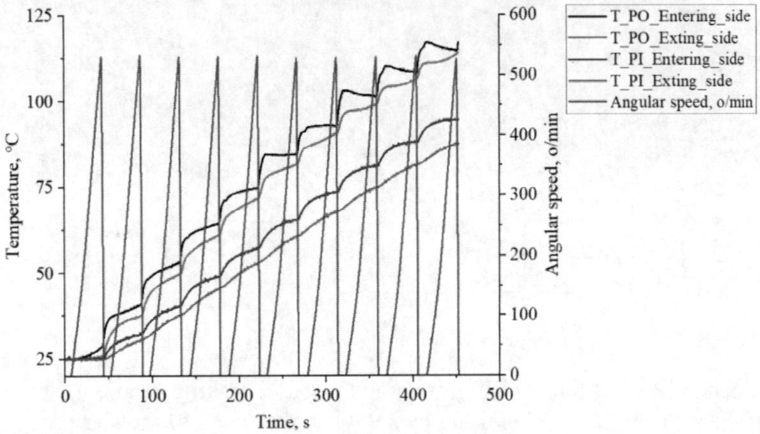

Figure 5. The temperature change on brake pads for the vehicle speed 60 km/h.

The same behaviour of the temperature change of brake pads, was recorded and for the case when the vehicle speed was 80 km/h, figure 6. However, for the case, when the vehicle speed was 80 km/h, temperatures were higher for about 50 °C, compare to the case when the vehicle speed was 60 km/h. Which is logic, because the same vehicle was simulated, that is, the vehicle with the same characteristics, only the speed was different.

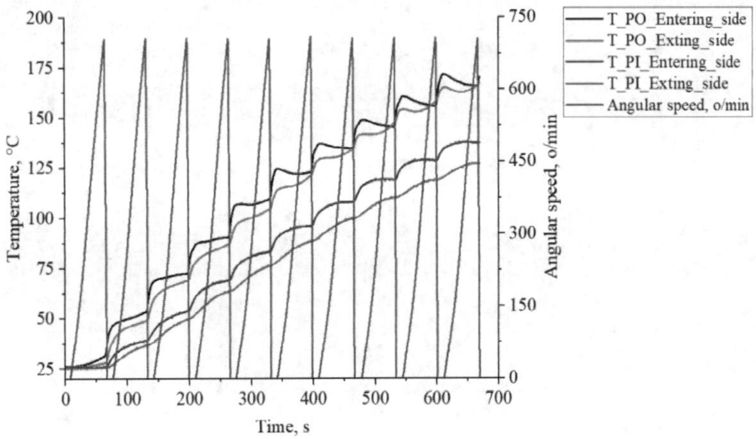

Figure 6. The temperature change on brake pads for the vehicle speed 80 km/h.

The highest temperatures appear on the external brake pad, and that on the entering side of the brake pad. By figures 7 and 8, is shown the dependence of the temperature of external brake pad, braking pressure, braking moment, angular speed and friction coefficient in the function of time, during the 10th braking cycle, for the cases when the vehicle speed was 60 km/h and 80 km/h.

IOP Conf. Series: Materials Science and Engineering 1271 (2022) 012024 doi:10.1088/1757-899X/1271/1/012024

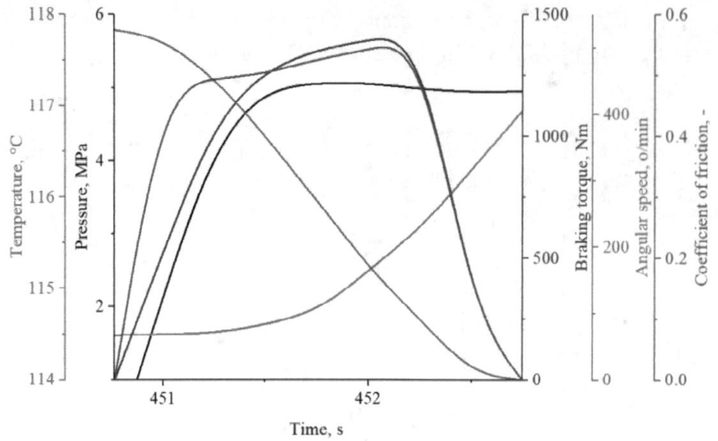

Figure 7. The dependence of temperature, braking pressure, braking moment, angular speed and friction coefficient during the 10th braking cycle, when the speed of the simulated vehicle was 60 km/h.

The value of the temperature does not change until the value of the pressure in the system don't achieves maximum value, figure 7. The shapes of the curves which describe the braking moment and friction coefficient almost the same, because one depends from another [14, 15]. The behaviour of the temperature on the entering side of the external brake pad, braking pressure, braking moment, angular speed and friction coefficient is the same for both cases, that is for case when the speed of the simulated vehicle was 60 km/h and 80 km/h.

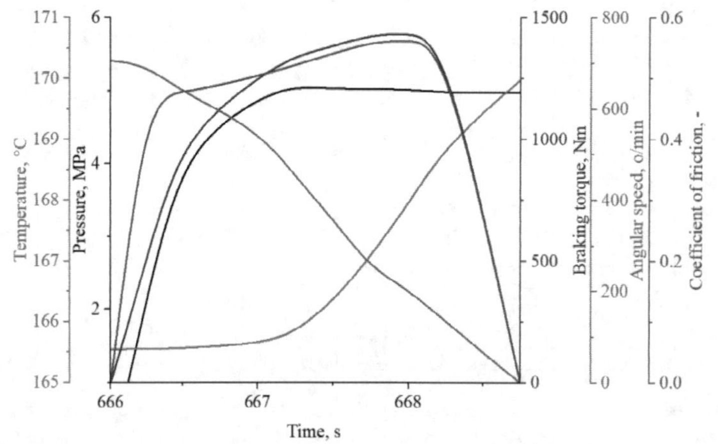

Figure 8. The dependence of temperature, braking pressure, braking moment, angular speed and friction coefficient during the 10th braking cycle, when the speed of the simulated vehicle was 80 km/h.

The temperature field of the brake disc for the 1st and 10th braking cycle is shown on the figure 9, for the case, when the speed of the simulated vehicle was 60 km/h. While by figure 10 is shown for the case when the speed of the simulated vehicle was 80 km/h. The final temperature on the entering side of the external brake pad for the case of 60 km/h was 116.95 °C, while for the case of 80 km/h was 169.98 °C. The speed increment of 20 km/h, cased the temperature increment of 45.34 %.

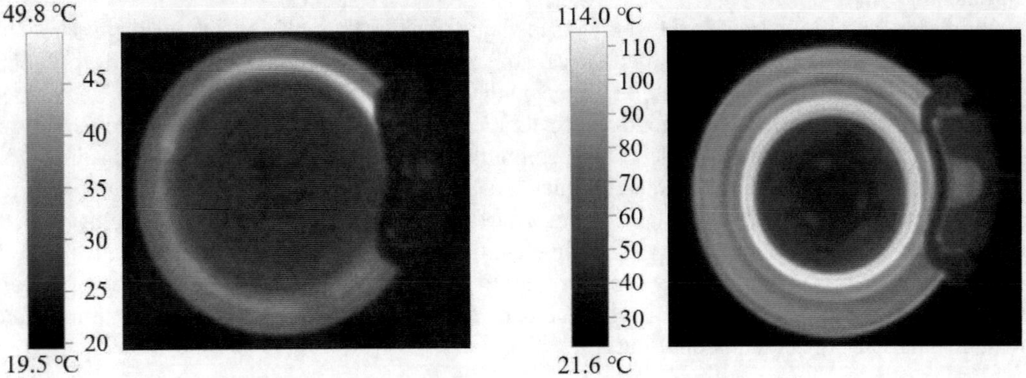

a) b)

Figure 9. The temperature field on the contact surface of the brake disc during the first and tenth braking cycle, for the case when the speed of the simulated vehicle was 60 km/h.

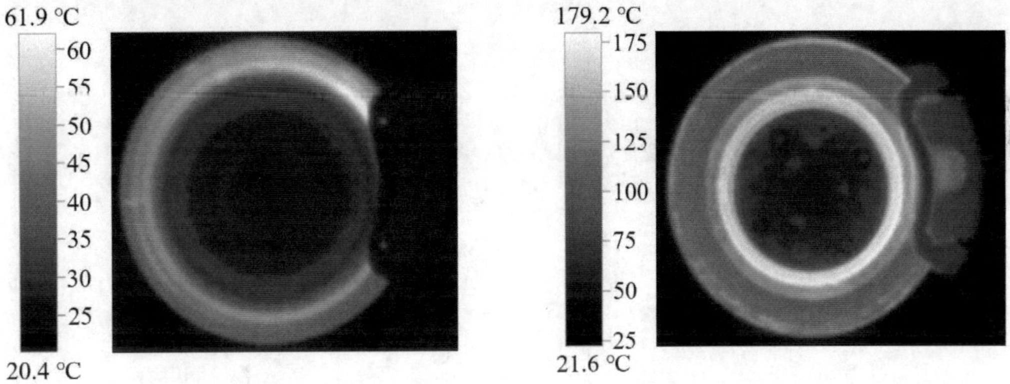

a) b)

Figure 10. The temperature field on the contact surface of the brake disc during the first and tenth braking cycle, for the case when the speed of the simulated vehicle was 80 km/h.

During the 1st braking cycle, by exit of the brake disc from the contact with the brake pad, are noticeable hot bands, on the contact surface of the brake disc, and that, one hot band for the case when the speed of the simulated vehicle was 60 km/h (figure 9.a), and two hot bands for the case when the speed of the simulated vehicle was 80 km/h (figure 10.a). The consequence of hot bands is the reduced contact between the brake disc and brake pads, which can result with the appearance of one or more hot bands, as in the research of Panier et al. [10]. The temperature of the brake disc, after ten consecutive braking cycles, in the case of the vehicle speed 60 km/h, was 114 °C, while for the case of the vehicle speed 80 km/h, was 179.2 °C. The increment of the vehicle speed for 20 km/h, after ten consecutive cycles of acceleration and stop, caused temperature increment of 57.19%. Also, at the end should emphasize, if the obtained temperatures overcome critical temperatures. The moment, when the temperature overcome the critical value, is the moment when the temperature of the brake disc overcome the value of 650 °C [8], which wasn't happened in this research. However, the optimal temperature, that is, the temperature when a best braking performance can be achieved (the greatest value of the friction coefficient), is when the temperature on the contact surface of the brake disc is around 150 °C [14, 16].

5. Conclusion

The exploitation conditions such is the vehicle speed and besides that, also and the vehicle parameters such is the vehicle mass, influence on the thermal stresses of the brake disc and brake pads. Of course, and the type of braking, in great amount influence on the thermal stresses. In this paper, were conducted two experimental tests, for ten consecutive accelerations and stops, for the same vehicle. The only difference was the speed of the vehicle at the beginning of braking process. The highest values of the temperature appeared in the case of greater vehicle speed. Besides that, on the basis of the type of the temperature field, which appears on the contact surface of the brake disc, comes the conclusion that in these regimes of the vehicle drive, appear the highest thermal stresses, which further influence on the mechanical stresses. Future researches should direct to the investigation of brake discs with different shape of ribs, as well as to brake discs and brake pads manufactured from alternative materials, under normal and extreme exploitation conditions.

Acknowledgment

This paper was realized within the framework of the project "The research of vehicle safety as part of a cybernetic system: Driver-Vehicle-Environment", ref. no. TR35041, funded by the Ministry of Education, Science and Technological Development of the Republic of Serbia.

References

[1] Vdovin A and Gigan GL 2020 *Energies* **13** 203
[2] Ahmed GMS and Algarni S 2018 *Materials* **11** 1211
[3] Saiz CB, Ingrassia T, Nigrelli V and Ricotta V 2015 *Frat. ed Integrità Strutt.* **9** 608-621
[4] García-León RA and Rojas EP 2017 *Ind. Syst. Eng.* **21** 71-96
[5] Kudal GB and Chopade MR 2016 *Int. J. Curr. Eng. Technol.* **5** 219-224
[6] Limpert R 1975 *SAE Technical Paper* 751014
[7] Talati F and Jalalifar S 2009 *Heat Mass Transf.* **45** 1047
[8] Breuer B and Bill KH 2012 *Bremsenhandbuch – Grundlagen, Komponenten, Systeme, Fahrdynamik* (Springer Fachmedien Wiesbaden)
[9] Yevtushenko AA and Grzes P 2015 *Int. Commun. Heat Mass Transf.* **62** 37-44
[10] Panier S, Dufrénoy P, Brunel J and Weichert D 2004 *J. Therm. Stresses* **28** 47-62
[11] Belhocine A and Bouchetara M 2013 *J. Appl. Res. Technol.* **11** 674-682
[12] Eriksson M, Bergman F and Jacobson S 2002 *Wear* **151** 26-36
[13] PGrzes 2019 *Adv. Mech. Eng.* **11** 1-13
[14] N Stojanovic 2021 *Identification of disc brake thermal strains for different braking operating parameters* (Kragujecvac: Faculty of Engineering)
[15] Stojanović N, Grujić I and Glišović J 2021 *15th International Conference on Accomplishments in Mechanical and Industrial Engineering* (Banja Luka, Repubika Srpska)
[16] Ahmed I, Leung P and Datta P (2000) *SAE Technical Paper* 2000-01-2778

IOP Conf. Series: Materials Science and Engineering 1271 (2022) 012025 doi:10.1088/1757-899X/1271/1/012025

Regenerative braking on electric vehicles: working principles and benefits of application

S Vasiljević[1,*], B Aleksandrović[1], J Glišović[2] and M Maslać[1]

[1] Academy of Professional Studies Šumadija, Department in Kragujevac, Kosovska 8, Kragujevac, Serbia

[2] University of Kragujevac Faculty of Engineering, Sestre Janjić 6, Kragujevac, Serbia

*Corresponding author vasiljevic.sasa036@gmail.com

Abstract. The application of electric vehicles leads to a change in the principle of operation and functioning of some systems in the vehicle, which also lead to a change in the concept of the vehicle itself. One of those systems that has a new concept, which differs from vehicles powered by IC engines, is the braking system. The previous function of the braking system was to stop the vehicle, i.e. to reduce the speed of the vehicle in a safe way. In the case of electric vehicles, the friction brakes were retained, with the addition of a regenerative braking system that has the role of replenishing the vehicle's batteries. The regenerative braking system has the role of converting the vehicle's kinetic energy into electrical energy that recharges the batteries. This system is already used today on full electric and hybrid vehicles, i.e. on vehicles powered by an electric motor. The benefits of regenerative braking are reflected on the fact that the vehicle batteries are recharged during braking, vehicle maintenance costs are reduced, the service life of discs and drum brakes on the vehicle is extended, brake non-exhaust emission is reduced, and heat energy emission is reduced, too.

1. Introduction

The number of electric vehicles and the interest in using of electric vehicles are increasing more and more. The fact is that the sale of electric vehicles has doubled in 2021 compared to 2020 [1]. In 2021, the number of sold electric vehicles (plug-in hybrid vehicles and battery electric vehicles) has increased by 108% compared to 2020, while compared to the year before 2020, that number has multiplied [2]. If we look at the number of vehicles sold during the period from 2018 to February 2022, a growing trend in the sale of electric vehicles is also noticeable [3]. The growing trend of sales of electric vehicles is particularly noticeable since July 2020. Based on the available data, it can be observed that the number of sales of electric vehicles is particularly dominant in the first two months of 2022. How big the sales of electric vehicles are today in the world is also shown by the number of vehicles sold according to the manufacturers. If we take into account the manufacturers of electric vehicles and their number of vehicles sold, it is noticeable that they have an increase in the number of sold plug-in hybrid vehicles as well as purely electric vehicles [3,4]. Based on data [5,6] from the manufacturers Tesla and Volkswagen Group BEV, in the first quarter of 2022 compared to the first quarter of 2021, the number of vehicles sold increased by several tens of percent.

Electric vehicles, i.e. battery-powered vehicles, use electricity from their batteries as fuel. The problem of electricity consumption, but also the lack of infrastructure [7], is one of the problems of using electric vehicles, which reduces the maximum range of vehicles and makes it impossible to use electric vehicles. As kinetic energy is converted into thermal energy during braking, engineers have devised a regenerative braking system. Regenerative braking, i.e. recuperation, is a principle in which when the vehicle slows down, i.e. braking the vehicle, the kinetic energy of the vehicle is converted into electrical energy, with which the batteries of both electric vehicles and hybrid vehicles are recharged.

The aim of this paper is to analyze the existing operating principles of the regenerative braking system in electric and hybrid vehicles. The paper presents the operation of this system from the aspect of charging the batteries and converting the kinetic energy of the vehicle into electrical energy, as well as the strategy of activating the regenerative braking system and friction brakes. The basic advantages of using this system on the vehicle are also presented, as well as a preview of the information that the driver receives about the operation of the system.

2. Basics of regenerative braking - recuperation

Regenerative braking, i.e. recuperation, is the only way to recharge the batteries of electric vehicles without any other connection than between the drive wheels and the electric motor, i.e. in hybrid vehicles between the electric motor and the IC engine [8]. Based on reference [9], regenerative braking has the role of increasing range, i.e. the number of kilometers traveled without additional battery charging. Based on reference [8], regenerative braking enables an increase in vehicle range by 10%. The regenerative braking system recharges the batteries when moving downhill, but also when braking. In the case of hybrid vehicles, the battery can be recharged in the mode when the IC engine is activated, producing excess power that is converted into electrical energy to recharge the batteries. Based on reference [9] regenerative braking can increase vehicle range by up to 30% depending on vehicle, terrain, temperature and other variables. When it comes to the type of vehicle and the energy saved by applying regenerative braking, figure 1 shows the percentage of electrical energy savings according to the type of vehicle. It is noticeable that heavier vehicles have a higher percentage of electrical energy using regenerative braking, compared to lighter vehicles such as electric bicycles [10].

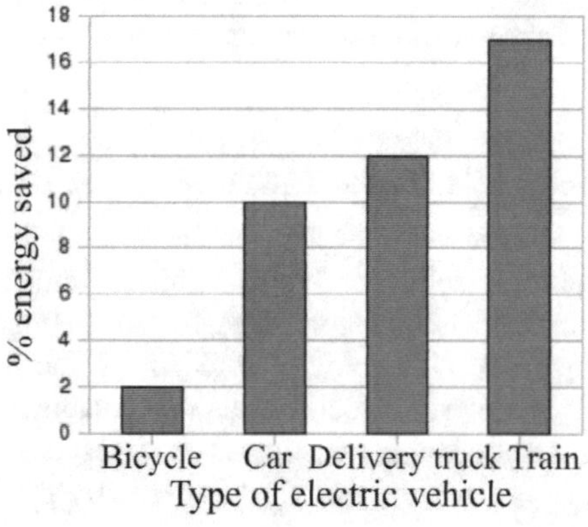

Figure 1. The percentage of saved energy using regenerative braking depending on the type of vehicle [10].

The application of regenerative braking works automatically, i.e. the driver, unless he is braking, has no effect on the operation of the system and control. However, the driver may see different signs on the dashboard that can inform the driver about recharging the batteries. Specifically, the driver can at any time have an insight into the flow of power, i.e. the flow of energy from the batteries to the wheels. The graphic, driver can have an insight into the flow of energy used to replenish the battery during regenerative braking. Figure 2a shows an example of energy monitoring, i.e. power flow on the driver's control monitor, while figure 2b shows a closer example of power monitoring where information is obtained about the operation of the entire power transmission system. In the case of hybrid vehicles, if the batteries are charged using the IC engine, information on battery charging is also obtained. When the regenerative braking system is activated, the arrows on the monitor change indicating that the batteries are being recharged.

a) b)
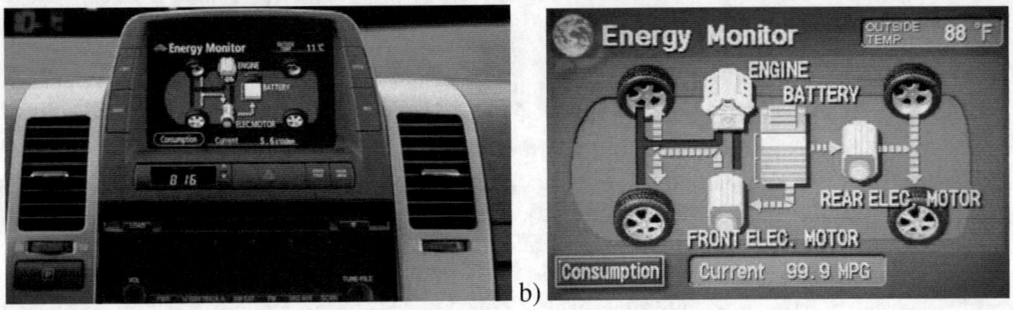

Figure 2. Power flow monitoring, a) display of power monitoring on the driver's control monitor, b) closer view of the appearance of power flow monitoring [11].

By monitoring the flow of power from the batteries to the wheels and vice versa, the driver or user of the vehicle can also receive information on the consumption of electricity and the energy used to recharge the battery. Figure 3 shows an example of data on the control panel on the monitoring of energy consumption and the energy with which the vehicle's battery was recharged. In this way, the driver has accurate insight into the use of energy, i.e. about the recovery of electricity.

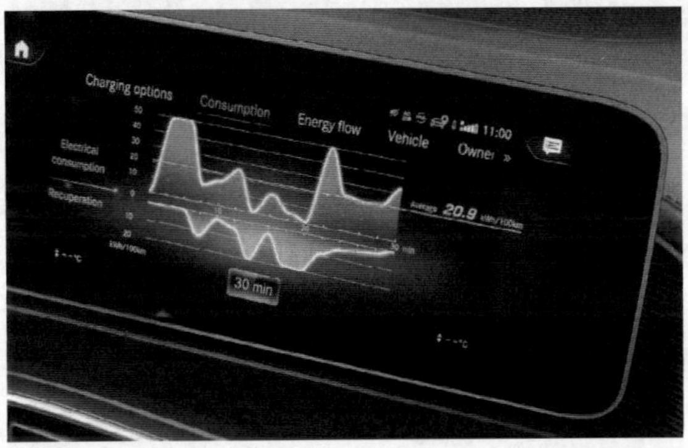

Figure 3. Information on electricity consumption and recovery [12].

The advantages of applying the regenerative braking system, according to the source [13], are the extension of the range of electric vehicles, the reduction of fuel consumption in the case of hybrid vehicles, the extension of the service life of the disc and brake pads of friction brakes, and thus the reduction of particle emissions (non-exhaust emission) caused by the wear of friction pairs [14]. The disadvantages of the regenerative braking system, according to [13], are lower effectiveness at lower speeds, potentially less force to stop the vehicle and the possibility of a different feeling when pressing the brake pedal compared to vehicles without a regenerative braking system.

3. The principle of operation of regenerative braking

Based on [15], there are several types of regenerative braking, i.e. they are typically mechanical, electrical and hydraulic. Mechanical regeneration (mechanical kinetic energy recovery system) is based on a flywheel and clutch system, in which the flywheel is connected to the driveline at the start of braking, and the momentum of the vehicle is used to accelerate the flywheel and thereby slow the vehicle down. The opposite situation is in the acceleration phase, where flywheel momentum is used to assist acceleration. Electrical energy recovery systems are based on a motor/generator powered by the kinetic energy of a moving vehicle. The electricity produced is used to charge the battery and then returned to the machine in the acceleration phase. Energy is transformed from kinetic energy into electrical energy and then into chemical energy in the regenerative braking phase.

Bearing in mind the nature of the electric vehicle and the regenerative braking observed in this case, the electrical system of the regenerative braking was analyzed bearing in mind the fact that in this case, the batteries are supplemented with electrical energy. A typical example of an electric-powered vehicle is shown in figure 4. The very principle of regenerative braking is based on the law of conservation of energy, which states that energy can neither be created nor destroyed, it can only be converted from one form of energy to another form [16]. When the vehicle is moving, the electric energy from the batteries is converted into the kinetic energy of the vehicle (i.e. the rotation of the wheels) by means of an electric motor (Figure 4a). On the other hand, when braking, i.e. when the brake is activated, the electric motor changes its direction, that is, it switches from the role of the converter to the role of the electric generator. Thus, the kinetic energy of the vehicle and through the rotation of the wheel, is converted into electrical energy by the engine, which works in the case of activation of the regenerative braking system in generator mode. In this way, the batteries are being recharged, which is shown in figure 4b.

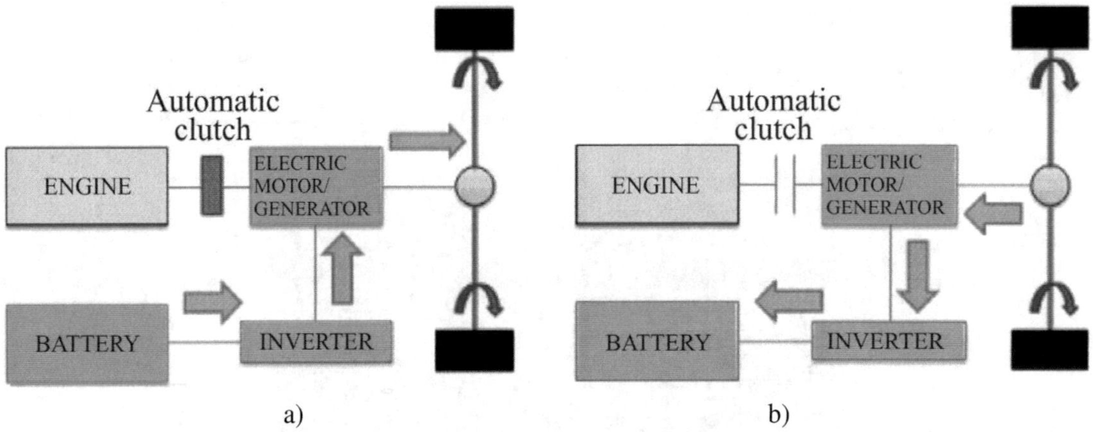

Figure 4. Vehicle a) acceleration process, b) regenerative braking process [17].

As there is a mechanical connection in the regenerative braking system between the various elements connected to the generator and there is a certain degree of efficiency of power transmission and generated energy. The losses in power transmission from the wheels to the electric motor or the generator in the case of regenerative braking are shows graphically in figure 5. There are also such losses in power transmission in the case of vehicle acceleration, but it is still necessary to point out the fact that there are losses in the case of regenerative braking, at least when it comes to mechanical connections between mechanical elements. Of course, as shown in figure 4b, there are certain losses due to natural phenomena, i.e. forces acting on the vehicle during the movement. It is important to point out that the forces acting on the vehicle in the case of braking in order to stop represent a positive phenomenon in order to shorten the distance and stopping time of the vehicle. But on the other hand, in the case of recuperation, they represent a negative phenomenon because they reduce the efficiency of the system.

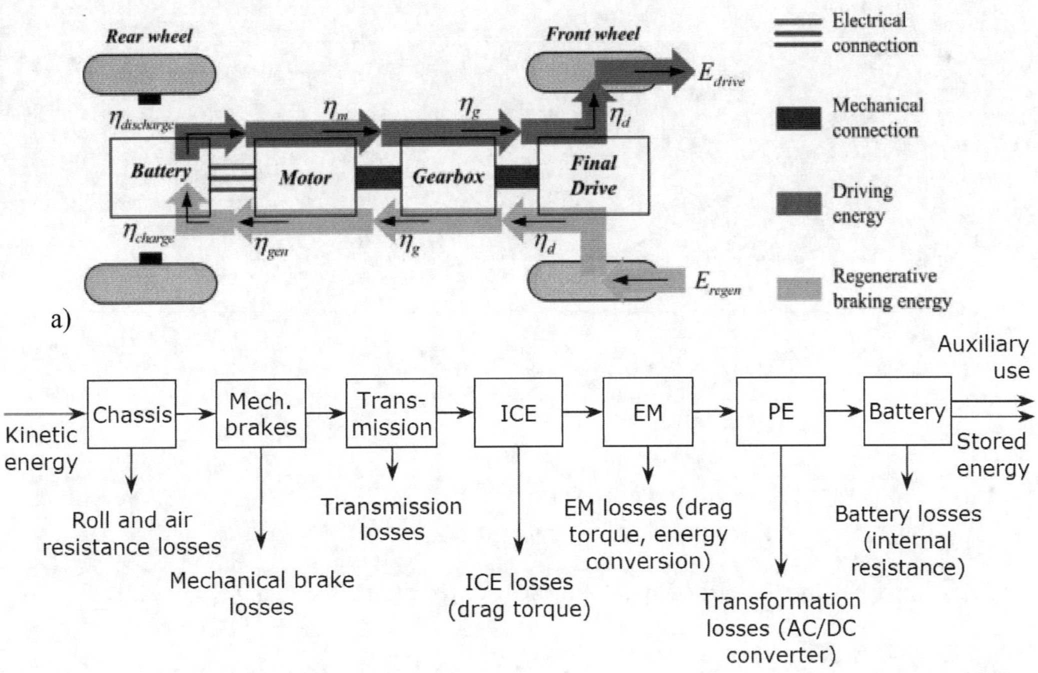

Figure 5. Potential energy losses in the regenerative braking system, where: a) graphic representation of the energy flow and utilization of elements, b) energy losses in the transmission system according to individual elements [18, 19].

During vehicle braking, the energy required to stop the vehicle must be divided between the friction brakes and the regenerative braking system, favoring the use of regenerative braking [20]. A similar operation principle and the strategy of reducing vehicle speed also applied to the application of retarders on the vehicle, i.e. retarders, which are used to reduce vehicle speed. The simplest retarder control was one level of operation (on or off), although today that system has also been developed to allow a combination with friction brakes. Problems with braking balance and vehicle stability were avoided. Combining regenerative braking with friction brakes is done so that operation of the brake and stopping is completely safe and does not affect the stability of the vehicle. Regenerative braking control must be integrated with all other electronic braking control systems such as Anti-lock braking system (ABS) and Electronic stability control (ESC). Based on this source, there are two types of regenerative braking:

- Category A: regenerative braking system that is not part of the service braking system, where regenerative braking is activated when the accelerator pedal is released,
- Category B: regenerative braking system that is part of the service braking system [20].

In the case of vehicles with a category B regenerative system, the service braking system may have only one control device. Vehicles equipped with both categories of regenerative braking (A and B) must not cause the effect of the service brake command and reduce the regenerative braking effect that would be created by releasing the throttle control. Knowing the existence of categories of regenerative braking is an important fact in considering the strategy of its management. It is thus possible for manufacturers to decide whether to apply regenerative braking when releasing the gas pedal to increase engine braking, i.e. to simulate engine braking, thus activating the recuperation system immediately, or to activate the entire system by pressing the brake pedal. It is certainly possible to combine both systems, i.e. recovery categories. An example of combining friction brakes and regenerative braking is shown in figure 6 [20].

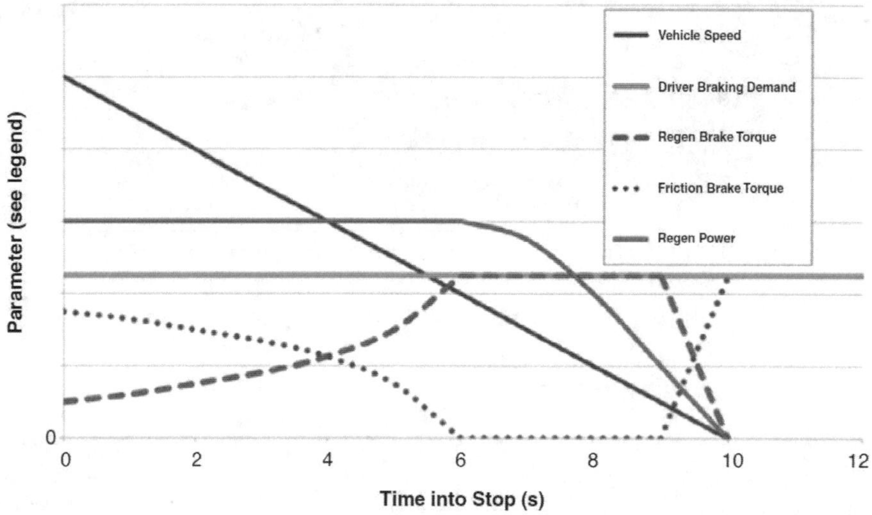

Figure 6. Brake system blending profile required during the braking process [20].

3.1 Regenerative braking and recovery strategies

Regenerative braking is an additional system that is connected to the braking system. Thus, the electric motor in generator mode creates resistance that slows down the vehicle, and on the other hand, the generator itself is started. Thus, when the brakes are activated, the regenerative braking system is first activated, and if greater braking force is required, the brakes on the vehicle itself are additionally activated. Illustrative activation of the regenerative braking system and brakes depending on the force on the brake pedal is shown in figure 7. This illustration shows the braking strategy, i.e. regenerative braking. Of course, there are different regenerative braking strategies depending on the vehicle manufacturer and the system.

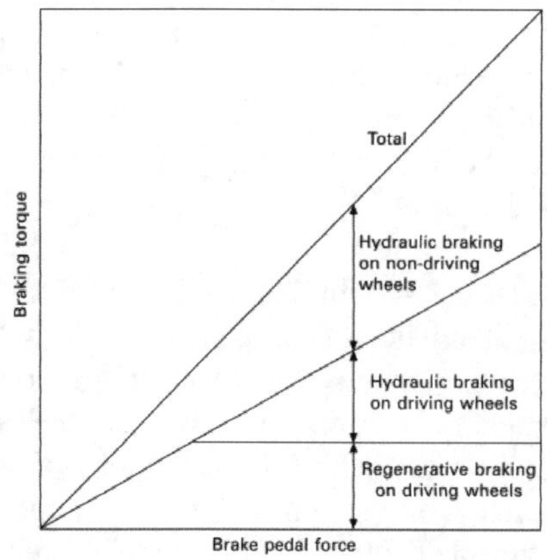

Figure 7. Regenerative braking strategy [21].

There are two basic regenerative braking strategies, which are shown in figure 8, and these two strategies are the parallel and series regenerative braking systems. Figure 8a shows a parallel regenerative braking strategy where it is noticeable that the friction brakes and the regenerative braking system are activated in parallel. On the other hand, figure 8b shows the serial regenerative braking strategy, where it is noticeable that in this case, the regenerative braking system is first activated, and then the friction brakes are activated.

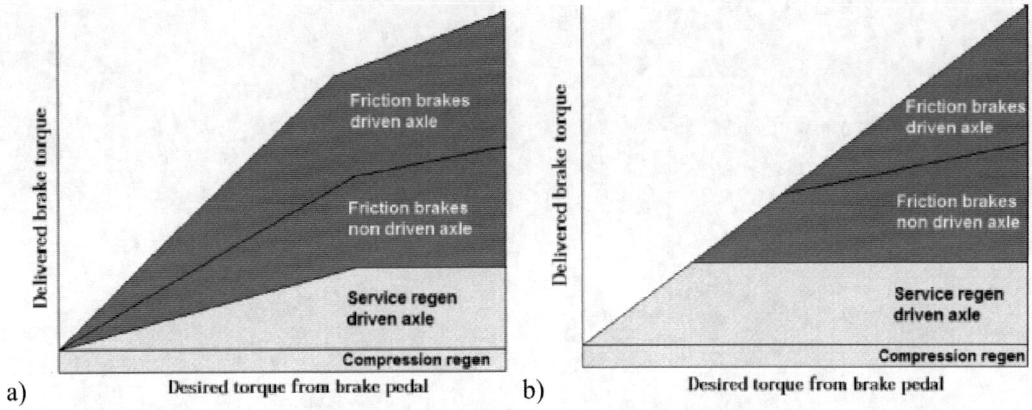

Figure 8. Types of regenerative braking, a) parallel strategy of regenerative braking, b) series strategy of regenerative braking [22].

In the research [23], three regenerative braking strategies were used; one of the three strategies is a parallel regenerative braking strategy. The other two strategies are serial regenerative braking strategies. It can be concluded from this that different variants of one type of strategy are derived, which differ according to the moment of activation of a specific friction brake on the vehicle. Applied strategies and variations are shown in figure 9.

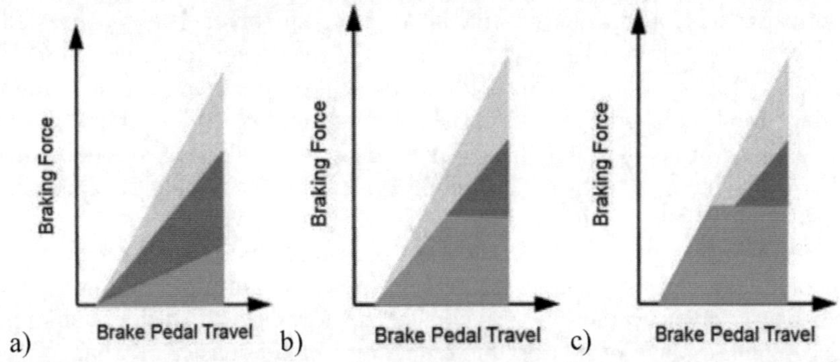

Figure 9. Different versions of regenerative braking strategies: a) parallel regenerative braking strategy, b) first version of serial regenerative braking strategy, c) second version of serial regenerative braking strategy [23].

As stated in the previous part of the paper, there are different strategies and variants of the regenerative braking strategies. In research [23], in addition to the variations of different strategies of regenerative braking, the effectiveness of different strategies was shown. The presented results are shown in figure 10. It is noticeable that the efficiency of regenerative braking depends on the initial braking speed. It can also be concluded that the series strategies of regenerative braking have a higher effectiveness compared to the parallel strategy of regenerative braking.

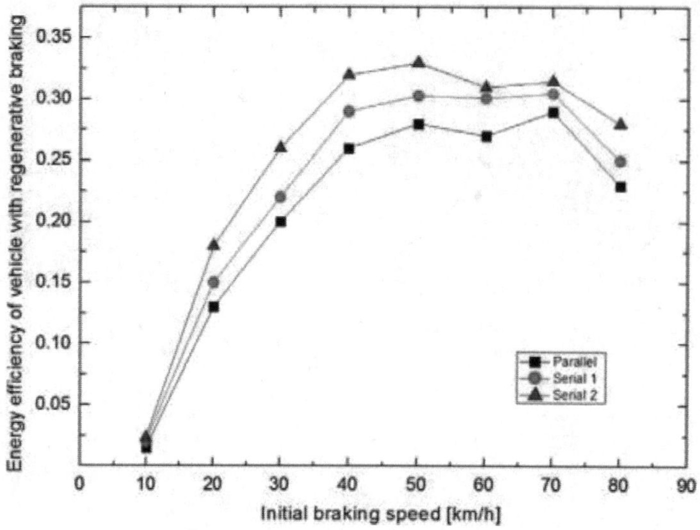

Figure 10. Regenerative braking efficiency depending on the regenerative braking strategy [23].

4. Conclusion

The range of electric-powered vehicles can be one of the problems in the application of such vehicles. In order to use the kinetic energy that the vehicle has when moving, a system of regenerative braking or recuperation has been designed on the vehicles. Using this system, instead of converting kinetic energy into thermal energy, using regenerative braking the kinetic energy is converted into electrical energy. In this way, the batteries of electric-powered vehicles are recharged and the range of the vehicle is increased. Regenerative braking systems present on the market today cannot use their full potential, i.e. they do not convert the entire kinetic energy into electricity, but only a part is used to charge the batteries, bearing in mind that friction brakes are also activated on the vehicle. The regenerative braking system is part of the braking system, i.e. it is only an addition to the friction brakes. The activation of the friction brakes and the regenerative braking system depends on the operation strategy, which depends on the force on the brake pedal. In practice, there are different regenerative braking strategies, which greatly affect efficiency. Furthermore, the effectiveness of this system depends on the category and type of vehicle. The advantages that this system provides are the extension of the vehicle's range of motion, the reduction of the intensity of wear of the brake friction pairs, and therefore, the reduction of particle emissions that occur during braking, and additional assistance when slowing down or stopping the vehicle. The regenerative braking system is already in use today on electric-powered vehicles, but the further developments of this system are aimed at increasing the efficiency of the regenerative braking system, but also the efficiency degree of the entire braking process.

References

[1] Global EV Outlook 2022, https://www.iea.org/data-and-statistics/data-product/global-ev-outlook-2022, (accessed 20 Jun 2022).
[2] Irle R, Global EV Sales for 2021, https://www.ev-volumes.com/, (accessed 20 Jun 2022).
[3] Kane M, Global Plug-In Electric Car Sales Doubled In February 2022, https://insideevs.com/news/578241/global-plugin-car-sales-february2022/, (accessed 22 Jun 2022).
[4] Richter F, Ford Announces Restructuring to Catch Up With EV Leaders, https://www.statista.com/chart/26981/top-electric-vehicle-makers/, (accessed 22 Jun 2022).
[5] Kane M, Volkswagen Group Global BEV Sales In Q1 2022: Almost 100,000, https://insideevs.com/news/580323/volkswagen-group-global-bev-sales-2022q1/, (accessed

22 Jun 2022).
[6] Kane M, Who Sells The Most All-Electric Cars: Tesla, Volkswagen, BYD?, https://insideevs.com/news/580346/who-sells-most-electric-cars-2022q1/, (accessed 28 Jun 2022).
[7] Li Y 2016 Infrastructure to Facilitate Usage of Electric Vehicles and its Impact *Transp. Res. Proc.* **14** 2537–43
[8] Ehsani M 2014 Conventional fuel/hybrid electric vehicles *Alternative Fuels and Advanced Vehicle Technologies for Improved Environmental Performance* ed R Folkson (Cambridge: Woodhead Publishing) pp 632–54.
[9] Loveday E, What Is Regenerative Braking And Why Is It Useful For Electric Cars?, https://insideevs.com/news/378161/regen-brake-electric-car/, (accessed 03 July 2022).
[10] Woodford C, https://www.explainthatstuff.com/how-regenerative-brakes-work.html, (accessed 08 July 2022).
[11] Atienza I, How Regenerative Braking Works, https://goflatoutph.com/2014/09/13/techno-logic-regenerative-braking/, (accessed 08 July 2022).
[12] Mercedes-Benz EQC, https://www.netcarshow.com/mercedes-benz/2020-eqc/#22, (accessed 08 July 2022).
[13] What is regenerative braking?, https://www.tiresplus.com/blog/brakes/what-is-regenerative-braking-in-electric-vehicles/, (accessed 10 July 2022).
[14] Grigoratos T and Martini G 2014 Brake wear particle emissions: a review *Environ. Sci. Pollut. Res.* **22** 2491 2501.
[15] Tawadros P, Zhang N and Boretti A 2014 Integration and performance of regenerative braking and energy recovery technologies in vehicles *Alternative Fuels and Advanced Vehicle Technologies for Improved Environmental Performance* ed R Folkson (Cambridge: Woodhead Publishing) pp 541–563.
[16] Boechler E, Campbell A, Hanania J, Jenden J and Donev J, Energy Education - Law of conservation of energy, https://energyeducation.ca/encyclopedia/Law_of_conservation_of_energy, (accessed 15 July 2022).
[17] Regenerative Braking, https://cecas.clemson.edu/cvel/auto/systems/regenerative_ braking.html, (accessed 15 July 2022).
[18] Lv C, Zhang J, Li Y and Yuan Y 2015 Mechanism analysis and evaluation methodology of regenerative braking contribution to energy efficiency improvement of electrified vehicles *Energy Convers. Manage* **92** 469–82
[19] Von Srbik M T and Martinez-Botas R F 2012 Vehicle optimisation for regenerative brake energy maximisation *Sustainable Vehicle Technologies* ed Gaydon and Warwickshire (Cambridge: Woodhead Publishing) pp 165–174.
[20] Day A and Bryant D 2022 Electronic Braking Systems *Braking of Road Vehicles* (Oxford, Cambridge: Butterworth-Heinemann is an imprint of Elsevier) pp 385-425.
[21] Chau K T 2014 Pure electric vehicles Alternative Fuels and Advanced Vehicle Technologies for Improved Environmental Performance ed R Folkson (Cambridge: Woodhead Publishing) pp 655–684.
[22] Varocky B J 2011 Benchmarking of Regenerative Braking for a Fully Electric Car, *TNO Automotive, Helmond & Technische Universiteit Eindhoven* (Eindhoven: University of Technology) Report No. D&C 2011.002.
[23] Qiu C and Wang G 2016 New evaluation methodology of regenerative braking contribution to energy efficiency improvement of electric vehicles *Energy Convers. Manage* **119** 389–98

IOP Conf. Series: Materials Science and Engineering 1271 (2022) 012026 doi:10.1088/1757-899X/1271/1/012026

Improving vehicle dynamics employing individual and coordinated sliding mode control in vehicle stability, active front wheel steering and active rear wheel steering systems in co-simulation environment

V Changoski[1], I Gjurkov[1], V Jordanoska[1]

[1] Faculty of Mechanical Engineering – Skopje, "Ss. Cyril and Methodius" University, Rugjer Boshkovikj 18, Skopje, Republic of N. Macedonia

E-mail: vasko.changoski@mf.edu.mk

Abstract. The goal for the future, safer and more sustainable transportation leads to implementation of many advanced systems in vehicles. In this paper, a vehicle model with stability control system that uses the braking system is regarded as a base active vehicle. Two alternative vehicle models are considered where this system is combined and integrated with active front wheel steering (AFS) and active rear wheel steering (ARS) system, separately. For the purpose of this research, 3D virtual vehicle models based on a B-segment vehicle were created in ADAMS/Car. As a research tool, a co-simulation approach between ADAMS/Car and MATLAB/Simulink was used. Several sliding mode controllers (SMC) have been proposed and implemented in MATLAB/Simulink in order to analyse the potential improvements of the vehicle dynamics due to the integration and coordination of these systems. In the same MATLAB/Simulink environment, a reference 2DOF nonlinear bicycle model was used. The vehicle models were simulated in driving scenarios based on standardized ISO 7401 test procedures. The scenarios include situations where the passive vehicle loses its stability or fails to complete the manoeuvre while the vehicles with stability control or integrated control systems successfully complete it.

1. Introduction

Implementation of advanced automated systems in the vehicles leads to improved vehicle dynamics, increased stability and safer road transportation. Many of these systems are part of the standard equipment in today's vehicles, but also according to the new European Regulation 2018/858 [1] the number of mandatory automated systems is expected to increase in the following years.

One of the most wide-spread systems and the one that is responsible for drastically improving vehicle stability by selective wheel braking is the system commercially known as Electronic Stability Program (ESP) or Vehicle Dynamic Control (VDC) [2], [3] and [4] but also as Vehicle Stability Control (VSC). For the purposes of this research the reference Vehicle Stability Control (VSC) will be used.

Beside the VSC system, the steering system of the vehicle represents wide research area which contributes into improving vehicle dynamics and stability. Authors in [5] proposed coordinated chassis control system which integrates differential braking, traction control and Active Front Steering (AFS) systems. The proposed control algorithm and tire-slip penalty function, enhances vehicle dynamics and allows easier minimization of tire dissipation. A fuzzy-logic coordinative cooperative controller is suggested in [6] where the combination of yaw control using selective wheel braking, AFS and active suspension normal force control (NFC) shows improved vehicle dynamics and handling over a vehicle with only VSC system.

In addition to the AFS system, a combination between AFS and Active Rear Steering (ARS) is presented in [7] where a Sliding Mode Controller (SMC) allows smaller values of the side-slip angle of the vehicle. The authors also suggest four-wheel steering (4WS) reference model which further improves the vehicle's performance. Ferrara in [8] presents an implementation of the SMC in longitudinal, lateral and vertical dynamic for two-wheel steering (2WS) and 4WS vehicles. Using more complex models and co-simulation methods, authors in [9] using MATALB/Simulink and ADAMS, analyzed the implementation of SMC controller for VDC, while in [10] a co-simulation approach between MATALB/Simulink and ADAMS is used to compare the usage of SMC controller and Linear Quadratic Regulator (LQR) for selective wheel braking in order to achieve improved vehicle dynamics. In the same paper, a comparison of more sudden and impulsive activation of the SMC controller over more continuous activation of the LQR controller is presented.

A co-simulation analysis is also conducted in [11] where slip-based traction controller is implemented in order to optimize the traction controller of an electric vehicle. In the paper, the SMC controllers are implemented in vehicle dynamics control, AFS and ARS systems. From the co-simulation method, by using 3D virtual models created in ADAMS/CAR as a test vehicles and non-linear bicycle models as reference models in MATALAB/Simulnik, more realistic and accurate results can be obtained. This research paper analyzes the importance of the implementation of appropriate reference model and it's influence on the output of the SMC controller. It also presents the advantages of implementation of coordinated systems by combining the SMC controllers. The results and advantages of the coordinated systems are presented through standardised test manouvers using the standard ISO 7401[12] using diagrammatic representation of the vehicle parameters.

2. Vehicle models

For the purpose of this co-simulation research and in order to analyze the implementation of the SMC controllers, several vehicle models were created. A virtual 3D vehicle model (figure 1) was created in ADAMS/CAR which is based on a B-segment vehicle. The basic model consists of conventional steering mechanism, independent front suspension (MacPherson suspension), rear semi-dependent twist beam axle suspension, power unit of 77kW, braking system, chassis and 195/65 R16 tire models from the library of ADAMS/CAR based on the Pacejka model 2002(Magic tire Formula). The parameters of the vehicle are presented in table 1.

Figure 1. 3D virtual vehicle model in ADAMS/CAR.

Table 1. Virtual vehicle model – parameters.

Mass:	1100 kg
Vehicle weight distribution (front/rear):	55/45 %
Length:	4040 mm
Width:	1734 mm
Wheelbase:	2493 mm
Front track width:	1508 mm
Rear track width:	1471 mm

The basic model of the vehicle is upgraded and alternative variations were created. The first alternative vehicle includes implementation of VSC that uses the brake system and is regarded as base "active" vehicle. Other models represent further upgrade of this model. In the second active vehicle model, the conventional steering mechanism is replaced with an AFS system, while in the third variant an ARS system is integrated into the base active vehicle.

In order to conduct the co-simulation research, reference models were created using MATLAB/Simulink. The basic reference model that was used represents a 2DOF bicycle model (figure 2). It is defined by its lateral velocity and yaw rate, while the longitudinal velocity is not constant. Its value is read on-line from the virtual 3D model. This model is used as reference model in the implementation of the VSC system alone and its integration with the AFS. The tire models are defined using the equation (1) and (2) of Pacejka model described in [13]:

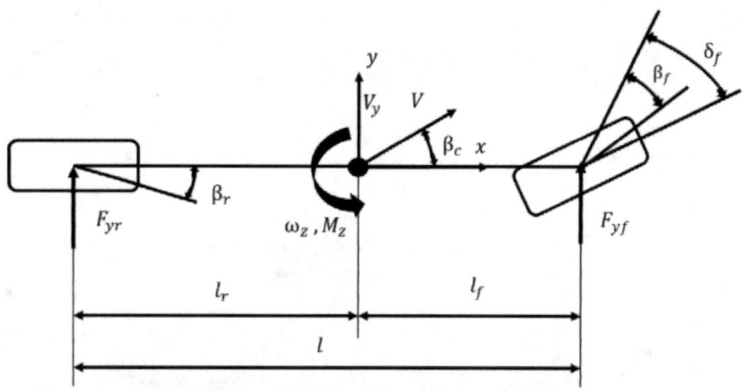

Figure 2. Bicycle vehicle model.

$$F_{yf} = Dsin\{Carctan[B\beta_f - E(B\beta_f - arctan(B\beta_f))]\} \quad (1)$$

$$F_{yr} = Dsin\{Carctan[B\beta_r - E(B\beta_r - arctan(B\beta_r))]\} \quad (2)$$

The parameters of the bicycle model and the parameters for the Magic Tire Formula for wet surface [14] are presented in table 2. The parameters for the Magic Tire Formula were chosen for driving manoeuvers performed/simulated on a wet surface with a friction coefficient of φ=0.4.

Table 2. Bicycle vehicle model – parameters.

Mass (m):	1100 kg
Wheelbase (l):	2493 mm
Distance from centre of mass to front axle (l_f):	1122 mm
Inertia radius (i):	1250 mm
Front wheel cornering stiffness ($K_{\alpha f}$):	77000 N/rad
Rear wheel cornering stiffness ($K_{\alpha r}$):	77000 N/rad
B	$C_{F\beta}/(CD)$
$C_{F\beta}$	$K_a sin\left\{(2\,atan\left(\frac{F_n}{F_{nom}}\right)\right\}$
C	1.2
D	φF_n
E	0

For the integration of the VSC and the ARS systems, this reference vehicle model wasn't suitable to satisfy the necessary requirements for the activation of the sliding mode controller and the reaction of the 3D vehicle model wasn't suitable and led to degradation of the performance and its stability (this was due to the fact that the reference model had slower reaction than the virtual 3D model; regarding this, a more detailed explanation will be presented in the next chapter). In order to overcome this, a 2DOF 4WS bicycle model was created. The control strategy for the rear wheels was derived from the linear bicycle mode [14], governed by the equations (3) and (4) with the requirement for a zero side-slip angle in both steady state and transient state.

Because a linear control strategy was used in nonlinear bicycle model, the reference slide slip angle did not have zero value, but it was significantly smaller than the reference value of the 2WS model.

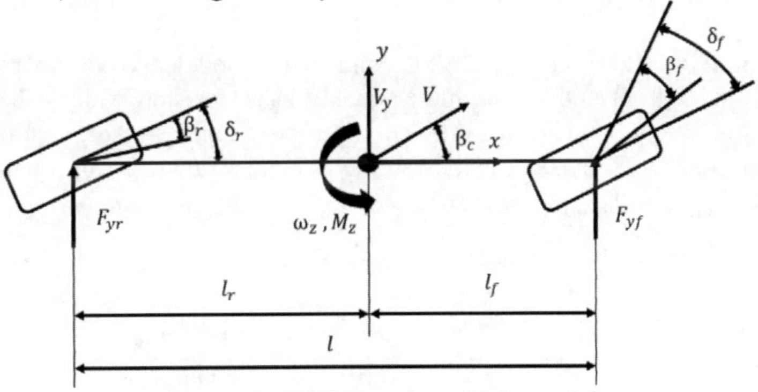

Figure 3. 4WS Bicycle vehicle model.

$$\dot{V}_y = \left(\frac{-K_{\alpha f}-K_{\alpha r}}{mV_x}\right)V_y + \left(\frac{K_{\alpha r}l_r-K_{\alpha f}l_f}{mV_x} - V_x\right)\omega_z + \frac{K_{\alpha f}}{m}\delta_f + \frac{K_{\alpha r}}{m}\delta_r \quad (3)$$

$$\dot{\omega}_z = \left(\frac{K_{\alpha r}l_r-K_{\alpha f}l_f}{I_zV_x}\right)V_y + \left(\frac{-K_{\alpha f}{l_f}^2-K_{\alpha r}{l_r}^2}{I_zV_x}\right)\omega_z + \frac{K_{\alpha f}l_f}{I_z}\delta_f - \frac{K_{\alpha r}l_r}{I_z}\delta_r \quad (4)$$

$$\delta_r = -\frac{K_{\alpha f}}{K_{\alpha r}}\delta_f + \frac{mV_x^2+K_{\alpha f}l_f-K_{\alpha r}l_r}{K_{\alpha r}V_x}\omega_z \quad (5)$$

3. Individual and coordinated Sliding Mode Controllers

As a control method in this paper, sliding mode control was used. Applying the theory of the Sliding Mode Control, firstly the sliding surface (s_i) is defined using equation (6). The error (e), that is observed represents the difference between the yaw rate of the 3D virtual model in ADAMS ω and the yaw rate of the reference model ω_z (7):

$$s_i = \dot{e} + \lambda_i e \tag{6}$$

$$e = \omega - \omega_{z.} \tag{7}$$

where λ_i is defined as a constant which multiplies the error and the sliding surface s is defined as $s = 0$ in stationary conditions.

For the controller output (u_i), different mathematical models are suggested in [15] such as using step and signum function, but for a more robust controller, a *tanh* function is suggested. This was used and validated in the design of vehicle stability controller in equation (8):

$$u_i = k_i tanh\left(\frac{s}{\phi_i}\right) + u_{eqi} \tag{8}$$

where k_i represents the maximum value of the signal that the controller can produce, ϕ_i represents the thickness of the boundary layer of the sliding surface and where u_{eqi} identifies the desired output signal of the controller in stationary conditions.

Due to the use of nonlinear bicycle model, the decrease of longitudinal speed during simulation and the fact that in the test manoeuvres the vehicles are brought to their limits, the response of the reference models is relatively slow. In order to create a signal from the controller that would improve vehicle response and settling time, a maximum value of the reference yaw rate is defined using equation (9):

$$\omega_{max} = \left|\frac{\varphi g}{V_x}\right| \tag{9}$$

where $\varphi = 0.4$ and longitudinal velocity V_x is obtained from the 3D virtual model. This suggests that the maximum yaw rate is not constant and depends on the current speed of the vehicle.

3.1 VSC controller

During the braking process of the wheels, the VSC controller is modeled to control and to activate both wheels on the same side (front and rear) simultaneously. The decision which wheels to be braked depends on two criteria: turning direction of the steering wheel and the steering character of the vehicle during the simulation process. This is defined using the reference yaw rate (ω_{ref}) and the yaw rate of the 3D virtual model (ω). The decision making process of the controller is presented in table 3.

Table 3. Decision making process of the controller [16].

ω_{ref}	$\omega - \omega_{ref}$	**Steering**	**Braked wheels**
$\omega_{ref} > 0$	≥ 0	Oversteer	Right wheels
	< 0	Understeer	Left wheels
$\omega_{ref} < 0$	> 0	Understeer	Right wheels
	≤ 0	Oversteer	Left wheels

The sliding mode controller for the VSC is defined using equations (6), (7) and (8). The equations are adapted for this control strategy and presented in equations (10), (11) and (12):

$$s_1 = \dot{e} + \lambda_1 e \tag{10}$$
$$u_1 = k_1 tanh\left(\frac{s_1}{\phi_1}\right) + u_{eq1} \tag{11}$$

where $k_1 = 6$, while $\phi_1 = 3$, $\lambda_1 = 30$ and $u_{eq1} = 0$.

The final form of the control strategy is presented in equation (12):

$$u_1 = 6tanh\left(\frac{\dot{e}+30e}{3}\right) \tag{12}$$

where u_1 is defined as hydraulic brake pressure for each wheel and the coefficient k_1 is defined as the maximum hydraulic pressure of the front brakes that can be achieved by the system (6 MPa). On the other hand, the maximum hydraulic pressure for the rear brakes is 4 MPa. To achieve this, the signal of the controller is divided with appropriate proportion coefficient. The values for ϕ_1, λ_1 and u_{eq1} were adopted depending on the vehicle behaviour. Larger values of ϕ_1, which is the thickness of the boundary layer of the sliding surface, would make the controller more robust, but would decrease the effect of the vehicle stabilisation process. The value of λ_1 allows scaling of the error (e) and lowers the oscillations caused by the first derivative ($\dot{e}$), while u_{eq1} is the hydraulic pressure during stationary conditions when the brakes are not activated.

Due to the sensitivity of the controller, even the smallest difference between the reference yaw rate and the yaw rate of the virtual model generates output of the controller and application of the brakes. This would maybe significantly improve the vehicle safety, but it would also cause severe degradation of the vehicle performance and reliability. In order to avoid this, another controller is created, using the phase-plane method ($\beta - \dot{\beta}$). The control algorithm is defined as:

$$\left| C_1\beta + C_2\dot{\beta} \right| \leq 0.8 \tag{13}$$

where $C_1 = 2.41$ and $C_2 = 9.615$. These values are recommended in [16] and used in a similar model. The difference in this paper is the threshold limit where it is defined as 0.8. This value was defined during the research in order to make the VSC controller prevent vehicle destabilisation, but also not to allow frequent activation of the brakes.

When the condition in equation (13) is satisfied, the vehicle is in the stable region of the phase-plane ($\beta - \dot{\beta}$) where there is no danger of losing its stability. On the other hand, when the condition is not met and the vehicle tends to destabilise, this controller allows activation of the VSC controller and the application of the brakes in order to return the vehicle back in the stable region.

By implementing this strategy, the brakes would activate only when there is danger of vehicle destabilization. The defined VSC controller is used in both combinations of the coordinated VSC and AFS and VSC and ARS controllers.

3.2 Coordinated VSC and AFS controller

The upgrade from the previous controller is made by implementing the control strategy for the AFS. Again, the difference between the yaw rate and the yaw rate of the reference model is used as an error in the control strategy.

$$s_2 = \dot{e} + \lambda_2 e \tag{14}$$
$$u_2 = k_2 tanh\left(\frac{s_2}{\phi_2}\right) + u_{eq2} \tag{15}$$

where $k_2 = 0.174$, while $\phi_2 = 3$, $\lambda_2 = 2$ and $u_{eq2} = 0$.

The final form of the control strategy is presented in equation (16):

$$u_2 = 0.174tanh\left(\frac{\dot{e}+2e}{3}\right) \tag{16}$$

IOP Conf. Series: Materials Science and Engineering 1271 (2022) 012026 doi:10.1088/1757-899X/1271/1/012026

where u_2 represents the front wheels correction angle of the AFS system and the value for k_2 limits the AFS system the correction of the steering angle of the front wheels to $10°$ $(0.174\,rad)$. The values of the coefficients ϕ_2, λ_2 and u_{eq2} were adopted using several simulation trials, depending on the vehicle behaviour.

Unlike the previous controller where the activation needed is required to be executed momentarily and in short time interval, here the working principle of the system and the controller require continuous control of the system. It must be pointed out that this system is not working independently, but the main input for the front wheels steering angle is given by the driver, while the AFS controller helps in vehicle stabilisation.

3.3 Coordinated VSC and ARS controller

The last combination of coordinated control includes the ARS controller. In this case, unlike the coordinated controller between the VSC and AFS, the control of the steering angle of the rear wheels depends solely on the controller output. By taking into consideration the fact that the sliding mode control acts more aggressively than some other control strategies, even smallest error can cause the activation of the controller. Because of this, a first order filter is implemented where the output signal of the controller is limited to 5 Hz. If the filter is not used then the steering of the rear wheels can be very sudden and often would cause irregularities to the vehicle yaw movement and destabilisation.

$$s_3 = \dot{e} + \lambda_3 e \tag{17}$$

$$u_3 = k_3 tanh\left(\frac{s_3}{\phi_3}\right) + u_{eq3} \tag{18}$$

where $k_3 = 0.087$, while $\phi_3 = 10$, $\lambda_3 = 50$ and $u_{eq3} = 0$.

The final form of the control strategy is presented in equation (19):

$$u_3 = 0.087 tanh\left(\frac{\dot{e}+50e}{10}\right) \tag{19}$$

where u_3 is defined as the rear wheels steering angle and the value for k_3, limits the ARS system to correction of the steering angle of the rear wheels to 5° $(0.087\,rad)$. By means of several simulation trials, the values of the coefficients ϕ_3, λ_3 and u_{eq3} were determined. In this case, where the control signal must be continuous and any sudden reaction would cause degradation of the vehicle performance, the values of the coefficients are larger than in the previous controllers. The values of the coefficients ϕ_3 and λ_3 allow the controller to be more robust. The coefficient λ_3 attenuates the oscillations of the system caused by the first derivative of the error ($\dot{e}$), as well.

As mentioned before, in this case a 4WS reference bicycle model is used. By conducting the research, it was concluded that the basic reference model is forcing the ARS controller to act more frequently and to degrade the vehicle performance, thus requiring a more sophisticated reference model.

4. Simulated manoeuvres and result analysis

The benefits of using the proposed sliding mode controllers are presented in this section by conducting co-simulation between ADAMS/CAR and MATLAB/Simulink and result analysis. In the analysis, the vehicles are simulated in standardised test procedures on a wet surface with friction coefficient of φ=0.4 according to the standard ISO 7401 [12], for testing lateral transient response of the vehicles. Deliberately, within the manoeuvres, the passive vehicle is forced over its limits and loses its stability.

The results and the analysis are presented for the step-steer manoeuvre and the single lane change by using comparative diagrams between the passive and the alternative vehicles. The diagrams that are presented are: steering wheel angle, steered wheels angle, lateral acceleration, yaw rate, side slip angle, longitudinal velocity, trajectory of the vehicles and the braking torque vs time.

As input signals to the ADAMS 3D vehicle models, where VSC system is implemented, the braking hydraulic pressure of the 4 wheels are used. Beside these signals, in the vehicle with implemented VSC and AFS controllers (VSC+AFS vehicle) the corrective steering wheel angle for the AFS system is used, while as an input in the vehicle with implemented VSC and ARS controllers (VSC+ARS vehicle), the rotation of the rear pinion gear for the ARS system is used. The output signals of the ADAMS model, such as the steering wheel angle, side slip angle, and yaw rate are used in the Simulink models as inputs for the reference vehicles and for the control strategies.

4.1 Step-steer

The first manoeuvre that will be presented and analysed is step-steer. The initial velocity of the vehicles is 80 km/h, and in the first second of the manoeuvre the vehicle is traveling in a straight line. Afterwards, steering wheel angle is applied within 0.5s, when it reaches its maximum value (figure 4). The values of the steering wheel angle are determined in order to achieve 4 m/s^2 in steady-state condition (figure 6) as suggested in the standard ISO 7401. The sudden drop of the lateral acceleration in figure 6 are caused by the activation of the brakes. During the simulation, the vehicle is traveling in a left turn and when the right side brakes are firstly activated (figure 12-14) they force the vehicle to travel on a curved trajectory with larger radius, thus lowering the lateral acceleration. Next, the activation of the left side brakes (figure 12-14), causes the vehicle to return on its original trajectory which results in increase of the lateral acceleration approximately up to the previous values, before the control action. Because of the different strategies, this results in different front and rear steered wheels angles (figure 5). Due to the fact that the maximum values of the reference yaw rate are achieved, the 3D virtual model is forced to drive in a curved trajectory on its limits. By taking into consideration equation (8), the maximum reference yaw rate is increasing during the simulation, because the actual longitudinal velocity of the vehicle is decreasing.

This leads to constant correction of the controllers, until the longitudinal velocity of the vehicle is lowered enough and the vehicle will no longer be driving on its limits. Unlike the passive vehicle where it is obvious that loses its stability, that's not the case with the alternative vehicles. The VSC controller activates at the moment when the maximum yaw rate (figure 7) is achieved and stabilises the vehicle. It is interesting to analyse the VSC+AFS and VSC+ARS vehicles where the value of the side-slip angle is lower than the other vehicles (figure 8 and figure 9) and they tend to have lower values of the yaw rate after the activation of the VSC controller.

The last observation is explicitly visible in the VSC+ARS vehicle. The rear wheel steering forces faster reaction from the vehicle and after reaching steady-state conditions, the controller slowly turns the rear wheels to follow the maximum yaw rate, defined by the reference model. After reaching the limits, it is obvious that the VSC controller activates and decreases the yaw rate of the vehicle. Also, in this case, the VSC+ARS vehicle completes the manoeuvre with the highest longitudinal speed (figure 10).

Different timing of activation of the VSC controllers is presented in figures (figure 12, figure 13 and figure 14).

It must be pointed out that this is an open loop manoeuvre, and the driver doesn't make a correction with the steering wheel. With this taken into consideration and the fact that the vehicles are on their physical limits, this results in activation of the controllers to stabilise the vehicle, but afterwards the vehicles still attempt to achieve their limits. This would continue until the longitudinal velocity is decreased under a value which is not considered to be critical for the stability of the vehicle. To conclude, the VSC controller and its coordinated activation with the other controllers results in vehicle stabilisation and faster response when analysing the VSC+ARS vehicle. In real driving situation the driver would make a correction with the steering wheel and the vehicles would not be forced to attempt to achieve their limits again.

IOP Conf. Series: Materials Science and Engineering 1271 (2022) 012026 doi:10.1088/1757-899X/1271/1/012026

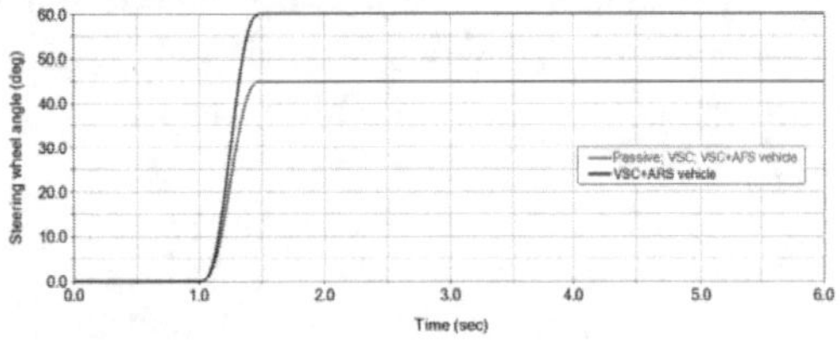

Figure 4. Steering wheel angle.

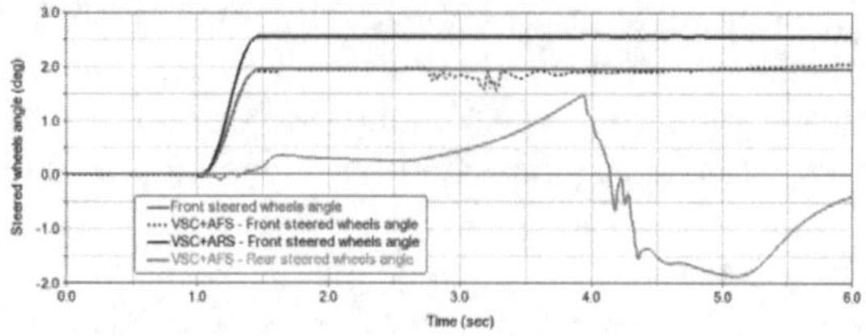

Figure 5. Front and rear steered wheels angle.

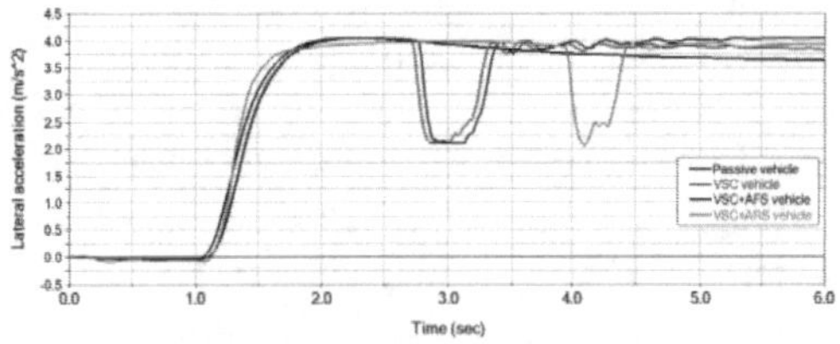

Figure 6. Lateral acceleration.

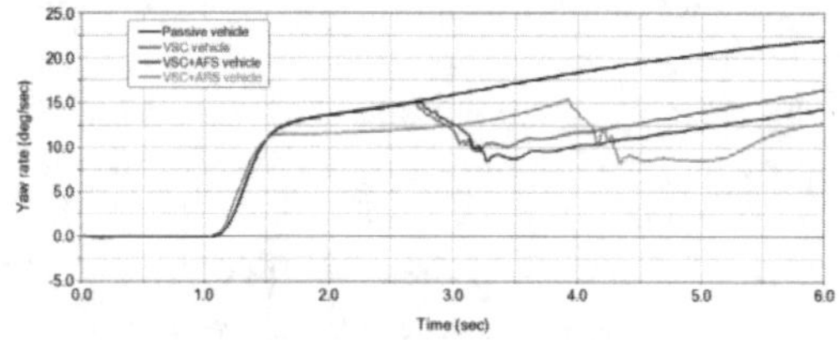

Figure 7. Yaw rate.

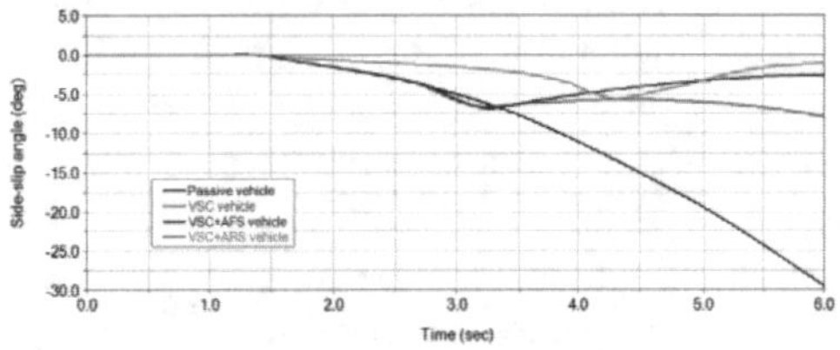

Figure 8. Side slip angle.

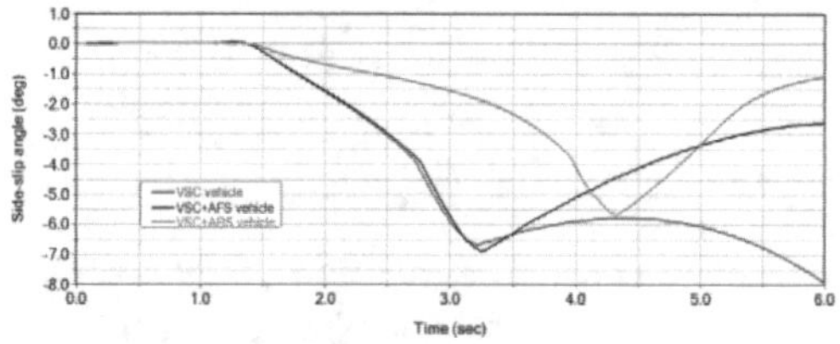

Figure 9. Side slip angle of the alternative vehicles.

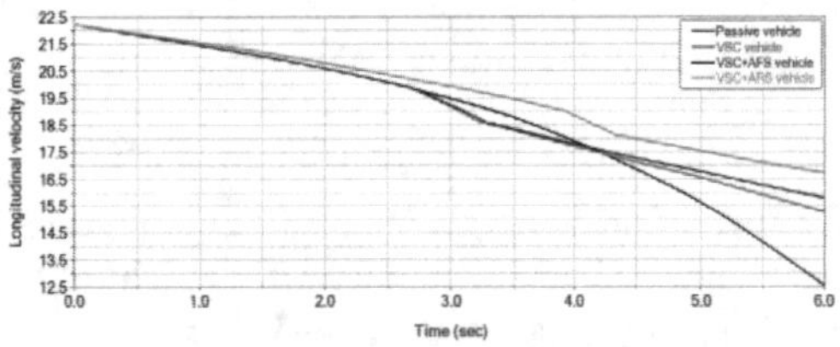

Figure 10. Longitudinal velocity.

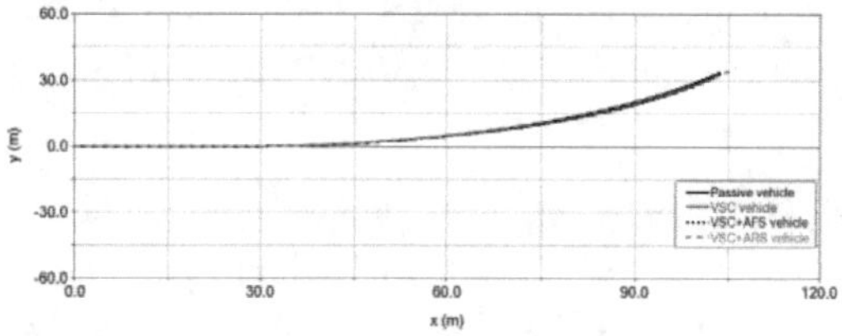

Figure 11. Trajectory.

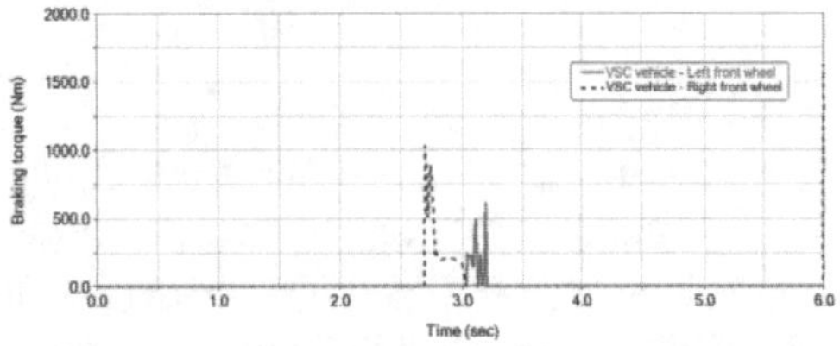

Figure 12. Braking torque – VSC vehicle.

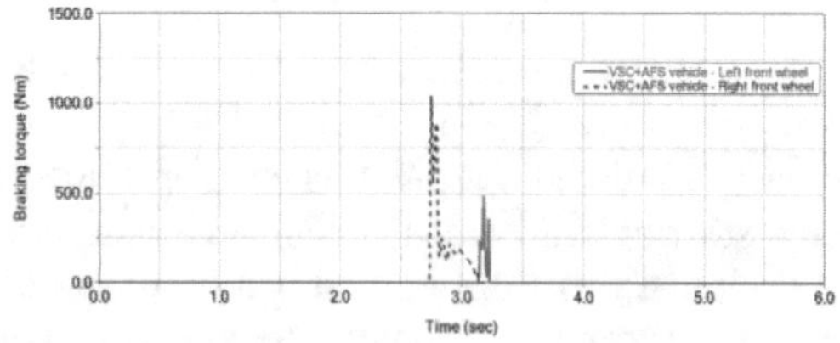

Figure 13. Braking torque – VSC+AFS vehicle.

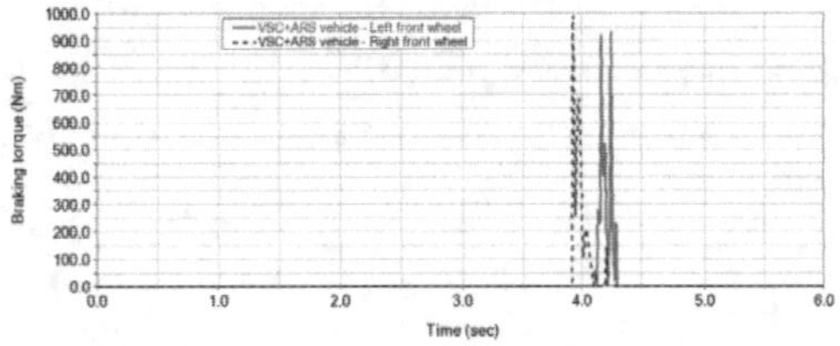

Figure 14. Braking torque – VSC+ARS vehicle.

IOP Conf. Series: Materials Science and Engineering 1271 (2022) 012026 doi:10.1088/1757-899X/1271/1/012026

4.2 Lane-change

In this simulation, the initial velocity of the vehicles is 80 km/h, the input signal form the driver represents a full sinusoid function of 0.5 Hz where the maximum steering wheel angle is 80° (figure 15) in order to achieve 4 m/s^2 of lateral acceleration in the first peak value. This approach is used for every vehicle, and depending on the vehicle variant, different steered wheels angles are applied (figure 16). The front steered wheels angle of the passive vehicle, the VSC and VSC+ARS vehicle depends exclusively on the driver input and physical parameters of the steering mechanism. On the other hand, the front steered wheels angle of the VSC+AFS vehicle is corrected due the action of the AFS system. Even though those corrections are relatively small, it can be observed that those corrections imply improvements in the vehicle dynamic behaviour.

In the results presented on figure 17, 18, 19 and 20 it can be observed that the passive vehicle fails to complete the manoeuvre and swerves out of control. That is explicitly visible on the side-slip angle diagram (figure 19) and of the trajectory diagram (figure 22).

On the other hand, every alternative vehicle with an active system successfully completes the manoeuvre. Based on the diagrams it can be concluded that the VSC system has the biggest impact on the vehicle stability because it stabilises the vehicle only with his solitary actuation.

Similar to the previous analysis, the sudden drop of the lateral acceleration (figure 17) can be observed in the VSC and VSC+AFS vehicles where the earlier activation of the brakes (figure 23 and 24) causes changes in the radius of the current curved trajectory. On the other hand, there is no sudden drop in the lateral acceleration values of the VSC+ARS vehicle. The rear steered wheels cause correction in the vehicle trajectory and keep the vehicle in the stable region, thus preventing earlier activation of the brakes (figure 25).

Further improvements of the vehicle dynamics and stability are observed in the vehicles in which the AFS and ARS systems are implemented on top of the VCS. That is obvious in the faster vehicle response and shorter settling time, lower values of the vehicle side-slip angle (figure 19 and 20) and smaller path width to complete the manoeuvre (figure 22).

Based on the results, it can be concluded that the combination of the VSC and ARS systems offers best improvements in the vehicle dynamics, compared to the other vehicles. This vehicle has fastest response and shorter settling time, smaller amplitudes and the longitudinal velocity of the vehicle is highest at the end of the manoeuvre (figure 21). This implies that this vehicle would successfully complete the manoeuvre in the shortest time and would least disturb driver's awareness of the vehicle traveling direction due to the smallest side-slip angle. It is interesting to observe the reaction of the VSC system on figure 23 which is activated more frequently, compared to the vehicles with coordinated control (figure 24 and figure 25). This is more visible in the VSC+ARS vehicle where the VSC activates later in manoeuvre, unlike the other vehicles. This results in larger longitudinal velocity during the manoeuvre and faster completion of the manoeuvre.

The advantages of using 4WS bicycle reference model can be observed bellow. The reference model forces a controller activation, which results in smaller values of the side slip angle, has fastest response time and less frequent activation of the VSC controller and requires minimal path width to complete the manoeuvre.

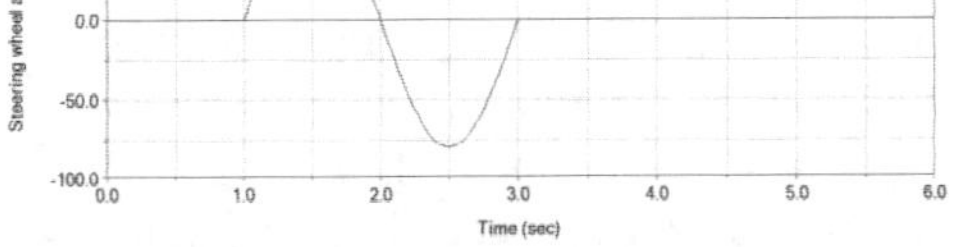

Figure 15. Steering wheel angle.

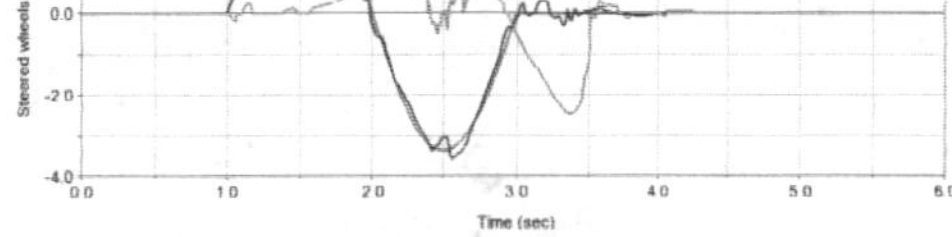

Figure 16. Front and rear steered wheels angle.

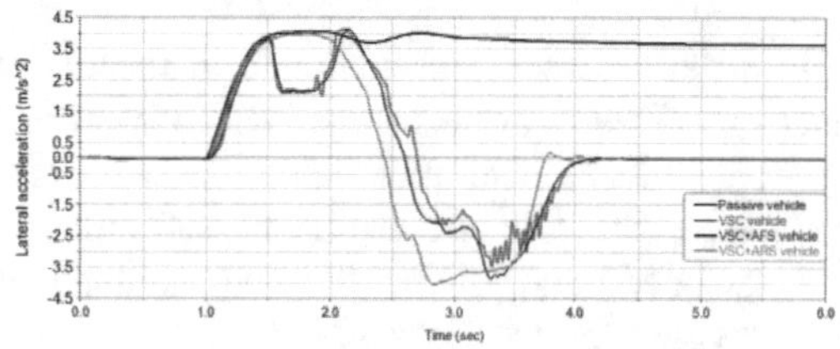

Figure 17. Lateral acceleration.

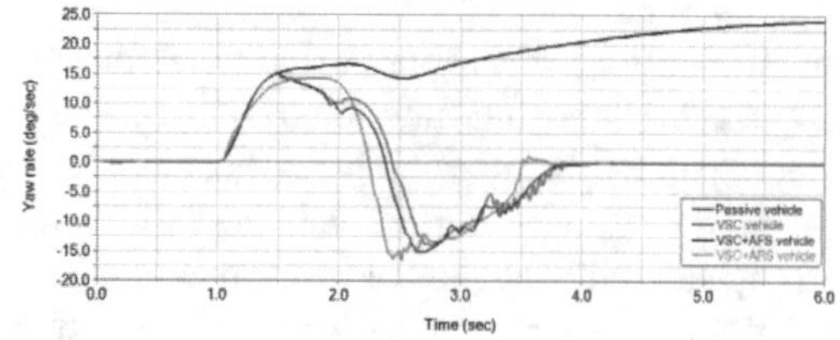

Figure 18. Yaw rate.

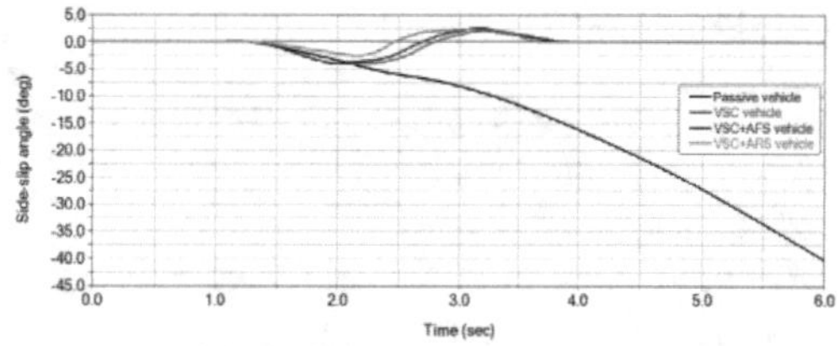

Figure 19. Side slip angle.

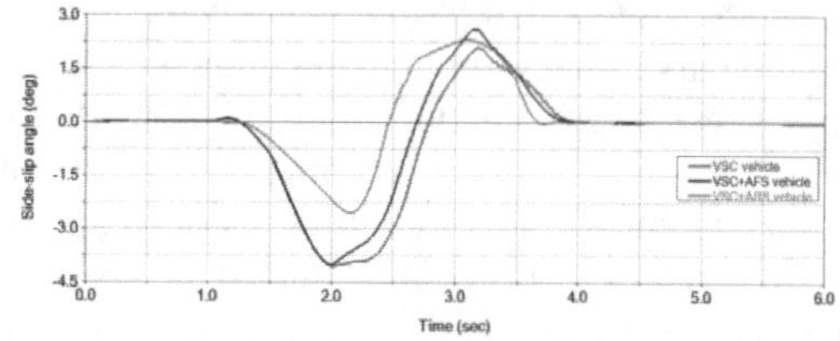

Figure 20. Side slip angle of the alternative vehicles.

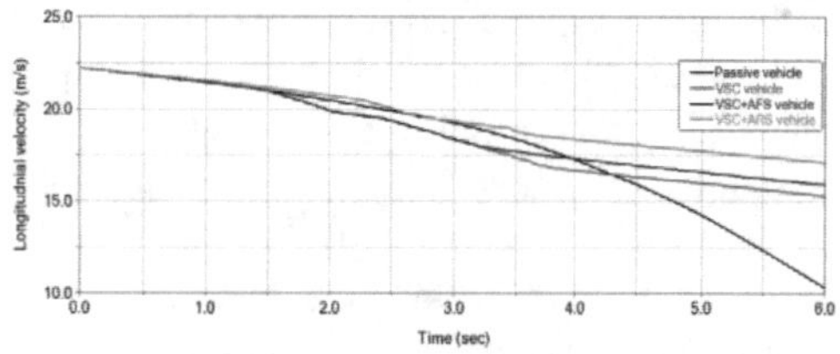

Figure 21. Longitudinal velocity.

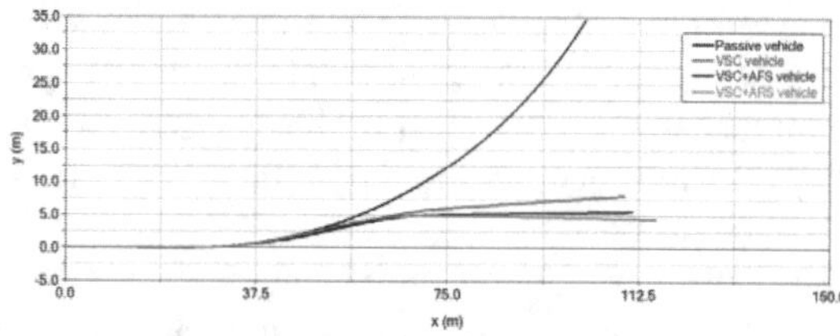

Figure 22. Trajectory.

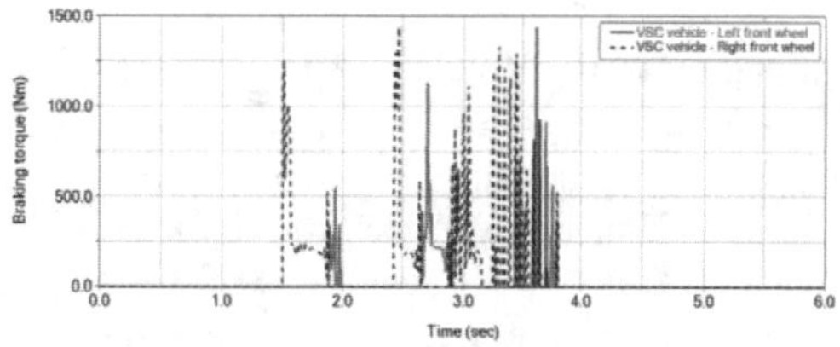

Figure 23. Braking torque – VSC vehicle.

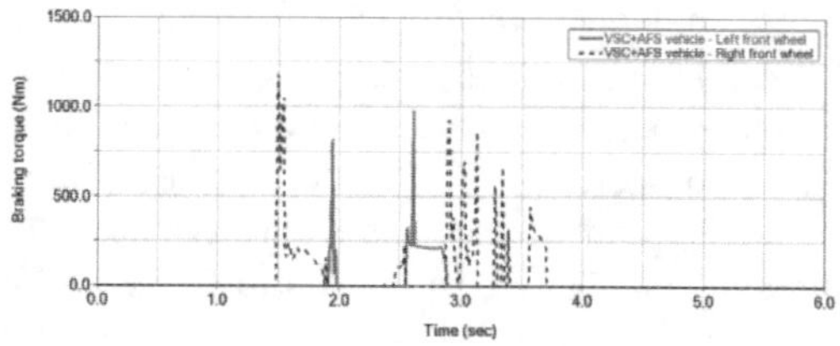

Figure 24. Braking torque – VSC+AFS vehicle.

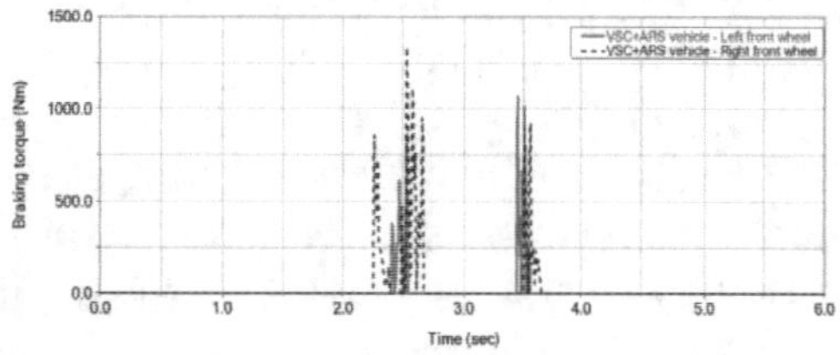

Figure 25. Braking torque – VSC+ARS vehicle.

5. Conclusions

By using co-simulation approach with a complex 3D vehicle model and nonlinear reference models, more accurate results were obtained that are closer to a real driving scenario. The implementation of the SMC controllers and their coordinated activation and action results in improvement in vehicle dynamics and stability. Although the VSC controller alone, has the biggest effect on the vehicle's stabilisation, it can be stated that the coordinative action of the controllers further improves the vehicle behaviour. This can be substantiated with the smaller amplitudes of the vehicle yaw rate and side-slip angle, the smallest required road space to complete the manoeuvre and also by the faster response from the vehicles.

Beside the co-simulation approach, the research showed the importance of using appropriate reference model. This is especially important when the reference vehicle has rear wheel steering. If a 2DOF bicycle model is used as reference model, it can lead to non-optimal controller output and unwanted vehicle behaviour. By implementing 2DOF 4WS bicycle model with a certain control strategy, improved controller output signal is produced enforced to follow the reference variable from the 2DOF 4WS model and achieve better vehicle dynamics and stability.

Nomenclature

B	Pacejka tire model stiffness factor
C	Pacejka tire model shape factor
$C_{F\beta}$	Pacejka tire model stiffness
C_1, C_2	$\beta - \dot{\beta}$ phase plane method coefficients
D	Pacejka tire model peak factor
E	Pacejka tire model curvature factor
e	Sliding mode error function
F_n	Vertical force
F_{nom}	Nominal vertical force
F_y	Lateral tire forces
F_{yf}	Front tire lateral force
F_{yr}	Rear tire lateral force
g	gravitational constant
I_z	Material moment of inertia
i	Inertia radius
k_i	Maximum value of the signal that the SMC controller can produce
K_a	Pacejka tire model change of stiffness with slip
$K_{\alpha f}$	Front wheel cornering stiffness
$K_{\alpha r}$	Rear wheel cornering stiffness
l	Wheelbase
l_f	Distance from the centre of mass to the front axle
l_r	Distance from the centre of mass to the rear axle
M_z	Yaw moment
m	Mass of the vehicle
s	Sliding mode surface function
u_i	SMC controller output
u_{eqi}	SMC desired output signal of the controller in stationary conditions
V_x	Longitudinal velocity
V_y	Lateral velocity
β_c	Side slip angle
β_f	Front tire side slip angle
β_r	Rear tire side slip angle
δ_f	Front steering wheels angle
δ_r	Rear steering wheels angle
λ_i	Constant which multiplies the error in SMC controller
ϕ_i	Thickness of the boundary layer of the sliding surface in SMC controller
φ	Road surface friction coefficient
ω	3D virtual model yaw rate
ω_{max}	Maximum value of the reference yaw rate

ω_{ref} Reference yaw rate
ω_z Bicycle vehicle model yaw rate

References

[1] https://eur-lex.europa.eu/legal-content/EN/TXT/?uri=CELEX%3A32018R0858
[2] Van Zanten A T 2000 Bosch ESP Systems: 5 Years of Experience *SAE Transactions* ***109*** 428–436
[3] Van Zanten A T, Erhardt R, Landesfeind K and Pfaff G 1998 VDC Systems Development and Perspective *SAE Transactions* ***107*** 424–444
[4] Van Zanten A T, Erhardt R and Pfaff G 1995 VDC The Vehicle Dynamics Control System of Bosch *SAE Transactions* ***104*** 1419–36
[5] Feng J, Chen S and Qi Z 2020 Coordinated Chassis Control of 4WD Vehicles Utilizing Differential Braking, Traction Distribution and Active Front Steering *IEEE Access* **8** 81055-68
[6] Jordanoska V, Danev D and Changoski V 2021 Evaluating Coordinated Cooperative Control of Three Active Car Systems Using Fuzzy-Logic *IOP Conference Series: Materials Science and Engineering* vol. 1190 no. 1 (IOP Publishing)
[7] Hiraoka T, Nishihara O and Kumamoto H 2004 Model following sliding mode control for active four-wheel steering vehicle *Review of Automotive Engineering* **25** 305-313.
[8] Ferrara A 2017 *Sliding Mode Control of Vehicle Dynamics* (London: The Institution of Engineering and Technology)
[9] Zhang S, Zhou S and Sun J 2009 Vehicle Dynamics Control Based on Sliding Mode Control Technology *2009 Chinese Control and Decision Conference* 2435-39
[10] Čangoski V, Đurkov I and Jordanoska V 2020 Implementation of advanced control methods for improved vehicle dynamics and safety using co-simulation approach *Mechanical engineering – scientific journal* **38(2)** 109-116
[11] Saha S and Amrr S M 2020 Design of slip-based traction control system for EV and validation using co-simulation between Adams and Matlab/Simulink *Simulation: Transactions of the Society for Modeling and Simulation International* **96(6)** 537-549
[12] https://www.iso.org/standard/54144.html
[13] Pacejka H 2012 *Tire and Vehicle Dynamics* (Butterworth-Heinemann)
[14] Ѓурков И 2015 Симулација на динамиката на возилата, (Скопје: Машински факултет – Скопје)
[15] Jalili-Kharaajoo M and Besharati F 2003 Sliding mode traction control of an electric vehicle with four separate wheel drives, *IEEE Conference on Emerging Technologies and Factory Automation* vol 2 pp 291-296
[16] Jin L, Xie X, Shen C, Wang F, Wang F, Ji S, Guan X and Xu J 2017 Study on electronic stability program control strategy based on the fuzzy logical and genetic optimization method *Advances in Mechanical Engineering* **9(5)** 1–13

Numerical simulation of an electromechanical brake concerning associated difficulties

A Poznić[1] and B Stojić[2]

[1] PhD assistant, University of Novi Sad, Faculty of Technical Sciences, Republic of Serbia, 21000 Novi Sad, Trg Dositeja Obradovića 6
[2] Associate professor, University of Novi Sad, Faculty of Technical Sciences, Republic of Serbia, 21000 Novi Sad, Trg Dositeja Obradovića 6

E-mail: alpoznic@uns.ac.rs.

Abstract. The biggest challenge in developing any type of brake (or any other actuator) is overcoming its inherent disadvantages. Brakes that use materials whose rheological properties change under the influence of an external magnetic field are called magnetorheological brakes. In addition to the mechanical issues, magnetorheological brakes are troubled with cooling, sealing, settling, fluids` gap size, etc. Furthermore, there are issues with brake's magnetic field density distribution and propagation. The poor distribution of the magnetic field density through the body of the magnetorheological brake is one of these inherent disadvantages. Magnetorheological's brake magnetic circuit modeling forms the basis for further development. This paper is part of ongoing research into the feasibility of using magnetorheological brakes in automotive applications. Certain points of the numerical simulation process are addressed. The research uses part of the results from previous numerical simulations, carried out using commercial finite element method software - COMSOL Multiphysics, AC/DC module, and results from the previous experiments.

(Some figures may appear in color and detail only in an electronic version)

1. Introduction

A magnetorheological brake - MRB is an electromechanical brake consisting of rotating and non-rotating elements, which can be made either from magnetic or non-magnetic materials. The working medium flows between the MRB stators and the rotors. Both magnetorheological fluids - MRFs and magnetorheological greases - MRGs are the best-known examples of working mediums, regularly used in the MRBs, [1, 2]. Using the control current in the excitation elements, the magnetic fields are generated inside the MRB's body. The working medium's resistance to flow changes under the influence of this magnetic field. The direction of the main magnetic field vector for each coil is directed through this MRB body toward the working medium's surface area. This induces a change in the stator-to-rotor rotational resistance, thus generating the braking torque. Sizes and shapes of all elements are directly correlated with the MRB type, torque requirements, and magnetic flux density distribution patterns.

The main issues with any MRB application are the working medium's viscosity level, solid-phase settling, and the overall braking torque value. There are several ways to overcome these issues. The first step is to use a high viscosity, low setting rate MR working medium. Secondly, if possible, to further reduce the MR gap size and increase the applied magnetic field acting on the MR working medium. The last one is to increase the total active area of the MR working medium by increasing the MRB's diameter and multiplying the number of working media layers.
The goal of this work was to simulate numerically magnetic field distribution through the MRG layers and the MRB. Working media layers are constant in gap size, shape, and height. A particular magnetic field distribution influence on each MRG layer was analyzed using a commercial Finite Element Method - FEM software. The simplification of the MRB FEM Model and the uniform magnetic flux distribution throughout all (MRG) layers were addressed in all the previous simulations as well. Although the results indicate a high matching level with results from previous simulations, there were numerous difficulties in this particular Model e.g.: the FEM Model size reduction, the right boundary conditions selection, and set-up, meshing process, and post-processing.

2. Magnetorheological brake

By combining some of the specifics found in other MRB types, a new hybrid type was introduced. Aside from all the advantages, some of the disadvantages were introduced as well.

2.1 Design

Compared to the previous [3], illustrated in figure 1, a) this iteration was corrected for some small design imperfections which have been subsequently noticed. A few small elements, whose shape did not affect the end results of the FEM simulation process, were simplified or even eliminated. The new Model had the diameter increased by a few percent. This design iteration also relied on a variant of the MRB [4, 5]. In [4], the Model had less evenly distributed magnetic flux and more modest torque capacity. The arrangement of magnetically conductive components coupled with excitation coils contributed to moderate magnetic flux distribution. The Model has twenty-four coil pairs, figure 1 b), thus forming a double-sided, multi-pole structure. The magnetic field is generated radially, on both sides of the brake, due to the excitation coils' orientation, figure 1 b). Each magnetic flux vector is directed from the center of the MRB body outwards, acting almost perpendicularly on the MRG layer, figure 1 c). The sheer number of individual coils and their positions contributed to more even magnetic field distribution but also increased the complexity of the Model, thus requiring more assets from FEM simulation settings.
Magnetically non-conductive housing, used to contain MRG, figure 1 a) - in color, reduced magnetic field dissipation and focused it (as much as possible) on the MBR's core and MRG areas.

2.2 Mathematical model

Although various improvements to the original design have been implemented, the braking torque mathematical model remained the same. The braking torque of the MRB is divided into two parts, induced and viscous, [3].

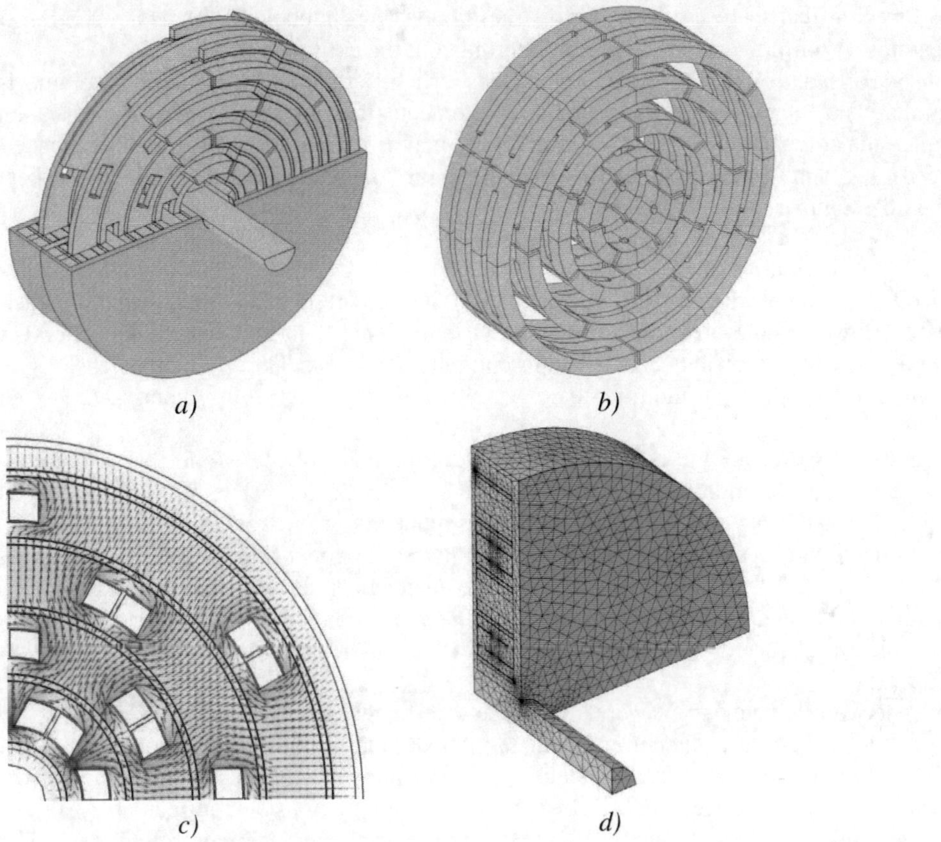

Figure 1. Illustrations of the multi-pole, multi-layer, double-sided magnetorheological brake Model, a) segmental cross-section, b) multi-pole coils arrangement, c) magnetic field density distribution pattern, d) meshed quarter Model.

3. Finite element method

3.1 Basics

Computers cannot handle continuous variations in the intensity of the magnetic field of a solid object, i.e. space, because they handle discrete numbers. This is forcing us to divide the proposed space (model and surrounding space) into many small - finite-size elements. In these small elements, the magnetic parameters have almost constant values. Here, the otherwise impossible-to-solve differential or integral equations transform into a large set of simple linear equations that can be handled by a computer.

Magnetic materials, such as soft iron and MRFs, usually have more complex properties than dielectrics. The properties of the magnetic materials may vary with their orientation and may depend on their previous exposure to magnetic fields [6].

To reduce the size of the model, in this commercial FEM software - Comsol Multiphysics, two types of symmetry were used: *Sector symmetry* and *Reflection symmetry* relative to the midplane, perpendicular to the axis of the Model. In order to use the Sector symmetry, *Periodic Boundary Conditions* must be used on the sides of the chosen sector, figure 1 d), of the model. The Reflection symmetry forces the normal component of the magnetic field to be zero at the midplane.

A highly non-linear B-H curve node was used to describe the magnetic behavior of both the soft-iron core and the MRG layers in simulation. Hysteresis effects in these materials were neglected. It was assumed that the coils were made of small-diameter wire and have multiple turns. Using the

assumption that there are many turns, these coils were modeled with *Homogenized Multi-turn coil* features. A uniform current distribution throughout the coils was applied.
The next step in the modeling process is mesh generation – the solution volume is divided into numerous elements and vertices. The set of elements that comprise the division of a solution volume is called the computational mesh. Solving for the magnetic field in a MRG was the most computationally costly segment of the Model but the overall amount of mesh elements needed to be kept at the lowest possible number.

3.2 Model definition

Due to the complexity of the MRB geometry and numerous coils, a 3D space dimension option was used. However, the previous FEM model simulation [3] had approximately 4.000.000 tetrahedral elements, which contributed to the high computational cost. The 3D models easily grow and require more computing power and time to solve. In fact, computing requirements grow exponentially with model size.
The aim is to reduce the complexity (and thus the cost) of the Model by exploiting its symmetry, removing the less important elements, and simplifying the geometry. In order to reduce the size of the Model, its symmetry was exploited, and it was reduced to a one-eighth of its size, figure 1, d).
The one-eighth section of the Model now has two orthogonal planes, both passing through the center of the coils and the center of the MRB. The magnetic field was mirrored across these two planes, sequentially using *The Magnetic Insulation Boundary* condition. This boundary condition, applied to the coils, interprets as a boundary through which excitation current can flow only in the perpendicular direction.
A spherical air boundary, several times the size of the Model, surrounded it. Every element of the Model was assigned with appropriate material nodes and boundary conditions.
Non-magnetic properties for: air, stator, and shaft nodes were selected from the software's database. Rotor - construction steel and MRG were defined with previously acquired data, [6].
Large Model parts, e.g. Stator, shaft, etc., were meshed using User-controlled mesh. The MRG layers, rotor, and their small radii were all meshed in the *Free tetrahedral* with various element sizes. Special attention was placed on the curvatures and the narrow regions of the MRG segments of the brake. The curvature radii were multiplied by the curvature factor parameter, which in turn gives the maximum allowed element size along a specific boundary. The *Resolution* of narrow regions' parameter controls the number of elements created in narrow regions. These parameters greatly improved the Model's mesh quality, which was at the threshold of 0.1, which is considered a satisfactory value.
To calculate the current in the coils, a *Coil Geometry Analysis study* step was added.
The solver was stationary but non-linear.

4. Conclusion

In this paper, some aspects of a novel magnetorheological brake design have been considered. The design is characterized by the combined use of both magnetic and non-magnetic materials. The materials with different magnetic properties shaped the unique distribution of magnetic flux through the magnetorheological brake's body, especially through magnetorheological grease layers. Some difficulties with the finite element method simulation process were briefly discussed here. For this purpose, the magnetorheological brake finite element Model was made using commercial finite element type software. To reduce the computation cost, the Model was simplified to a one-eighth the size of the original Model.
The accent was placed on the Model's symmetries. Specific boundary conditions needed to be implemented both for magnetic field propagation and for excitation current flow. To simplify the Model an assumption needed to be introduced: the coils have a constant cross-sectional area along their circular paths. This assumption is not completely accurate and needs to be addressed in future studies. The basic geometry and coil properties were the same as for the previous model.

A combination of different materials types contributed to the magnetic flux intensity increase in specific, small areas of the core and its better propagation through layers of the magnetorheological grease. The proposed multi-pole multi-T-rotor magnetorheological brake design shows great potential, but there is room for improvement.
This paper is a part of ongoing magnetorheological brake topology optimization.

5. Acknowledgments

This paper has been supported by the Ministry of Education, Science and Technological Development through project no. 451-03-68/2022-14/ 200156 “Innovative scientific and artistic research from the FTS domain”.

6. References

[1] Ahamed R, Choi S B and Ferdaus M M. 2018 A state of art on magneto-rheological materials and their potential applications. *J intell mater syst struct*, **29** 10, 2051-2095

[2] Mohamad N, Mazlan S, Sabino U, Choi S-B and Nordin M F 2016 The field-dependent rheological properties of magnetorheological grease based on carbonyl-iron-particles. *Smart mater struct*, **25** 10, 1-10

[3] Poznic A and Stojic B 2022 *Proc. Machine and Industrial design in Mechanical Engineering KOD 2021. part of: Mechan. Machine Science*, **109** 531–538

[4] Poznic A and Stojic B 2020 *Proc. The 8th International congress motor vehicles and motors 2020*, 1–8.

[5] Yaojung S and Quang-Anh N 2013 Development of a multi-pole magnetorheological brake. *Smart mater struct*, **22** 6.

[6] Poznic A, Miloradovic D and Zelic A 2015 *Proc. The 3rd International conference and workshop mechatronics in practice and education – MECHEDU 2015* 130-133.

IOP Conf. Series: Materials Science and Engineering 1271 (2022) 012028 doi:10.1088/1757-899X/1271/1/012028

Computational assessment of light experimental hybrid vehicle braking efficiency

B Stojić[1], A Poznić[1]

[1]University of Novi Sad, Faculty of Technical Sciences, Department for Mechanization and Design Engineering, Trg Dositeja Obradovića 6, 21000 Novi Sad - Serbia

E-mail: bstojic@uns.ac.rs

Abstract. This paper deals with braking performance of light parallel electric hybrid vehicle conceived as platform for educational and research work with students. As next step in vehicle design improvement, braking system modification was proposed. Using these data, in this paper braking performance on grounds with both high and low adhesion was assessed. Calculation results shown that braking efficiency is better for high than for low adhesion ground, which is result of design criteria applied. In both cases application of the regenerative braking impairs braking efficiency. This is consequence of insufficient braking force being applied at the rear axle as a result of system design.

1. Introduction

Braking performance of the road vehicles, in the sense of being able to deliver the maximum achievable braking rate in critical situations, depends on the road adhesion utilization of all vehicle wheels engaged with the road during braking. Different driving and road conditions require different proportioning of the braking forces between axles, or even individual wheels of the same axle. Sophisticated design solutions for braking system and its control were developed and upgraded in the last several decades, including systems that can use regenerative braking, to provide for optimum braking performance while maintaining vehicle handling, stability and energy efficiency. International regulations (e.g. [1-5]) prescribe not only required braking performance of the vehicle, but also functions and performances of braking control systems that also enable realization of several more advanced functions.

As opposed to contemporary series-production vehicles featuring complex braking systems and advanced devices, this paper deals with simple vehicle prototype developed by the group of students of Faculty of technical sciences – University of Novi Sad, Serbia, to serve as further educational and research platform. Vehicle is equipped with conventional hydraulic braking system without any means of affecting brake force distribution between front and rear, i.e. with linear proportioning. At the same time vehicle has in-wheel electric motors at the front axle, capable of applying regenerative braking torque. Acting on the front wheels only, this additional braking torque will affect braking force distribution and therefore capacity for road adhesion utilization during critical braking. Aim of this work is to assess the magnitude of this effect on the ground with both strong and weak adhesion with tyres, under certain assumptions that will be introduced subsequently.

2. Basic vehicle description

Vehicle is shown in the figure 1. Its overall design and the main features were introduced in [6]. Vehicle is made with tubular frame, with parallel hybrid powertrain configuration consisting of the front axle with hub electric motors and internal combustion engine (ICE) driven rear axle with gearbox and chain-driven final drive. There is conventional hydraulic braking system with friction brakes, with possibility of adding effect of regenerative braking on the front wheels. Wheelbase of the vehicle is $l = 2.2$ m.

Figure 1. Light experimental vehicle.

3. Assumptions for analysis

In order to enable student team more efficient and simple embarking on the initial phase of practical work under budget and organizational constraints, it was decided to realize first version of the vehicle prototype using parts that were already available or most straightforward to obtain. As well as other vehicle systems and assemblies, braking system hence consists of mixed parts originating from different vehicles including motorcycles, all-terrain vehicles (ATV) and passenger cars. Obviously, engineering rules for appropriate brake proportioning based on conditions for adhesion utilization during braking were not observed. Task of proper choice of braking system components was queued for the next phase of project development. As the first step, engineering parameters for vehicle design improvement were determined in bachelor thesis in mechanical engineering [7], according to which the following vehicle data were determined or adopted, assuming the case of lightly loaded vehicle:

$G_F = 3500$ N – static load at the front axle

$G_R = 3300$ N – static load at the rear axle

$h_{CG} = 0.5$ m – centre of gravity height

Tyre size: 205/55R18

$r_W = 0.34$ m – rolling radius

$B = 2.6$ – ratio between braking force at the front and rear axle respectively ($B = F_{BF} / F_{BR}$)

This last parameter (B) was determined by dealing with choice of braking system components from passenger cars, readily available on the market, taking into account all relevant design parameters of the vehicle. Observing requirement from UN Regulation for braking of passenger cars R13-H [1] and targeting adopted condition that simultaneous lockup of the front and rear wheels should occur on the ground with maximum coefficient of adhesion value of $\varphi_{MAX} = 0.9$, whereby wheel lock sequence for grounds with adhesion lower than that is that front wheels have to lock first, proposed braking system configuration was obtained. System design features constant brake torque distribution between front and rear axle, which is determined by the factor *B*.

Electric motors used on the vehicle's front axle were dealt with in the master thesis in electrical engineering [8]. It was assessed that each electric motor can generate permanent mechanical driving power of approximately 2 kW. Motors were produced in China, whereby neither manufacturer nor exact specifications are known. Since insufficient data were available in order to assess electric motors performance in view of regenerative braking, simplification will be introduced by adopting assumption that the vehicle is capable of achieving deceleration rate $z = 0.2$ g on the level ground when using only regenerative braking. Neglecting air and rolling resistance, braking torque M_{BW} that has to be developed at the single front wheel can then be calculated accordingly:

$$M_{BW} = \frac{1}{2} \cdot r_W \cdot G \cdot z = \frac{1}{2} \cdot 0.34 \cdot 6800 \cdot 0.2 \approx 230 \text{ Nm}$$

Here, $G = G_F + G_R = 6800$ N is vehicle weight. For this value of braking torque, instant braking power of the single electric machine at the beginning of the braking from 60 km/h will be:

$$P_B = \frac{M_B}{r_W} \cdot v = \frac{230}{0.34} \cdot \frac{\frac{60}{3.6}}{1000} \approx 11 \text{ kW}$$

Regenerative braking force at the front axle will be:

$$F_{BF,Rec} = 2 \cdot \frac{M_{BW}}{r_W} = 2 \cdot \frac{230}{0.34} = 1353\ N$$

4. Braking efficiency calculations

Braking efficiency η_B represents the ratio between realized deceleration rate of the vehicle z and maximum possible deceleration rate z_{MAX} if adhesion between tyres and the road is fully exploited at all wheels. It can easily be shown (e.g. [9]) that the latter is equal to the maximum value of coefficient of adhesion between tire and road, i.e. it applies:

$$z_{MAX} = \varphi_{MAX} \Rightarrow \eta_B = \frac{z}{\varphi_{MAX}} \tag{1}$$

In the Annex 5 of the UN Regulation for braking of passenger cars R13H [1], dealing with distribution of braking among the axles of vehicles, requirements for vehicle behaviour are defined for roads with maximum value of coefficient of adhesion between tire and road between $\varphi_{MAX} = 0.2$ and $\varphi_{MAX} = 0.8$. In this paper, these two boundary values are selected to assess braking efficiency on the road with high and low adhesion respectively. It is important to note that the following considerations apply for the case when adhesion utilisation is greater at the front axle than on the rear (i.e. front wheels lock first), otherwise some of the conditions used subsequently would have to be rewritten in order to obtain proper results. It has been ascertained that this requirement is fulfilled for all cases and data used in this paper, though the presentation of this procedure is omitted herein.

4.1. *Case 1: Braking efficiency when braking only with friction brakes*

For calculation of the braking efficiency it is assumed that at the front axle adhesion between tyres and the ground is fully utilized, while at the rear axle, observing prescribed wheel lock sequence, adhesion utilization is below its physically possible maximum. It follows:

$$F_{BF} = \varphi_{MAX} \cdot G_{F,dyn} \tag{2}$$

$$F_{BR} = \frac{F_{BF}}{B} = \frac{\varphi_{MAX} \cdot G_{F,dyn}}{B} \tag{3}$$

Here, $G_{F,dyn}$ represent dynamic front axle load during braking under dynamic influences. Equation of motion for braked vehicle on the level ground, neglecting air and rolling resistance, can now be written as:

$$G \cdot z = F_{BF} + F_{BR} = \varphi_{MAX} \cdot G_{F,dyn} \cdot \left(1 + \frac{1}{B}\right) \tag{4}$$

Dynamic front axle load $G_{F,dyn}$ is:

$$G_{F,dyn} = G_F + G \cdot \frac{h_{CG}}{l} \cdot z \tag{5}$$

Combining two expressions and solving for deceleration rate z yields:

$$z = \frac{\varphi_{MAX} \cdot \frac{G_F}{G} \cdot \left(1 + \frac{1}{B}\right)}{1 - \varphi_{MAX} \cdot \frac{h_{CG}}{l} \cdot \left(1 + \frac{1}{B}\right)} \tag{6}$$

4.2. *Case 2: Braking efficiency when braking with both friction and regenerative brakes*
It is assumed for this case that regenerative braking is fully applied to adopted extent of regenerative braking force value calculated above ($F_{BF,Rec} = 1353$ N), along with frictional braking system in a way that combination of friction and regenerative braking at the front axle yields full utilization of available adhesion. Hence, appropriate conditions apply as carried out subsequently.

$$F_{BF} = \varphi_{MAX} \cdot G_{F,dyn} = F_{BF,Rec} + F_{BF,Fr} \tag{7}$$

Overall braking force at the front axle consists of regenerative ($F_{BF,Rec}$) and friction ($F_{BF,Fr}$) parts. Braking system design gives:

$$F_{BR} = \frac{F_{BF,Fr}}{B} = \frac{\varphi_{MAX} \cdot G_{F,dyn} - F_{BF,Rec}}{B} \tag{8}$$

Since no electric recuperation devices are present at the rear axle, rear braking force F_{BR} is result of purely friction mechanism. Using given conditions and solving for deceleration rate z, in this case it is obtained from the following expression:

$$z = \frac{\varphi_{MAX} \cdot \frac{G_F}{G} \cdot \left(1 + \frac{1}{B}\right) - \frac{F_{BF,Rec}}{B \cdot G}}{1 - \varphi_{MAX} \cdot \frac{h_{CG}}{l} \cdot \left(1 + \frac{1}{B}\right)} \tag{9}$$

It can be seen that, putting $F_{BF,Rec} = 0$, expression (9) reduces to (6).

5. Calculation results

Using expressions developed and conditions adopted above, braking efficiency values are obtained as summarized in the Table 1.

Table 1. Calculation results for braking efficiency.

	Only friction braking	Friction and regenerative braking
High adhesion 0.8	0.95	0.82
Low adhesion 0.2	0.76	0.35

6. Conclusions

Regarding braking using only friction brakes, on the high adhesion ground relatively high braking efficiency is achieved. This was expected, because brake proportioning was designed so as to obtain good adhesion utilisation in such conditions, which is why real brake force distribution will be more close to the ideal one. On the ground with low adhesion, wheel lock will happen at significantly lower deceleration rate, so that the ratio between front and rear brake force will be more distant from the ideal point and efficiency will therefore be lower.

As expected, braking with simultaneous applying of both frictional and regenerative braking will feature lower efficiency. This is because when adhesion at the front wheels is fully utilized, front friction brakes share only a portion of overall braking force, so that, due to linear proportioning, brake force at the rear wheels will be less than without regenerative braking, leading to decreased deceleration rate and therefore lower efficiency. This is especially distinguishable in case of low adhesion, where even pure regenerative braking would lead to wheel lock, i.e. by full adhesion utilization at the front wheels, friction braking system does not even take a part in the process, which means that the braking force at the rear axle will also remain zero, leading to particularly low braking efficiency.

Acknowledgment
This research has been supported by the Ministry of Education, Science and Technological Development through project no. 451-03-68/2022-14/ 200156 “Innovative scientific and artistic research from the FTS domain”.

References
[1] UNECE Regulation 13-H: Uniform provisions concerning the approval of passenger cars with regard to braking, Economic Commission for Europe of the United Nations
[2] UNECE Regulation 13: Uniform provisions concerning the approval of vehicles of categories M, N and O with regard to braking, Economic Commission for Europe of the United Nations
[3] UNECE Regulation 131: Uniform provisions concerning the approval of motor vehicles with regard to the Advanced Emergency Braking Systems (AEBS), Economic Commission for Europe of the United Nations
[4] UNECE Regulation 139: Uniform provisions concerning the approval of passenger cars with regard to Brake Assist Systems (BAS), Economic Commission for Europe of the United Nations
[5] UNECE Regulation 140: Uniform provisions concerning the approval of passenger cars with regard to Electronic Stability Control (ESC) Systems, Economic Commission for Europe of the United Nations
[6] Stojić B., Ružić D., Ivanović Z, Adžić E., Ilanković N.: Light hybrid-electric vehicle as educational and research platform, 7. International Congress Motor Vehicles & Motors 2018, Faculty of Engineering Kragujevac, Kragujevac, 2018., ISBN 978-86-6335-055-7, pp 377 – 382
[7] Popović M.: Izbor komponenata kočnog sistema lakog drumskog vozila, BSc Thesis, Faculty of technical sciences Novi Sad, Serbia, 2021
[8] Gabrić T.: Projektovanje i ispitivanje sinhronog motora sa stalnim magnetima za primenu u električnim vozilima, MSc Thesis, Faculty of technical sciences Novi Sad, Serbia, 2017
[9] Mitschke M., Wallentowitz H.: Dynamik der Kraftfahrzeuge, Springer Verlag Berlin Heidelberg New York, 2004

IOP Conf. Series: Materials Science and Engineering 1271 (2022) 012029 doi:10.1088/1757-899X/1271/1/012029

Improving ride comfort by optimising suspension system of an electric street sweeper

V Changoski[1], I Mircheski[1], D Danev[1]

[1] Faculty of Mechanical Engineering – Skopje, "Ss. Cyril and Methodius" University, Rugjer Boshkovikj 18, Skopje, Republic of N. Macedonia

E-mail: vasko.changoski@mf.edu.mk

Abstract. The opinion of the operators of the working machines in today's world is much more valued than in the past. Their comfort, safety and satisfaction during the working hours can be crucial in the decision making process of buying new working machine. Therefore, in this research paper the vertical dynamics and ride comfort of an electric street sweeper would be analysed by using a multibody dynamic model. The model is created using MSC.ADAMS, based on a real electric street sweeper. In order to test the ride comfort of the existing model, a road profile is modelled, according to the standard ISO 8608. The virtual tests are conducted while the vehicle is traveling with maximum velocity of 40 km/h, achieved during transit and lower velocities which are achieved during the operating hours of the machine. An optimization of the suspension system of the real electric sweeper is made, by targeting its components location and positioning, stiffness and damping characteristics. This results in improved ride comfort and smaller root mean square (RMS) values of vertical acceleration of the driver seat. These improvements would influence on operator's better comfort and health. Results are presented using comparative diagrams of the original and optimised model.

1. Introduction

The idea for creation of a more sustainable transportation is not just connected to the passenger vehicles but also to the working machines. Therefore, a devotion exist among the companies to create safer, greener and more comfortable machines. Operators opinion and feelings about the machine and it's comfort may be a deciding factor when buying new working machine. Because of that, many companies invest in research of the ride comfort of their products. In [1] an analysis is made on an operator's seat of a dump truck using virtual model created in MSC.ADAMS based on the standard ISO 5982:2001. Using multibody dynamics, authors in [2] achieved optimization of the driver's body vibrations of a wheel loader by improving the suspension of the machine. Reference [3] focuses on creation of optimization NSGA-II algorithm for development of lightweight dump truck carriage. Authors in [4] propose analytical cascading methodology for industrial application for optimization of chassis and suspension system of a lightweight bus. Using ADAMS in [5] a simulation and analysis of a scissors mechanism for high dump truck is made. MSC.ADAMS is also used for creation of a multibody model of a ultra large dump truck in [6] where authors propose a methodology for a road design, depending on the dynamic forces of the trucks. Another research is conducted using ADAMS, which is focused on improving drivers comfort is presented in [7] where the focus of the optimization is on the suspension system of the trucks cabin.

Beside ADAMS, co-simulation approach is presented in [8] where ADAMS, Simulink and AEMSim are used for improving ride comfort and handling stability of an articulated dump truck. An optimization of the front suspension of a mine truck is conducted in [9] using ADAMS/Car where the target is the toe and camber angles and the position of the elements. In [10] an optimization of the ride comfort is made, using intelligent algorithm and nonlinear damping models in co-simulation environment. Road profile is modeled in Simulink according to the standard ISO 8608 while a virtual model in ADAMS is

IOP Conf. Series: Materials Science and Engineering 1271 (2022) 012029 doi:10.1088/1757-899X/1271/1/012029

tested on the previously created road profile. Standards ISO 8608 and ISO 2631 are used in [11] to propose evaluation method for analysis of the ride comfort of passengers in commercial vehicles where the results are presented in comparison to a real measurements.

In this paper an optimization of the ride comfort is made by targeting the suspension system of an electric street sweeper, using standard ISO 8608 [12]. C-level road profile is created in order to analyze the RMS value of the vertical acceleration of the driver's seat on which the driver is exposed during his working hours. The research was conducted in collaboration between the Faculty of Mechanical Engineering in Skopje and the company Green Machines DOO Veles.

2. Multi-body sweeper modelling

The multi-body model, created in MSC.ADAMS is based on a real electric street sweeper GM500ze (figure 1) from the company Green Machines. GM500ze is a vacuum sweeper for gathering the street waste, with container capacity of $1m^3$ and maximum weight of 2750 kg. The front and rear suspension represents elastic dependable system with conventional coil springs and hydraulic dampers.

Figure 1. GM500ze – electric street sweeper [13].

Renders from the virtual model in ADAMS are presented in figure 2. The model consists of rigid parts connected with joints which replicate the real kinematics of the electric sweeper. Contact between the model and the road is achieved using 155/70 R12 tires based on the Fiala tire model. It is important to mention that the brushes are not involved in the ADAMS model and the analysis of the ride comfort is focused solely on the suspension system of the machine. This was done because the design study and the optimisation were conducted when the machine is in transit mode, traveling with its maximum speed of 40 km/h. In this case the RMS values of the driver's seat vertical acceleration have the highest value, thus reducing the ride comfort. The contact between the brushes and the road and the vibrations caused by the vacuum suction system have undoubtedly negative effect on the ride comfort, but its influence and effects have not been tested yet. In order to achieve this, a testing procedure should be conducted in order to measure their effects on the ride comfort and afterwards, those results should be implemented in the ADAMS model in the future. The parameters of the virtual model are presented in table 1.

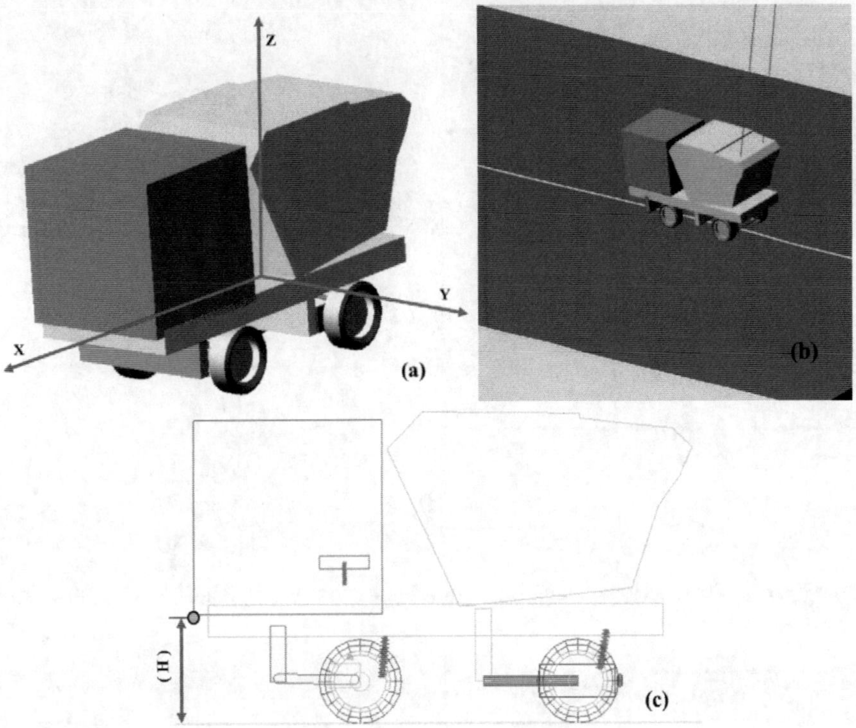

Figure 2. Multibody model of the electric street sweeper (a) Isometric view, (b) Sweeper during simulation tests (c) Schematic view of the sweeper.

Table 1. Multibody sweeper model parameters.

Mass:	2750 kg
Vehicle weight distribution (front/rear axle):	44/56 %
Length:	4040 mm
Width:	1200 mm
Wheelbase:	1380 mm

Schematics of the elastic front and rear suspension are presented in figure 3 and 4. Front suspension consists of two coil springs and dampers and two longitudinal links, connected to the chassis of the machine. The rear axle represents driven axle, where beside the two coil springs and dampers, also possess one longitudinal link with more complex form and one Panhard rod which form the connection between the axle and the chassis. All connections between the chassis and the axles are spherical joints. The springs and dampers of the model are described using linear mathematical interpretation of the elements obtained from the real model, while due to lack of information about the suspension of the driver's seat, it's parameters are obtained from proposed values in [14]. Characteristics of the suspension system are presented in table 2.

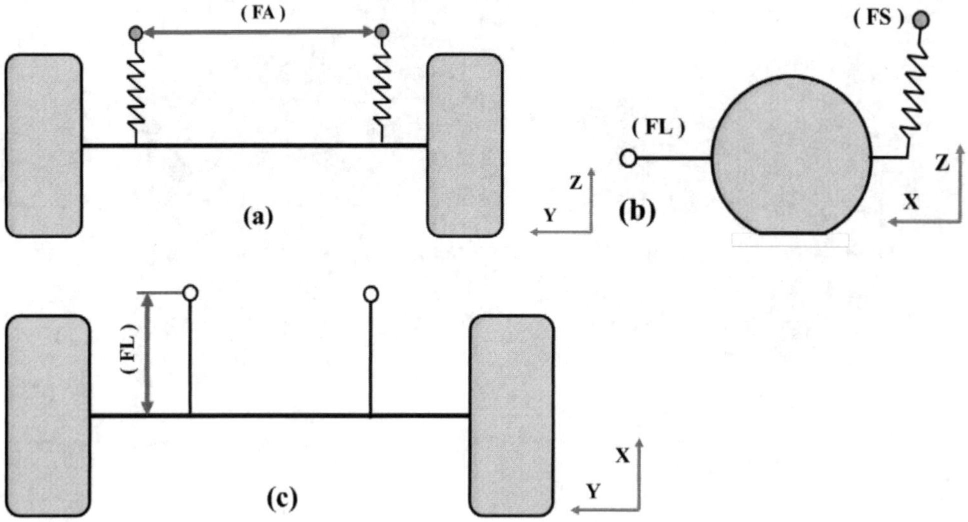

Figure 3. GM500ze – Front axle suspension system, (a) Front view, (b) Side view and (c) Plan view.

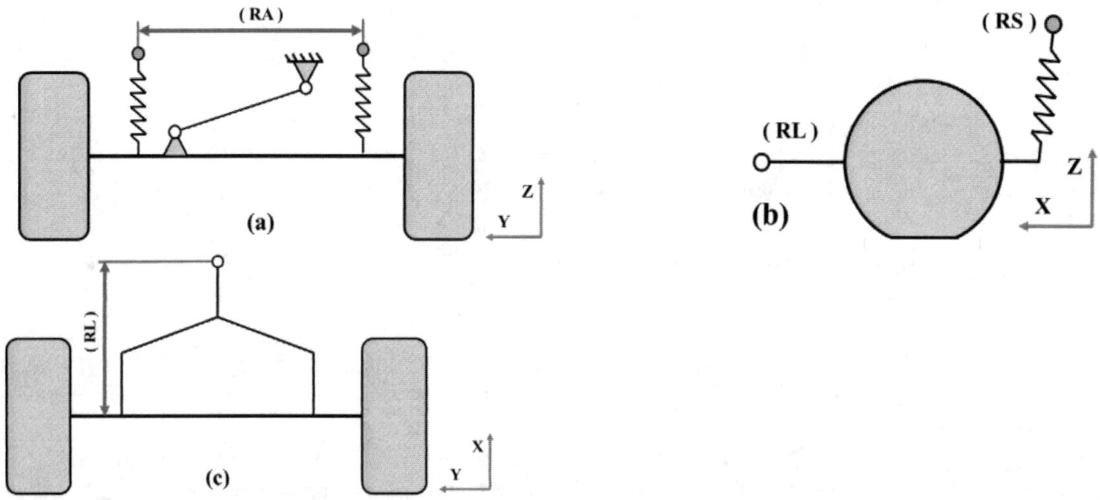

Figure 4. GM500ze – Rear axle suspension system, (a) Front view, (b) Side view and (c) Plan view.

Table 2. Multibody sweeper model suspension characteristics.

Front spring stiffness coefficient:	130 N/mm
Front damping coefficient:	25 Ns/mm
Rear spring stiffness coefficient:	155 N/mm
Rear damping coefficient:	25 Ns/mm
Seat stiffness coefficient:	29.9 N/mm
Seat damping coefficient:	0.78 Ns/mm

3. Design studies of the suspension system

The idea of the design studies is to try to make minimal changes of the suspension system which would initiate minimal changes of the chassis, attachment elements, springs and dampers. This analysis in conducted in order to achieve desirable improvement of the ride comfort of the machine without drastically increasing the production cost or delay of the machine. The possibility of repositioning the attachment points between the springs and dampers with the axles is not analysed because any change in the axles would result in procurement need of new or modified axles, leading to increase of the research budget and spendings. Also, the seat replacement is not subject to the design studies because the seat is specially designed for the GM500ze. Beside the suppling cost, the replacement of the seat would lead to additional changes in the cabin itself.

Taking into consideration the previous statement, several control variables were chosen in order to optimise the ride comfort, such as: stiffness of the springs, damping coefficients of the shock absorbers, attachment points between the chassis and the axles. The variables and their analysed range are presented in table 3.

The range of the variables is defined by two factors: limited construction space of the machine and similar commercially available elements. Due to the large number of elements and subsystems of the machine, the construction space for making any larger changes of the construction is very limited. In addition to that, and to avoid larger changes, springs and shock absorbers with similar dimensions but with different characteristics are analysed.

Table 3. Design study variables and their range.

Springs stiffness coefficients:	$\pm 30\%$
Shock absorbers damping coefficients:	$\pm 30\%$
Position of point (FS) in x direction:	$\pm 120\ mm$
Distance (FA) in y direction:	$\pm 200\ mm$
Position of point (FL) in x direction:	$\pm 100\ mm$
Position of point (RS) in x direction:	$\pm 100\ mm$
Distance (RA) in y direction:	$-150\ mm/+200\ mm$
Position of point (RL) in x direction:	$\pm 100\ mm$

Several design studies are made in order to investigate the impact on the ride comfort. During the simulations the machine is traveling by 40 km/h on a C-level ISO ROAD. The value of the velocity is chosen because it is planned to increase the current transit speed of the machines from 25 km/h to 40 km/h in the future. The road profile is created using equation 1, described in [10]:

$$\dot{q}(t) = -2\pi v n_q q(t) + 2\pi n_0 \sqrt{G_q(n_0) v}\,\omega(t) \tag{1}$$

where v is the vehicle speed (m/s); n_q represents the cutoff space frequency; $n_q = 0.011\ m^{-1}$; $q(t)$ is the road excitation; $n_0 = 0.1^{-1}$ represents the spatial reference frequency $G_q(n_0) = 16x10^{-6}$; is the road excitation coefficient; and $\omega(t)$ is Gauss white noise signal with zero mean [10, 14].

The C-lever ISO road profile is presented in figure 5.

IOP Conf. Series: Materials Science and Engineering 1271 (2022) 012029 doi:10.1088/1757-899X/1271/1/012029

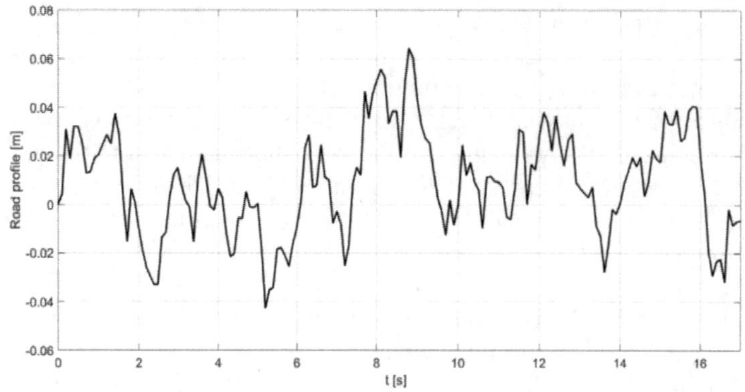

Figure 5. C-Level road profile according to the standard ISO 8608.

In order to analyse the ride comfort of the machine, the RMS value of the vertical acceleration of the driver's seat is used as reference value. Beside this parameter, the vertical distance (H) (figure 2) between the front edge of the diver's cabin and the road is analysed. This value is important in order to maintain the contact between the brushes and the road and to avoid collision between the front part of machine and the road during intensified braking. During the optimisation, the vertical distance (H) must be observed constantly, because if we focus solely on the ride comfort the primary task of the machine (sweeping) may be disturbed. In figures 6-19 the results of the design studies are presented, where the changes of RMS values of the vertical acceleration and the vertical distance (H) are shown, depending on the modification of the suspension system.

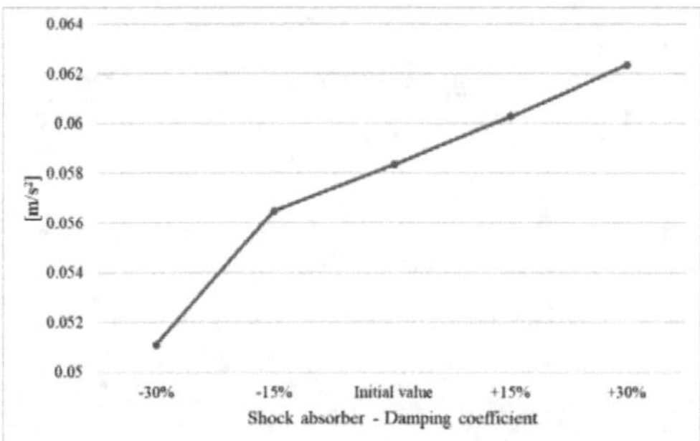

Figure 6. Changes of RMS value of the driver's seat vertical acceleration in relation to the damping coefficient of the shock absorbers.

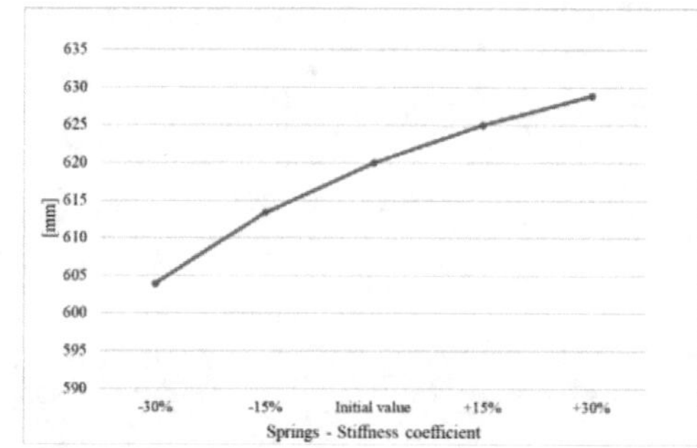

Figure 7. Changes of vertical distance (H) in relation to the stiffness coefficient of the springs.

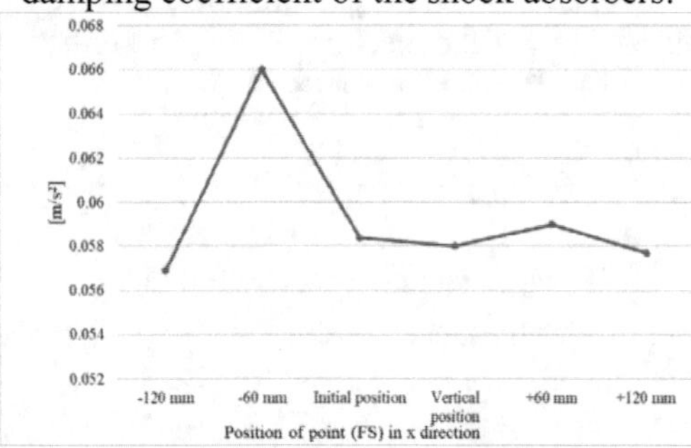

Figure 8. Changes of RMS value of the driver's seat vertical acceleration in relation to the position of point (FS) in x direction.

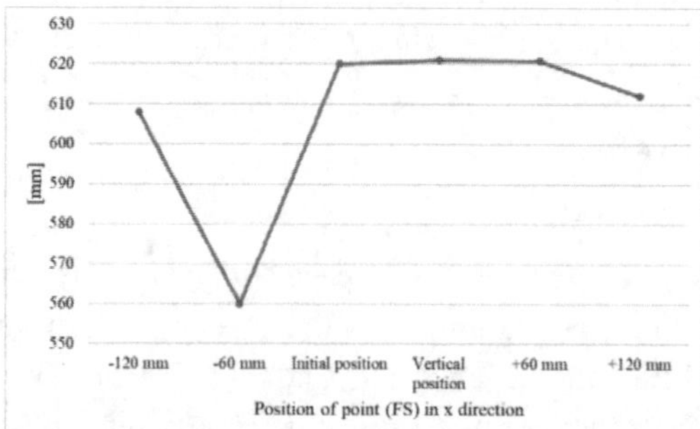

Figure 9. Changes of vertical distance (H) in relation to the position of point (FS) in x direction.

IOP Conf. Series: Materials Science and Engineering 1271 (2022) 012029 doi:10.1088/1757-899X/1271/1/012029

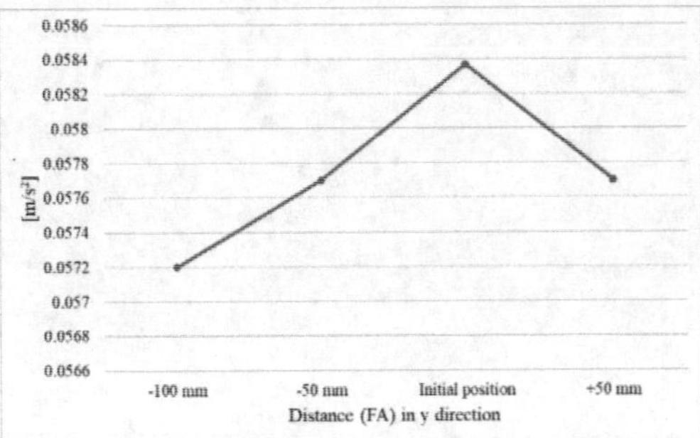

Figure 10. Changes of RMS value of the driver's seat vertical acceleration in relation to the distance (FA) in y direction.

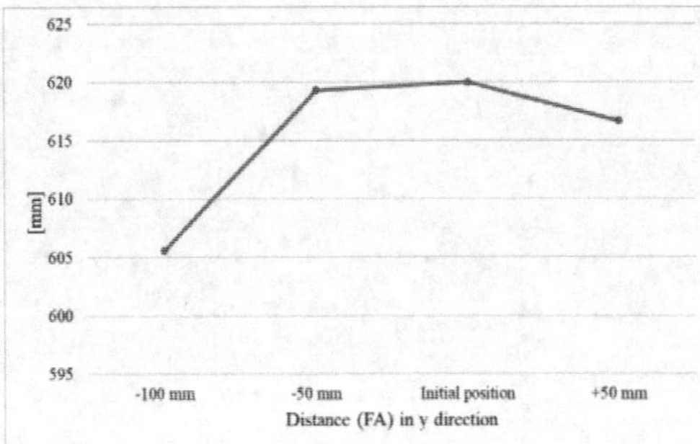

Figure 11. Changes of vertical distance (H) in relation to the distance (FA) in y direction.

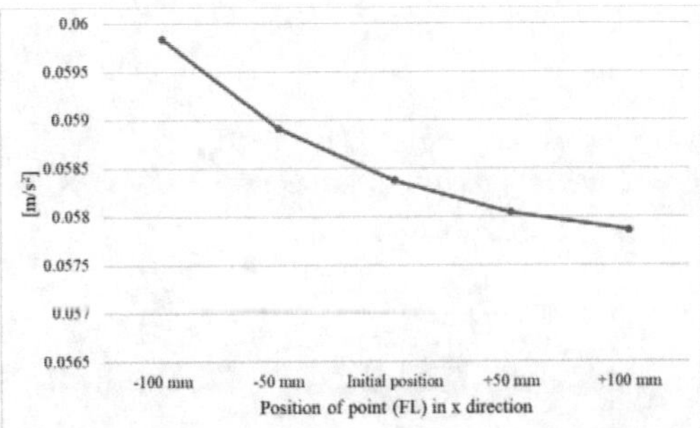

Figure 12. Changes of RMS value of the driver's seat vertical acceleration in relation to the position of point (FL) in x direction.

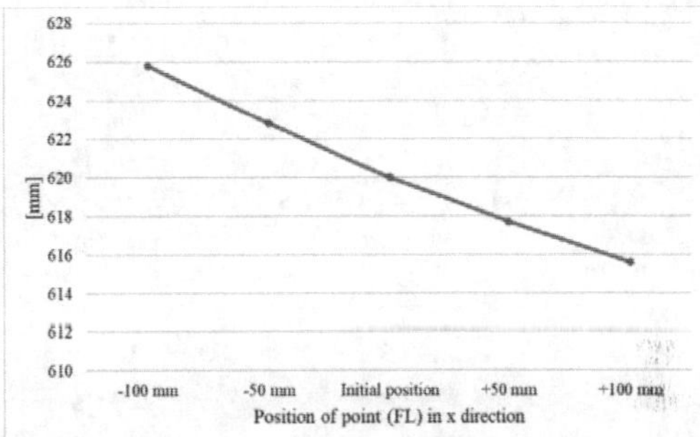

Figure 13. Changes of vertical distance (H) in relation to the position of point (FL) in x direction.

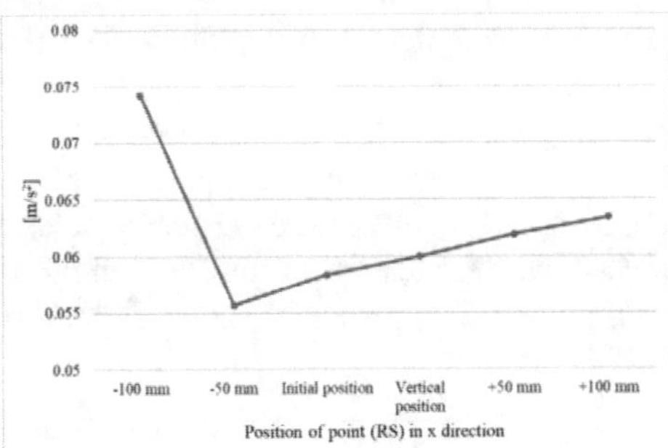

Figure 14. Changes of RMS value of the driver's seat vertical acceleration in relation to position of point (RS) in x direction.

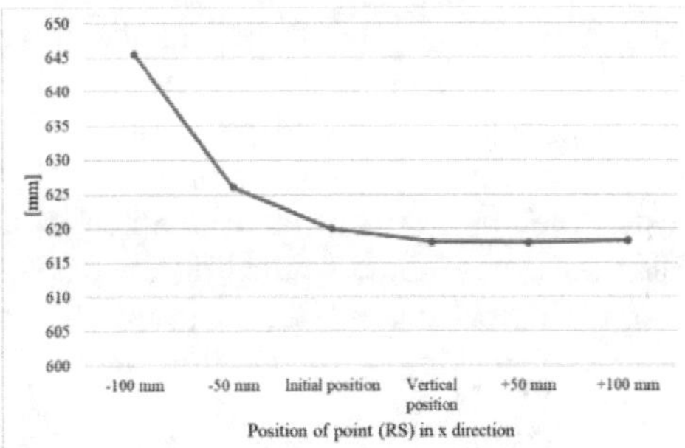

Figure 15. Changes of vertical distance (H) in relation to position of point (RS) in x direction.

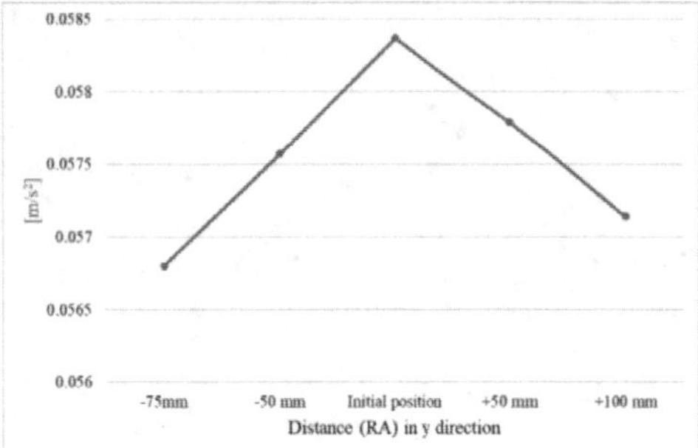

Figure 16. Changes of RMS value of the driver's seat vertical acceleration in relation to the distance (RA) in x direction.

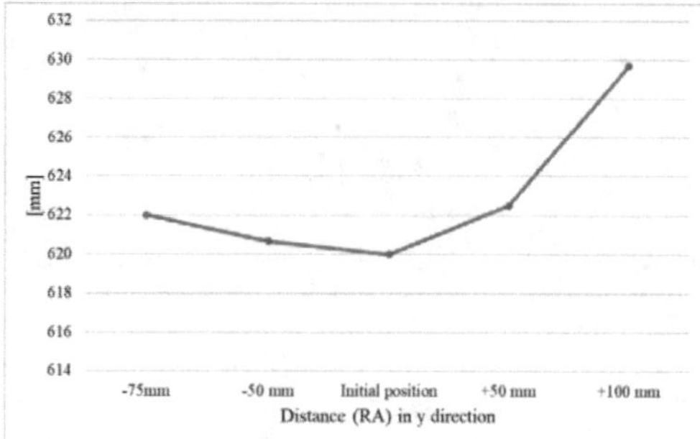

Figure 17. Changes of vertical distance (H) in relation to the distance (RA) in x direction.

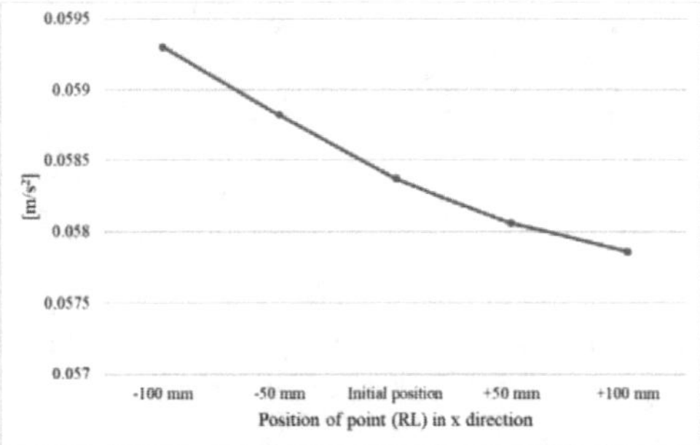

Figure 18. Changes of RMS value of the driver's seat vertical acceleration in relation to position of point (RL) in x direction.

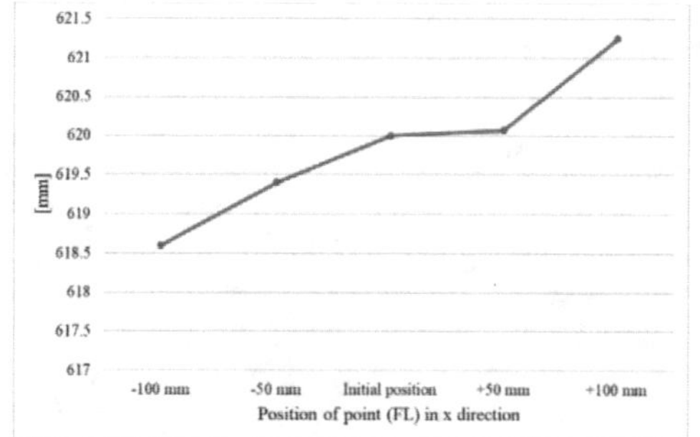

Figure 19. Changes of vertical distance (H) in relation to position of point (RL) in x direction.

After conducting the design studies, the biggest improvement in the decreasing of the RMS value of the driver's seat vertical acceleration by 12.45% can be obtained by decreasing the value of the damping coefficient of the shock absorbers. Also, significant decreasing of the RMS value by 4.59% can be observed by changing the position of point (RS) not more than 50 mm in the negative part of the x axis. The rest of the improvements are presented in table 4. These improvements in the vertical acceleration also have influence on the vertical distance (H) of the front edge of the driver's cabin. In table 5 it can be observed that the changes of the suspension system from table 4, increase and decrease the vertical distance (H) depending on the modification.

Table 4. Deduction of RMS value of vertical acceleration.

Shock absorbers damping coefficients:	12.45%
Position of point (FS) in x direction:	2.52%
Distance (FA) in y direction:	2%
Position of point (FL) in x direction:	0.87%
Position of point (RS) in x direction:	4.59%
Distance (RA) in y direction:	2.1%
Position of point (RL) in x direction:	0.87%

Table 5. Change of vertical distance (H).

Springs stiffness coefficients:	$-16.1\ mm$
Position of point (FS) in x direction:	$-12\ mm$
Distance (FA) in y direction:	$-14.4\ mm$
Position of point (FL) in x direction:	$-4.4\ mm$
Position of point (RS) in x direction:	$+6\ mm$
Distance (RA) in y direction:	$+9.7\ mm$
Position of point (RL) in x direction:	$+1.25\ mm$

4. Optimisation of the machine's ride comfort

By utilizing the influence of the certain elements on the ride comfort and by trying to maintain the same vertical distance or to apply small change that won't influence the contact between the brushes and the road, an optimisation is made. The optimisation in ADAMS was made by taking into consideration the previous values and the their influence on the ride comfort, and combining them in order to further improve the ride comfort. Several iterations were made in which there was significant improvement in the ride comfort, but that contributed to lowering the front part of the vehicle (more than 10 mm), worsening the contact between the brushes and the road, but also endangering the front part of the machine colliding with the surface during intensified braking. Therefore, those iterations were excluded from the analysis and the optimised solution is presented in table 6 where all changes of the suspension system are presented. The values of the variables that are not presented had not been changed during the optimisation.

A comparison between the original and the optimised model is presented in figure 20. The analysis is conducted by simulating the vehicle model on three different velocities: 12, 25 and 40 km/h. 12 km/h represents the speed of the vehicle during working period when it is cleaning the streets (sweeping regime), 25 km/h is the current transit speed when the machine is traveling from one location to another and 40 km/h is planned to be transit speed in the future improved model. Improvement of the ride comfort by 18.96%, while traveling with 40 km/h, is gained from the optimization while the vertical distance (H) is changed from the original 620 mm to 624 mm. These changes would allow better ride comfort, while increasing the vertical distance (H) by 4 mm will not represent a problem because a hydraulic system responsible for the brushes is capable of lowering them in order to achieve the contact between them and the road. As mention before, the problem occurs when that distance is smaller than on the original machine, because in that case beside the worsen contact between the road and the brushes, there is possibility of collision between the road and the front part of the machine.

It must be pointed out that during the sweeping regime when the machine is traveling with 12 km/h, the real RMS values would be higher because, as mentioned before, the brushes and the effects of the vacuum cleaning system are not implemented in the model due to lack of real measured data. But even if we take that into consideration, the improvements of the optimised suspension would still have influence on lowering the RMS value of the vertical acceleration of the driver's seat compared to the current machine, thus improving the ride comfort of the machine.

Table 6. Optimised variables values.

Front springs stiffness coefficient:	$+19.2\%$
Shock absorbers damping coefficients:	-30%
Position of point (FS) in x direction:	$-61mm$
Distance (FA) in y direction:	$-100\ mm$
Position of point (RS) in x direction:	$-50\ mm$
Distance (RA) in y direction:	$-50\ mm$

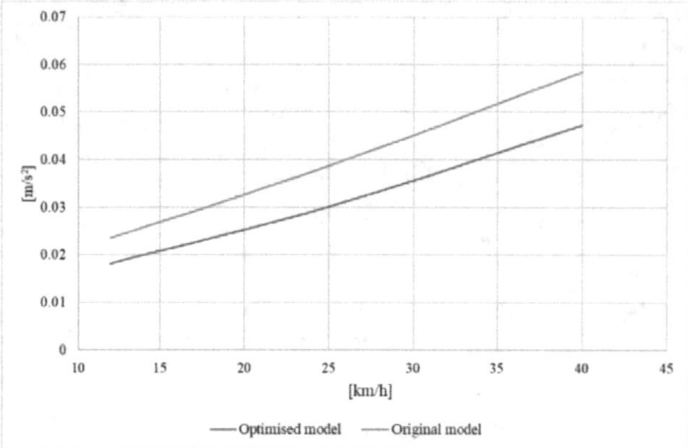

Figure 20. Comparison of the RMS values of the vertical acceleration of the driver's seat between the original and the optimised model.

5. Conclusions

In this research, the optimisation had shown that with relatively small changes in the geometry of the suspension system and its elements an improved ride comfort can be achieved. That would result in improved operator's comfort and health and would make the machine more desirable for the byers. All of these changes would result in modifications of the current chassis and systems but those modifications will not drastically increase the production cost of the machines or delay their shipment. The study had also shown that even further optimisation can be achieved in order to improve the ride comfort, but in that case, larger investments and modifications must be made which would question the value and the production profit of the machines. Also, it was concluded that beside the main focus of improving the ride comfort, a detail analysis and observation must be made on the main functionality of the machine. That was presented by analysis of the distance (H) between the cabin and the ground in order to maintain a satisfactory contact between the brushes and the ground and not endangering the safety of the machine while braking. If the contact is worsen then the main function of the machine, could be lost. Because of this beside the RMS values of the vertical acceleration of the driver's seat, the vertical distance between the cabin and the road is used as a reference value that should not suffer larger changes.

Implementation of semi-active or active suspension system and driver seat with improved characteristics and analysis of the vacuum suction system's influence on the ride comfort may be a task for future work for development of a newer generation or larger machine.

Acknowledgments

We would like to express our deep gratitude to the company Green Machines DOO Veles from Republic of N. Macedonia, it's owner professor dr. Kocho Angjushev and the engineers from the company for their collaboration and financial supporting. Our grateful thanks are also extended to our coworkers and teaching assistants Anita Vasileva, Blagoja Nestorovski and Elena Angeleska form the Faculty of Mechanical Engineering in Skopje.

References

[1] Frimpong S, Galecki G and Chang Z 2011 Dump truck operator vibration control in high-impact shovel loading operations *International Journal of Mining Reclamation and Environment* pp. 213-225

[2] Rehnberg A and Drugge L 2008 Ride comfort simulation of a wheel loader with suspended axles. *International Journal of Vehicle Systems Modelling and Testing* **3**

[3] Jiang R, Ci S, Liu D, Cheng X and Pan Z 2021A Hybrid Multi-Objective Optimization Method Based on NSGA-II Algorithm and Entropy Weighted TOPSIS for Lightweight Design of Dump Truck Carriage *Machines* **9(8):**156

[4] Kang N, Kokkolaras M, Papalambros P Y *et al.* 2014 Optimal design of commercial vehicle systems using analytical target cascading *Struct Multidisc Optim* **50,** 1103–1114 (2014)

[5] Li X R and Yuan D X 2011 Virtual Prototype Design and Dynamic Simulation of High-Dump Truck Based on ADAMS *In Advanced Materials Research* **Vols. 189–193** pp. 1882–1885 (Trans Tech Publications, Ltd.)

[6] Kansake B, Frimpong S and Ali D 2019 Multi-body dynamic modelling of ultra-large dump truck-haul road interactions towards haul road design integrity *International Journal of Mining, Reclamation and Environment* **34** 1-23

[7] Jiang C, Zhu Q and Li H 2019 Simulation and Optimization of Truck Cab Suspension System Based on ADAMS *IOP Conference Series: Earth and Environmental Science* **242** 032039

[8] Xueyan Z and Xiaxu H 2017 Multi-Objective Optimization on Ride Comfort and Handling Stability of the Articulated Dump Truck *4th International Conference on Information Science and Control Engineering (ICISCE)* pp. 1028-1032

[9] Zhang J, Pei J and Zhang J 2016 The Optimization Study of the Front Suspension Structure Parameters for the Mine Truck Based on Adams/Car In *Proc International Conference on Vehicle, Mechanical and Electrical Engineering* pp. 1-9

[10] He S, Chen K, Xu E, Wang W and Jiang Z 2019 Commercial Vehicle Ride Comfort Optimization Based on Intelligent Algorithms and Nonlinear Damping *Shock and Vibration* pp.1-16

[11] Žuraulis V, Pečeliūnas R and Jakutis G 2014 Semi-Active Suspension Influence on Comfort Sensation of a Vehicle Occupant *Agricultural Engineering* **46** pp. 116-124

[12] https://www.iso.org/standard/71202.html

[13] https://greenmachines.com/500ze/

[14] Micke M and Valentovic H 2009 Dinamika na motornite vozila, Tabernakul Skopje

IOP Conf. Series: Materials Science and Engineering 1271 (2022) 012030 doi:10.1088/1757-899X/1271/1/012030

Vehicle Suspension System with Integrated Inerter - Extended Analysis

Nikola Nešić[1*], Julijana Simonović[2], Milan Blagojević[1], Saša Milojević[3], Srđan Jović[1]

[1] Faculty of Technical Sciences, University of Priština, Kosovska Mitrovica, Serbia
[2] Faculty of Mechanical Engineering, University of Niš, Niš, Serbia
[3] Faculty of Engineering Sciences, University of Kragujevac, Kragujevac, Serbia
[*] e-mail: nikola.nesic@pr.ac.rs

Abstract. Inerter devices are used in Formula 1 vehicles for a decade to reduce vertical and rolling displacement at the corner of the sharp racing curves and provide higher tire grip on racing challenges. To improve dynamics Formula SAE Car, several topological designs for suspension system are already proposed and published [1]. Based on our passive suspension system for quarter-car model, we carried out analysis for different parameters and under the impact from the road disturbance and demonstrated when integrated inerter device can increase ride comfort and decrease vertical displacement. Proposed solution system with integrated inerter mechanism can improve vehicle suspension systems development and have effects on dynamics and stability of a vehicle.

1. Introduction

Inerter devices are used in Formula 1 vehicles (figure 1) for a decade to reduce vertical and rolling displacement at the corner of the sharp racing curves and provide higher tire grip on racing challenges. Its application to Formula SAE race car (SAE – Society of Automotive Engineers) has not been studied enough. Some pioneering work in this field has been done by Tran et al. [1-3]. In general engineering practice, suspension systems with passive, semi-active and active elements have been used to improve vehicle ride comfort [4,5]. Active suspension systems are the most investigated in the literature [6], but in order to increase robustness of the system, passive suspension systems are the most desirable [7]. Smith [8] introduced inerter elements including practical realization and Chen et al. [9] studied different mechanical network topologies with springs, dampers and inerter elements. In our study, we investigated passive suspension systems which can improve rolling stability by reducing vertical displacement and increase drive comfort. Our passive suspension system containing springs, dampers and inerter elements and discusses its influence on behavior of both the sprung and un-sprung mass of the vehicle subjected to road disturbance. The vertical displacement of vehicle body can be reduced by optimizing the springs, dampers and inerter elements which represent parameters of tire and suspension.

2. Physical model

The force produced by inerter is:

$$F = b(\ddot{z}_2 - \ddot{z}_1), \tag{1}$$

where b is the inertance, and $\ddot{z}_1$, $\ddot{z}_2$ are accelerations on two terminals (figure 2). For inerter model presented in figure 2, the constant b can be obtained as:

$$b = m\alpha_1\alpha_2 \tag{2}$$

with

$$\alpha_1 = \frac{\gamma}{r_3}, \ \alpha_2 = \frac{r_2}{r_1}. \tag{3}$$

Another parameters are: r_1 - the radius of the rack pinion, r_2 - the radius of the gear wheel, r_3 - the radius of the flywheel pinion, γ - the radius of gyration of the flywheel, m - the mass of the flywheel.

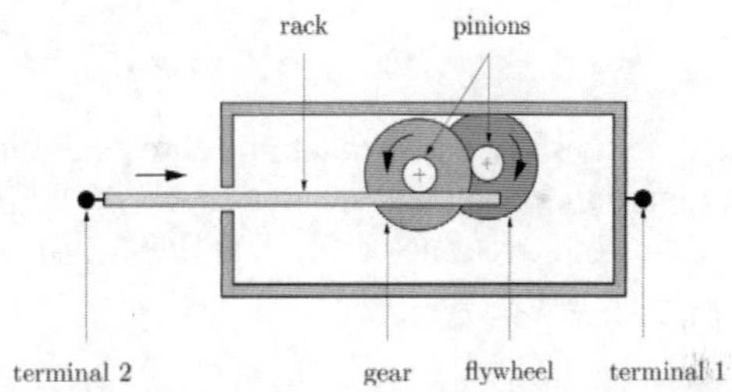

Figure 1. Ferrari inerter fitted on F10 racer, 2010. **Figure 2.** Inerter model proposed by Smith [8]

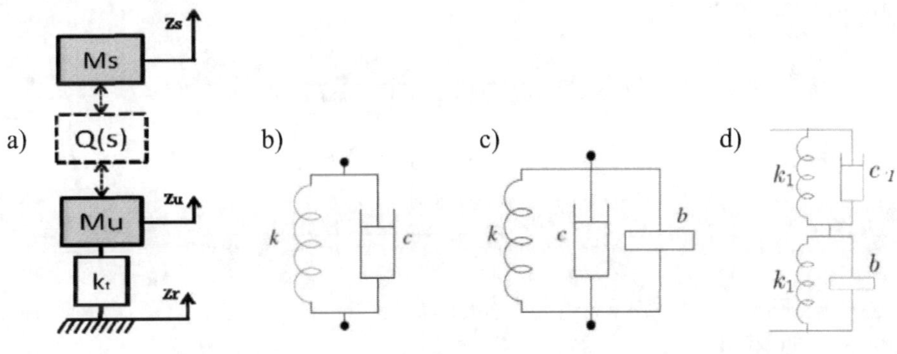

Figure 3. The quarter-car model with suspension function block Q(s) in Laplace domain (a), and different arhitectures for Q block: b) basic, c) parallel, d) series

In the figure 3a is presented quarter car model composed of two masses. k_t is tire stiffness. M_u is mass of the wheel (unsprung mass) and M_s is mass of the chassis (sprung mass). Function block $Q(s)$ in figure 3a is suspension function in Laplace domain and can have one of the forms presented in figure 3b, 3c or 3d. Constants k, k_1 are spring stiffnesses, c is damping and b is inertance.

3. Mathematical model

Equations of motion for the basic quarter-car model in time-domain are:

$$M_s\ddot{z}_s(t) = F_k(t) + F_c(t) \tag{4}$$
$$M_u\ddot{z}_u(t) = F_{kt}(t) - (F_k(t) + F_c(t)) \tag{5}$$

where forces due to suspension spring stiffness (F_k), suspension damper (F_c) and tire stiffness (F_{kt}) are defined as:

$$F_k(t) = k(z_u(t) - z_s(t)) \tag{6}$$
$$F_c(t) = c(\dot{z}_u(t) - \dot{z}_s(t)) \tag{7}$$
$$F_{kt}(t) = k_t(z_r(t) - z_u(t)) \tag{8}$$

z_r, z_u, z_s are displacement due to road roughness, displacement of unsprung mass (wheal) and displacement of the sprung mass (chassis), respectively, in time domain. Over dots are time derivatives. Variables Z_r, Z_u, Z_s represent displacements z_r, z_u, z_s, but in Laplace transformed domain.

The equations of motion (4,5) can be transformed in the Laplace domain leading to:

$$M_s s^2 Z_s = -sQ(s)(Z_s - Z_u) \tag{9}$$
$$M_u s^2 Z_u = sQ(s)(Z_s - Z_u) + k_t(Z_r - Z_u) \tag{10}$$

where $Q(s)$ is suspension function represented in Laplace transformed domain. From equations (9,10) are computed relevant transfer functions in the following text and equations, because they are suitable for studying our dynamic problem. Influence of the road disturbance z_r to the displacement of the sprung mass z_s is described as

$$T_{z_r \to z_s} = T_{Z_r \to Z_s} = \frac{k_t Q(s)}{M_s s(M_u s^2 + k_t) + ((M_u + M_s)s^2 + k_t)Q(s)} \tag{11}$$

Influence of the road disturbance z_r to the acceleration of the sprung mass $\ddot{z}_s$ is described with the following transfer function:

$$T_{z_r \to \ddot{z}_s} = s^2 T_{Z_r \to Z_s} = \frac{s^2 k_t Q(s)}{M_s s(M_u s^2 + k_t) + ((M_u + M_s)s^2 + k_t)Q(s)} \tag{12}$$

Influence of the road disturbance z_r to the displacement of the unsprung mass z_u is described as

$$T_{z_r \to z_u} = T_{Z_r \to Z_u} = \frac{k_t(M_s s + Q(s))}{(M_u s^2 + k_t)(M_s s + Q(s)) + M_s s^2 Q(s)} \tag{13}$$

Influence of the road disturbance z_r to the acceleration of the unsprung mass $\ddot{z}_u$ is described with the following transfer function:

$$T_{z_r \to \ddot{z}_u} = s^2 T_{Z_r \to Z_u} = \frac{s^2 k_t(M_s s + Q(s))}{(M_u s^2 + k_t)(M_s s + Q(s)) + M_s s^2 Q(s)} \tag{14}$$

The conventional suspension function represented in Laplace transformed domain:

$$Q(s) = Y_k + Y_c = \frac{k}{s} + c \tag{15}$$

The parallel suspension function represented in Laplace transformed domain:

$$Q(s) = Y_k + Y_c + Y_b = \frac{k}{s} + c + bs \tag{16}$$

The series suspension function represented in Laplace transformed domain:

$$Q(s) = \left(\frac{1}{Y_c+Y_{k_1}} + \frac{1}{Y_b+Y_{k_2}}\right)^{-1} = \left(\frac{1}{c+\frac{k_1}{s}} + \frac{1}{bs+\frac{k_2}{s}}\right)^{-1} \quad (17)$$

where Y_k, Y_c, Y_b are stiffness, damping and inertance in Laplace domain. In our special case two springs have equal stiffness leading to $Y_{k1} = Y_{k2}$. Ride comfort is defined as:

$$J_1 = E\left[\ddot{Z}_s^2(t)\right] = r.m.s \text{ body vertical acceleration} \quad (18)$$

4. Numerical results

With data from the table 1 are obtained results presented on figures 4-7. In the figure 4 are presented displacement and acceleration response of sprung body mass in 3 different configurations presented on the figure 3. Since the parallel configuration gave the best results, it is used in the further analysis. figures 6 – 7. present displacement and acceleration response of sprung mass for 4 different values of inertance parameter (0,20,40,60 kg). Figure 8. presents maximal displacement and maximal acceleration of sprung mass and Figure 9. ride comfort as a function of inertance. This diagram can serve for engineering design purposes.

Table 1. Used parameters for Formula SAE quarter-car.

Symbols	Parameters	Values
M_s	Mass of body	63 kg
M_u	Mass of tire	12 kg
k	Stiffness coefficient	24000 N/m
k_1	Stiffness coefficient	48000 N/m
c	Damping coefficient	1200 Ns/m
c_1	Damping coefficient	2400 Ns/m
b	Mass of inertance	20 kg
k_t	Stiffness coefficient of tyre	70000 N/m
H_0	Road disturbance hump	0.05 m

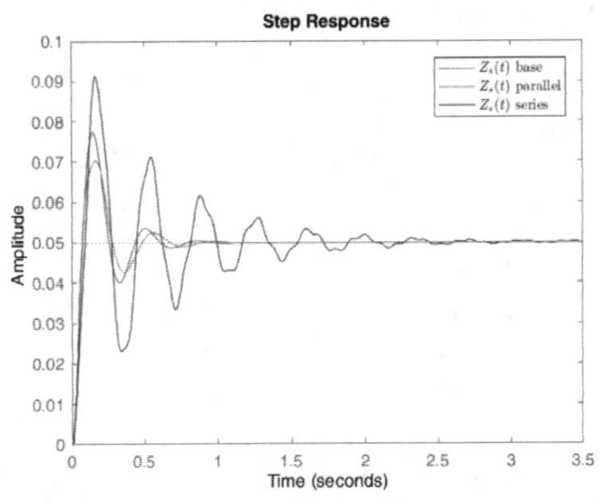

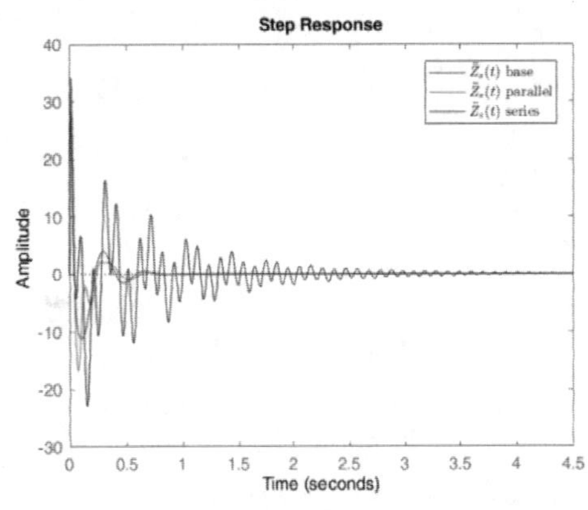

Figure 4. Time response diagrams for a) displacement and b) acceleration of sprung mass for 3 different quarter-car models subjected to road disturbance hump

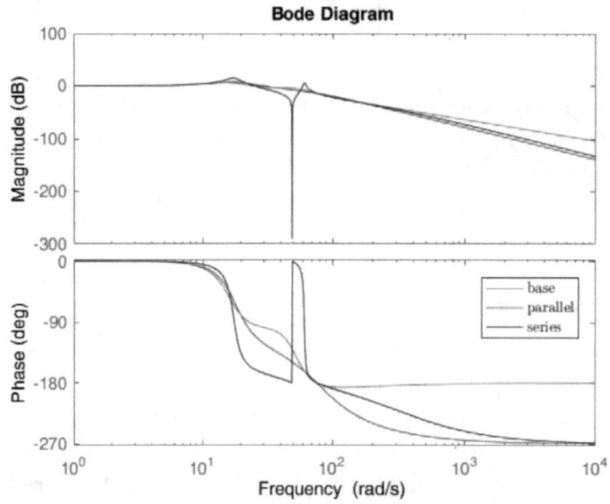

Figure 5. Bode diagrams for 3 different quarter car models

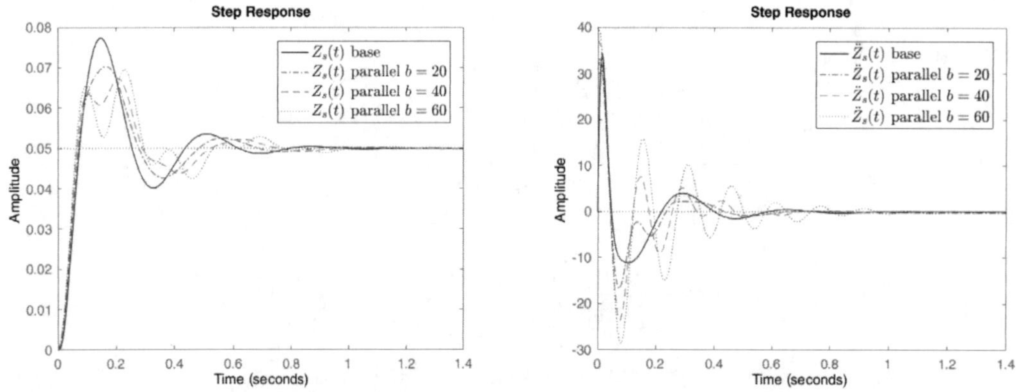

Figure 6. Time response of sprung mass of quarter-car model for different values of inertance paramer in parallel configuration: a) displacement and b) acceleration

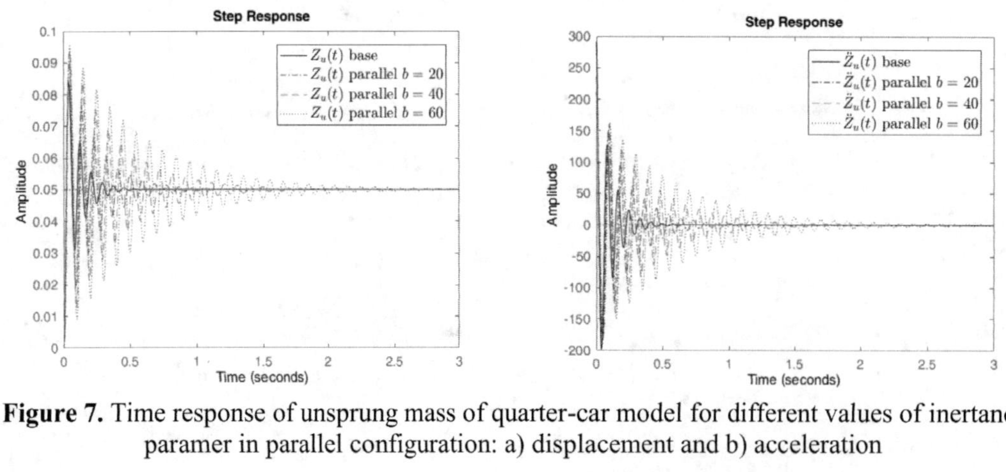

Figure 7. Time response of unsprung mass of quarter-car model for different values of inertance paramer in parallel configuration: a) displacement and b) acceleration

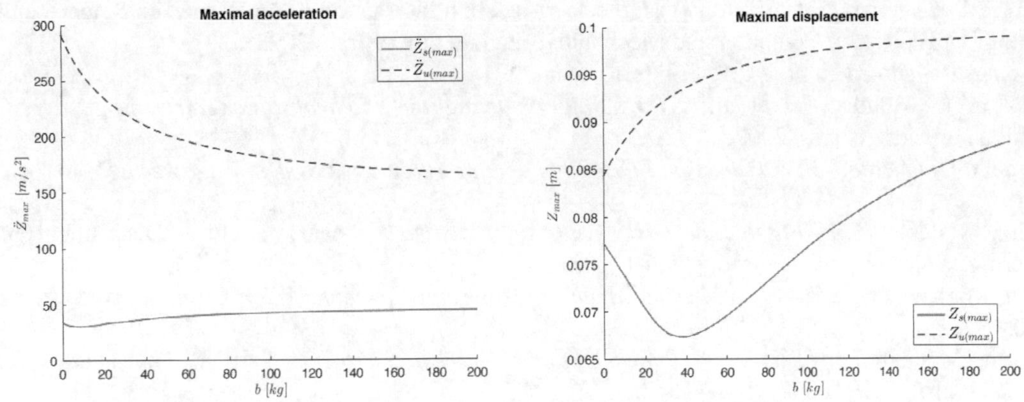

Figure 8. Parallel quarter car model: a) Maximal acceleration, b) maximal displacement

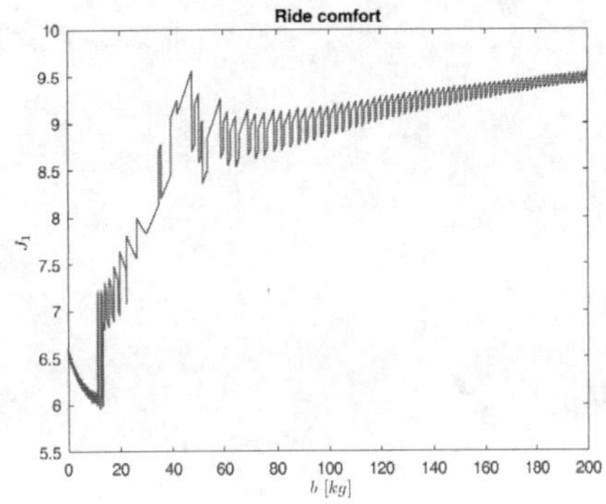

Figure 9. Parallel quarter car model: ride comfort as a function of inertance b

5. Conclusion

This paper demonstrated that quarter-car Formula SAE passive suspension model with inerter can reduce the oscillation, comparing to conventional spring and damper model. Influence of inertance on dynamic behavior is also presented which can serve for an optimal suspension design. Based on our passive suspension system for quarter-car model, we carried out analysis for different parameters and under the impact from the road disturbance and demonstrated when integrated inerter device can increase ride comfort and decrease vertical displacement. Therefore, proposed solution system with integrated inerter mechanism can improve vehicle suspension systems development and have effects on dynamics and stability of a vehicle. For more profound investigation, this study can be extended with half-car and full car model.

References

[1] Tran, Thanh-Tung, et al., *"Integrated inerter design and application to optimal vehicle suspension system."* International Journal of Computer-Aided technologies (IJCAx) Vol 1, 2014.

[2] Tran, Thanh-Tung, et al., *"1B23 Verification and optimization of Formula SAE suspension employing Inerter mechanism"*, The Proceedings of the 12th International Conference on the Motion and Vibration Control, The Japan Society of Mechanical Engineers, 2014.

[3] Tran, Thanh-Tung, and Hiroshi Hasegawa. *"Advanced passive suspension with Inerter devices and optimization design for vehicle oscillation."* International Journal of Mechanical Engineering and Robotics Research 4.4 (2015): 354.

[4] Pourghavam, Masoud, et al. *"Mathematical Modeling of a Formula SAE Student Racecar Passive*

Suspension System Using Two Different Methods." The sixth International Conference on Science and Technology of Electrical, Computer and Mechanical Engineering of Iran, 2022.
[5] Bosch, *Automotive Handbook*, SAE International, 2018.
[6] Kulakowski, Bohdan T., et al., *"Dynamic modeling and control of engineering systems"*, Cambridge University Press, 2007.
[7] Marcello Chiaberge, "*New Trends and Developments in Automotive System Engineering*", InTech, 2011.
[8] Smith, Malcolm C., *"Passive network synthesis revisited."* Plenary Lecture, Department of Engineering, University of Cambridge, UK (2003).
[9] Chen, Michael ZQ, et al. *"The Missing Mechanical Circuit Element."* IEEE Circuits and Systems Magazine 9.1.
[10] Smith, Malcolm C. *"Synthesis of mechanical networks: the inerter."* IEEE Transactions on automatic control 47.10 (2002): 1648-1662.

AUTHOR INDEX